Behavioral Modeling and Linearization of RF Power Amplifiers

For a complete listing of titles in the
Artech House Microwave Library,
turn to the back of this book.

Behavioral Modeling and Linearization of RF Power Amplifiers

John Wood

ARTECH HOUSE
BOSTON | LONDON
artechhouse.com

Library of Congress Cataloging-in-Publication Data
A catalog record for this book is available from the U.S. Library of Congress.

British Library Cataloguing in Publication Data
A catalogue record for this book is available from the British Library.

Cover design by Adam Renvoize

ISBN 13: 978-1-60807-120-3

685 Canton Street
Norwood, MA 02062

10 9 8 7 6 5 4 3 2

For my desk buddy, Chinagirl

Contents

Preface

This book describes some of the currently popular techniques for the behavioral or system-level modeling of RF power amplifiers (PAs), to use in the context of linearization of these amplifiers. And by linearization, the focus here is on the digital pre-distortion (DPD) methods that are pre-eminent in the wireless infrastructure PA industry. Generally, such PAs are relatively narrowband devices, and much of the DPD work to date has been carried out in that context. As the demand for more and more data continues to increase, the signal bandwidths increase, and the digital modulations schemes employed to satisfy this demand make the design and realization of both the PA and the DPD system much more difficult. Linear operation of the PA is necessary to cope with modern modulation techniques such as LTE, and DPD is essential to enable the PA or transmitter to meet the linearity specifications.

I believe that PA/transmitter and DPD design is about to enter an era where significant technical challenges will need to be faced and solved. For those companies that can solve these technical challenges, the rewards in terms of market share and revenue could be vast. The critical challenges include: wider bandwidths, and even multi-band operation of the transmitter; multi-protocol radios where different modulation schemes are employed on different channels in the same band; higher frequency operation; and realizing and using next-generation, high-efficiency PA architectures. Such challenges provide ample opportunities for new approaches and clever designs, where understanding of the nonlinear behavior and dynamic properties of the PA and transmitter system will be key enablers.

In this book, I have tried to provide a basic background and framework to the nonlinear dynamical modeling of a PA, starting from linear systems. These nonlinear dynamical modeling techniques can then be used directly to design the DPD linearization algorithm. I have focused on Volterra series methods as these are currently in wide use in practical DPD systems, and they provide a good foundation for the understanding of how to handle the nonlinear behavior and the dynamic or time-dependent effects in the transmitter. These approaches can be extended to other nonlinear basis functions. I have also included a chapter on so-called *memory effects*: these are the transient time-dependent behaviors displayed by the PA, but this expression is often used as a catch-all for any unspecified dynamic effects, or even any unexplained nonlinear behavior. As such, the term may be more confusing than illuminating, so I hope to provide some clarification here.

There are also three chapters on aspects of pre-distortion in transmitters. Crest factor reduction (CFR) and DPD are each described in some detail, these techniques being the workhorses of a typical linearizer. There is also a description of RF pre-distortion, an analog-based technique that is beginning to find favor in several applications such as small cell transmitters. Finally, as a coda, there is a brief description of frequency-domain modeling. While there are many applications and requirements for models of wide bandwidth amplifiers and subsystems, for which models in the frequency domain are very appropriate, this is not the approach that is typically used for pre-distortion in the digital baseband, as described in this book.

Who should buy this book? You, of course! Joking aside, the intention is that this book can be used as a basic reference for nonlinear modeling and DPD by practicing engineers, and also as an introduction to these topics for graduate students and newcomers to the field. Nevertheless, I have not laden this book with references; there must be over a thousand papers and articles on behavioral modeling and linearization (and I have read many of them), and an exhaustive set of references would be overwhelming, and of limited value. Instead, I have included some citations that are most relevant and illustrative of the points that are being made in the text, and a bibliography, to guide the reader to the current and classical works on the subject, and to those authors who have produced important contributions to modeling and DPD over time. From these points of information, it is possible to produce a personal literature survey quite easily, targeting those authors and topics of particular interest, using web-based search engines such as Google and IEEE Xplore. What a huge change from how I did my first searches as a fresh new PhD student: tracking down the reference in the dusty volumes of the bound yearbooks of the learned journals, often in the gloomy stacks in the basement of the university library, then carrying the weighty tome up to the photocopier and obtaining my own paper copy. After reading and digesting the paper, I then had to search for the important references in it, following the same procedure over again. And yes, the bound volumes did always seem to have a thin coating of powder; I just assumed it was dust.

I shall take this opportunity to acknowledge most gratefully the contributions of my guest authors, without whom this book would have been much less informative, and a good deal shorter. First, thanks to Gayle Collins of MaXentric Technologies for her contributions to the content of Chapter 4, in analyzing and describing the nonlinear behavior of PAs. Gayle has experience in nonlinear modeling and system identification, and has applied these insights to RF and microwave PA design. Neil Braithwaite contributed Chapter 8, on crest factor reduction techniques: Neil has worked on DPD and CFR algorithms using advanced signal processing techniques for several years, at both Powerwave and now at Qualcomm, and has valuable insights in these topics. Frédéric Roger provided Chapter 9 on RF pre-distortion, outlining the approaches and techniques that he used in designing and building these novel RFIC pre-distorters when working as lead designer for Scintera Technologies.

I would also like to thank the staff of Artech House, particularly Mark Walsh and Samantha Ronan for their encouragement, support, and almost endless patience during the production of this book.

John Wood
March 14th, 2014

Chapter 1

Introduction

Voice and data communications by wireless are now taken for granted, with connectivity virtually ubiquitous. The quality of service of these communications links is necessarily very high, as the users demand essentially perfect transmission and reception of voice and data; any errors are noticed and are a source of irritation. The infrastructure that supports this wide connectivity and nearly error-free delivery of information is complex, costly, and continually being improved.

The coverage area of the wireless signal depends on the power of the signal transmitter and the signal frequency. The signal propagation distance is proportional to the transmitter power, and falls with increasing frequency. Wireless communications signals commonly fall into the radio frequency (RF) range of a few hundred MHz to a few GHz, because these frequencies have sufficient propagation range to cover several square miles at the power levels that are practically attainable for mobile systems. Higher frequencies are attenuated by topography, buildings, and weather and atmospheric factors, including rainfall, whereas lower frequencies, while having long range, require physically large antennas that are somewhat inconvenient to carry around.

The data rates required to support the information flow in popular applications such as audio and video streaming are very high. This has led to the adoption of complex digital modulation schemes to make the best use of the available channel bandwidth in the wireless systems: the *spectral efficiency*, or number of data symbols per hertz of bandwidth, of the digital coding and modulation schemes is high. One consequence of using a spectrally-efficient modulation is that the dynamic range of the signal may be quite high. This is generally measured in terms of the ratio of the peak signal power to the average power of the modulated signal, the *peak-to-average power ratio* (PAPR). While in itself a large PAPR is not such a problem for signal transmission, it can have an impact on the efficiency of the RF power amplifier in the transmitter, as will be outlined shortly, in Section 1.2.

While the demand for more data, video, and audio, continues to increase, and wireless mobility remains paramount, the average revenue per user for these services has not increased in real terms. The costs remain low because of an increased number of users, and also by reducing the operating costs of the transmitter base-stations. The RF power amplifier (PA) is the main consumer of the electricity supplied to the base-stations. There have been significant and successful efforts in recent years to improve the energy efficiency of the RF PA in the base-station, particularly for the spectrally-efficient modulation signals. This effort continues, spurred by cost and environmental concerns of energy usage.

RF power amplifiers are at their most energy efficient when operated at their highest output power, in compression, close to saturation. This also happens to be where the power amplifier produces most distortion. The amplified signal must be free from error, within limits specified by international regulating bodies, to provide the best quality of service for the customer. The raw linearity of the efficient RF PA in compression is sufficiently poor that some form of error correction is essential. Linearization using analog techniques has been popular for some time; nowadays, digital pre-distortion (DPD) techniques are becoming the dominant linearization method.

In this book, we shall describe the mathematical methods and practical implementations of linearization techniques for RF power amplifiers for mobile communications. This presentation shall include a review of RF power amplifier design for high efficiency operation. We shall then discuss the mathematical approaches to modeling nonlinear dynamical systems, which can be applied in the context of modeling the PA for identification in a pre-distortion system. Finally, we shall describe typical approaches to linearization and digital pre-distortion that are used in practice.

But first, in this chapter, I shall provide a review of the digital modulation signals, and their characteristics, used in mobile wireless communications. This is followed by an outline of some of the main features of RF power amplifiers and the push for higher efficiency, the regulatory environment, and the need for linearization. The concept of behavioral modeling of the nonlinear system, that is, the RF PA, will be introduced, and some historical approaches will be outlined briefly, providing the context for modern approaches to linearization using DPD.

1.1 A BRIEF HISTORY OF MOBILE WIRELESS COMMUNICATIONS

My early memories of listening to the 'Billy Cotton Band Show' on the BBC Light Programme at Sunday lunchtimes on the 'wireless' were probably my first introduction to radio communications. I was quite fascinated by how these glowing tubes in our radio could create the music hall show from out of the air. Of course, this wasn't exactly mobile, but that would come later. The Billy Cotton show was broadcast using amplitude modulation (AM) of the radio signal: the music or audio signal is impressed on the sinusoidal RF carrier as changes in its instantaneous amplitude. While not the ultimate in quality, more of which a little later, the modulation for transmission and demodulation on reception could be achieved straightforwardly with simple components and circuitry.

The BBC would also broadcast some programs using frequency modulation (FM) of the radio signal. In this modulation method, the audio is impressed onto the carrier as changes in the instantaneous frequency, but the amplitude of this modulated RF signal remains constant. This *constant envelope* nature of the RF signal offers quite an advantage to the RF PA designer: the power amplifier in the transmitter can be operated at high power, in compression, and hence high energy efficiency, keeping the operating costs low. Further, because any amplitude modulation can be rejected by the receiver, amplitude distortion is not a problem. Thus, FM was originally used for the highest quality radio transmissions, since AM suffered from unwanted amplitude modulation distortions arising from the nonlinear gain characteristics of the RF PA.

The typical illustrations of AM and FM waveforms in the time domain are shown in Figure 1.1. The action of the relatively low-frequency modulating signal on the sinusoidal carrier can be clearly seen. Another way of representing these analogue modulations is in polar form on a *phasor* diagram (Figure 1.2). In the phasor diagram, the directed graph represents the modulating signal, and is assumed to rotate around the center at the RF carrier rate. In other words, the RF carrier action is a rotation at a constant speed, and can be neglected without any loss of information or clarity. This signal has a magnitude of A, and an arbitrary phase Φ. Amplitude modulation is described by changing the length of the phasor at the modulation rate, such that $A = A(t)$, but the angle or phase remains constant. In practice, it is difficult to achieve pure amplitude modulation, as the variation in signal level invariably leads to some (usually unwanted) phase modulation as well. If instead we keep the magnitude constant, but change the phase as a function of time $\Phi(t)$, we have phase modulation; the position of the end of the signal moves around the radius. For frequency modulation, the length or amplitude of the phasor is again kept constant, but its position on the radius will change with time at the modulation rate; this looks similar to phase modulation on the phasor diagram. The AM and FM modulating techniques outlined here are *analog* modulations. The phasor representation will be used later for describing the *constellations* of discrete data points created by *digital* modulation schemes.

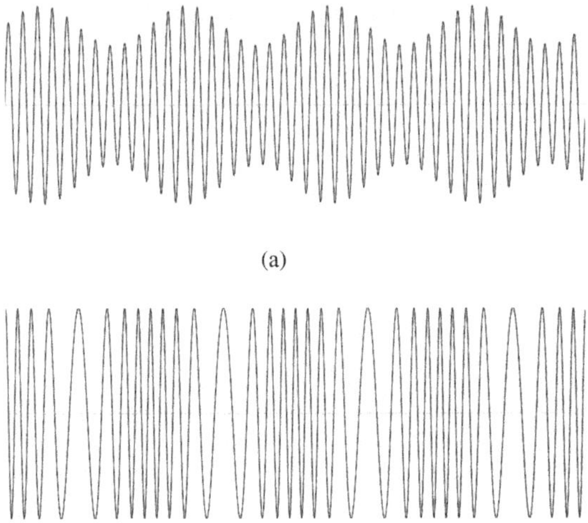

(a)

(b)

Figure 1.1 Amplitude- and frequency-modulated sinewaves in the time domain.

The radio and, later, TV broadcast transmissions using the analog modulations described above were made from fixed locations. As radio circuitry progressed from tube through transistors to integrated circuits, the size and power demands of these radios became small enough to enable the transmission and reception of radio signals on convenient and compact mobile platforms.

One of the earliest examples of personal mobile radio communications is the 'walkie-talkie' from World War II. This was invented in 1940 by an engineering team at the Galvin Manufacturing Company, which became Motorola. The team included Dan Noble, who devised the analog frequency modulation scheme used by this radio transmitter/receiver; he later became Director of Research at Motorola. The walkie-talkie radio used vacuum tubes for the RF transmission and reception, and was carried in a backpack; the (slightly) more convenient hand-held version of the radio was nicknamed the 'handie-talkie,' though this had lower power and a much more limited range.

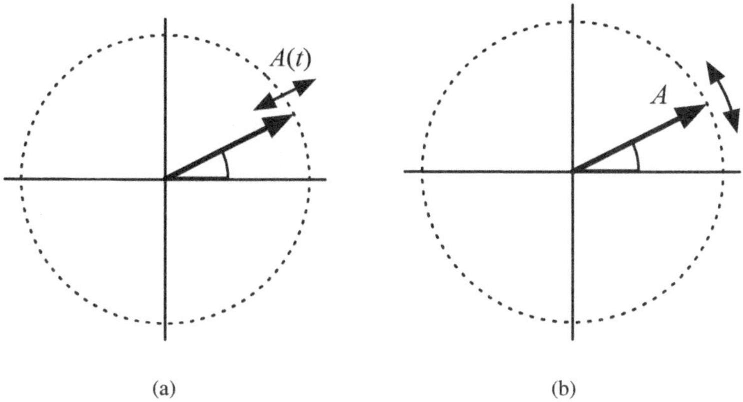

Figure 1.2 (a) Amplitude and (b) frequency modulation illustrated using phasors.

In the post-war years, Motorola continued to develop the walkie-talkie concept for Public Service radio, using solid-state components and circuits to reduce the size and cost of these radio transceivers. Motorola has a strong presence in Public Service radio to this day, both in the United States and overseas. These radios again used analog frequency modulation for the transmission of the radio signals. The constant envelope FM signal meant that relatively inexpensive power amplifiers could be used, running in compression for high efficiency to conserve battery life, without causing degradation of the signal quality resulting from unintended amplitude modulation due to the nonlinear gain characteristic at high power, in the compression region.

The channel bandwidth for the Public Service signal was initially chosen to be 25 kHz. This permitted the use of relatively simple filters and provided good voice quality. As use became more widespread, the channel was reduced to 12.5 kHz, and eventually to 6.25 kHz, to accommodate the increased number of users. This bandwidth is still wide enough for speech communications, but further reductions would compromise quality. Further expansion of the system required a different modulation approach: digital modulation techniques were applied.

The pre-history of mobile telephones follows a similar story. The early mobile telephone systems were based on analog FM for the signal transmission. Again, the constant envelope of the FM signal permitted the use of relatively inexpensive and simple-to-design power amplifiers in the handsets and base-stations: Class AB PAs running in compression at high efficiency, with minimal impact from amplitude modulation-induced distortions. The channel bandwidth, or peak FM (frequency) deviation, was around 6–8 kHz depending on the particular system. Each user was assigned a specific channel in the available part of the frequency spectrum, allowing many users to access the network simultaneously; this mode of channel assignment is known as frequency division multiple access (FDMA). As the number of users (subscribers) increased with the popularity of mobile communications, the network became unable to accommodate the number of simultaneous users. And as with the Public Service systems, the next generation of mobile communications systems used digital modulation techniques.

1.1.1 Introduction to Digital Modulation

There are many excellent texts on digital coding and modulation. Here, the goal is simply to outline the properties of the digital modulation schemes commonly found in mobile wireless communications systems, and then go on to indicate some of the ramifications for linear RF power amplifier design.

In digital coding of an analog signal, we generally think in terms of 'ones and zeros' to represent the signal in the digital domain. These ones and zeros are the individual digital bits in the digital signal. The information in the digital stream is contained in *symbols*, comprising one or more bits. The types of digital modulation of a signal are similar to the analog modulations that we met earlier:

amplitude shift keying (ASK) – information is encoded in the values of the amplitude of the pulse;

phase shift keying (PSK) – information is encoded in steps of the signal phase;

frequency shift keying (FSK) – information is encoded in the values of the signal frequency deviation.

The idea of keying dates from the days of telegraphy where the codes were entered manually by tapping on a switch or key.

The actual encoding of the symbols as modulation of the signal, while represented as ones and zeros, is in terms of discrete steps of the the modulation format. This is commonly represented as *in-phase* I and *quadrature* Q components of the modulating or information signal in the phasor diagram format, as indicated in Figure 1.3. The magnitude and phase of the signal are converted into rectangular form. The digital phase and amplitude modulation are represented as discrete IQ points in the phasor diagram; these sets of points are known as constellations and are characteristic of the various modulation formats.

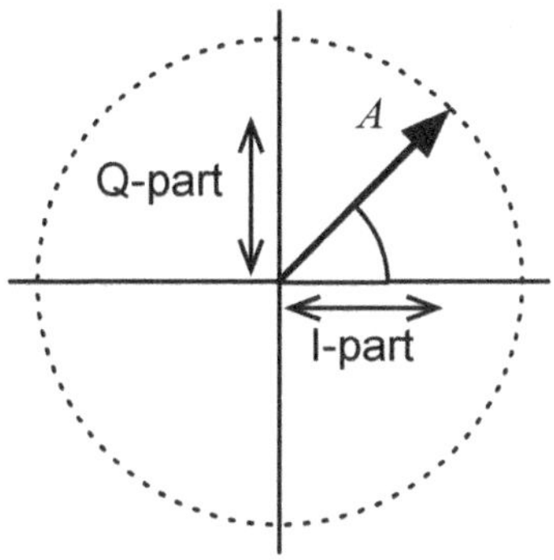

Figure 1.3 The representation of the signal in IQ format on the phasor diagram.

The IQ format reflects the way in which the digital communications signal is modulated and demodulated in practice. For transmission, the separate I and Q components of the data stream are mixed with the local oscillator (LO) at the carrier frequency; a 90° phase shift is placed in one of the LO paths to the mixers, creating a 90° separation between the modulated signals. These signals are therefore orthogonal to each other and can be added directly without interference to create the modulated communications signal. The reception is the reverse of this process, creating the separate I and Q data signals, which can then be decoded into symbols representing the analog information signal. The up-converters and down-converters are called IQ modulators and demodulators, respectively. Digital phase and amplitude (de)modulation can be accomplished easily using IQ (de)modulators; the circuits comprise the local oscillator, two mixers, a 90° phase shifter, and a summing amplifier, as indicated in Figure 1.4. Using an amplitude modulator and phase modulator to do this is much more complex.

The second-generation, digitally-modulated mobile telephony (2G) and Public Service systems all used time-division multiple access (TDMA) to accommodate the increased number of users or subscribers. This means that each user in a given frequency channel is allocated a short amount of time (a *time slot*) for their signal in each frame of the transmission, with a typical frame being 10 milliseconds. This gives the appearance of continuous communication for the users. Changing from an FDMA allocation to TDMA can be accomplished by using either phase- or frequency-shift keying modulation to encode the signal. Different approaches were used in various parts of the world.

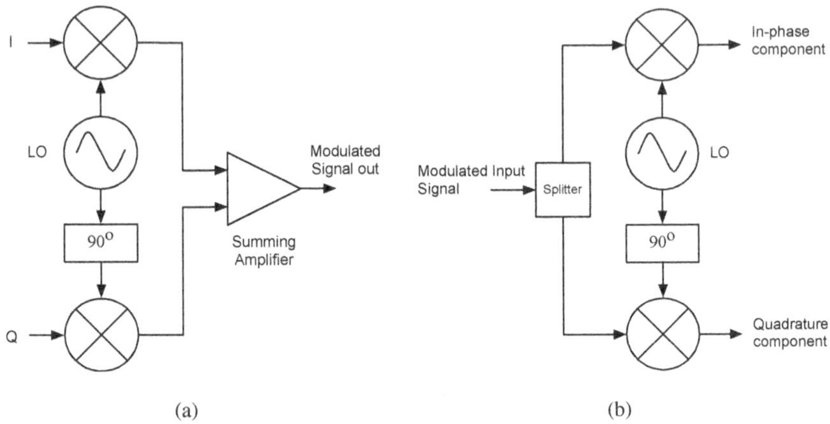

Figure 1.4 Simple block diagram circuit representations of (a) an IQ modulator and (b) a demodulator.

Phase-shift keying modulation was adopted for mobile telephony in the United States (North American digital cellular, NADC) and Japan (Pacific digital cellular, PDC), and for Public Service systems (trans-European trunk network, TETRA). This modulation technique is also used for many other communications systems. The simplest PSK coding is binary PSK, in which there are only two phase states, and hence only one bit per symbol. Quadrature PSK uses two bits per symbol, to create four phase states. The NADC, PDC, and TETRA digital modulations are specially designed QPSK formats, where there are two QPSK constellations, offset by 45°, and the signal moves from a location on one constellation to a location on the other (offset) location on each symbol clock cycle – differential mode operation; this format is known as $\pi/4$ differential QPSK ($\pi/4$-DQPSK). The QPSK and $\pi/4$-DQPSK constellations are shown in Figure 1.5.

The state transitions in the QPSK modulation are seen to go through the origin, whereas the transitions in the $\pi/4$-DQPSK modulation have been constructed to avoid this. When the signal goes through the origin, the signal magnitude is zero (and its phase is, strictly, indeterminate), and the RF PA must have no output for an instant. Switching high-power RF amplifiers off and on rapidly is quite difficult, and leads to distortions in the signal. The $\pi/4$-DQPSK modulation is more friendly to the PA in this respect, and delivers better distortion performance. It can also be seen from Figure 1.5 that the power level in the transmitted signal is not constant as the signal moves from symbol to symbol: the signal has a varying envelope power

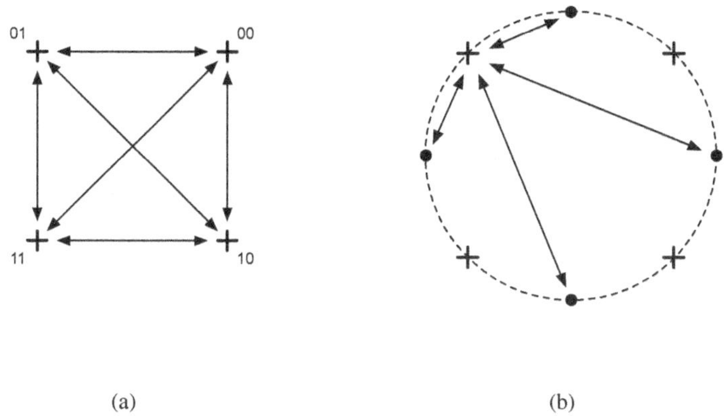

Figure 1.5 Constellation patterns for (a) quadrature PSK and (b) $\pi/4$ differential QPSK, showing also the transitions between symbol states.

and hence a finite, nonzero peak-to-average power ratio (described in more detail in Section 1.1.3).

In Europe, a frequency-shift keying approach was used. In particular, a special case of binary FSK (BPSK) was chosen. In BPSK, the two frequency states are symmetrical about the nominal (but not transmitted) carrier frequency, and so the frequency spacing between these two symbol states is twice the nominal *frequency deviation* of the modulation, Δf. The frequency shift between the two symbol states will result in a phase advance or lag, and so by sampling the phase, the frequency shift can be detected. Using an IQ demodulator, we can detect quite easily phase shifts that are of multiples of 90°: for even-numbered symbols, the I-channel signal polarity (positive or negative voltage representing phase advance or retardation) represents the data, and for odd-numbered symbols, the Q-channel signal polarity provides the data. The minimum quadrature phase shift that can be detected is 90°; this condition is defined when the frequency deviation is

$$\Delta f = 0.5/(2 * T_S) \tag{1.1}$$

where T_S is the symbol time. After each symbol, the phase advances or retards by 90°. This form of BPSK modulation is the minimum quadrature-rotation frequency-shift keying, known as minimum shift keying (MSK). When used with a Gaussian lowpass filter (more on filters for digital modulation later), this modulation format is Gaussian MSK, or GMSK. The GMSK modulation method used for mobile telephony in Europe uses a bandwidth scaling factor of 0.3 for the Gaussian filter: this is the GSM standard, which is now found worldwide.

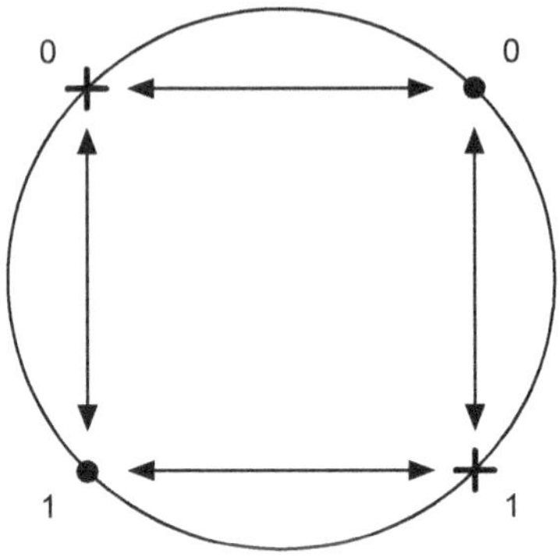

Figure 1.6 Gaussian minimum shift keying (GMSK) represented in the IQ domain, showing transitions between symbols and the constant envelope of the modulation.

The IQ pattern for the GSM signal is shown in Figure 1.6, where the quadrature transitions between the symbols are indicated. These transitions produce a constant envelope signal, which affords the benefits to the RF PA design outlined earlier for analog FM modulation. This GMSK modulation scheme is only one bit per symbol.

To increase the capacity of the GSM system, the GMSK modulation was changed to a Gaussian-filtered 8-PSK scheme with a $3\pi/8$ rotation, while maintaining the same channel bandwidths, symbol rate, and TDMA frame structure. This signal is 'enhanced data rate for GSM evolution,' known as EDGE. The constellation and transition vector diagram are shown in Figure 1.7. One outcome of this modulation is a finite, nonzero PAPR.

A few years after the introduction of the 2G communications standards outlined above, a different approach from TDMA and FDMA for mobile telecommunication was developed by Qualcomm, Inc. The technique is based in information theoretic principles, with each user transmitting a wideband signal covering the same frequency range and the network using spread spectrum techniques to distinguish these co-channel signals. The approach is known as CDMA (Code-division multiple access), which describes the user access method rather than the modulation scheme employed for the digital signal.

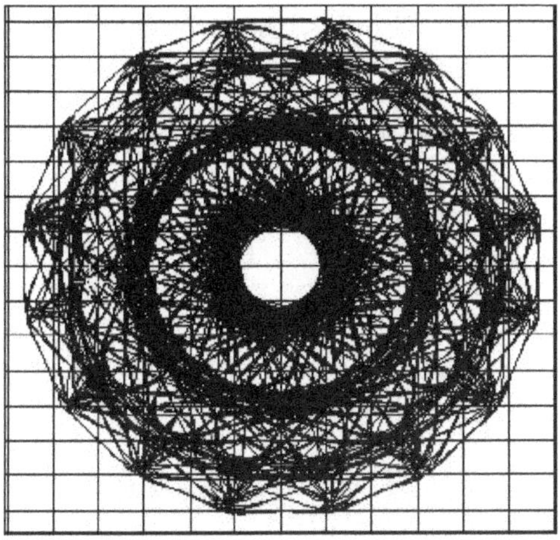

Figure 1.7 EDGE signal constellation points and the vector diagram showing the transitions between these symbol states. Some very attractive vector patterns can be derived for some of these digital modulations, although that is probably not the primary focus.

In the CDMA system, during the coding process, the data signal is multiplied (in the time domain) with a wide bandwidth digital spreading code to yield a wideband signal. At the receiver, the application of the same spreading code will retrieve the original information signal. The advantage of this approach is the rejection of interference signals: several orthogonal codes can be used for the encoding, enabling many signals to occupy the same frequency band without interfering with one another. As these other signals in the receiver bandwidth will not have the same encoding sequence, when the receiver's spreading code is applied, these interferers will decode to give a null signal.

The coding of the data signal occurs prior to the application of the modulation. The modulation scheme used in CDMA is QPSK, one bit per symbol, with a relatively low data rate of about 14 kbit/s. The modulation bandwidth is 1.2288 MHz due to the spreading code, with 20 frequency channels available. This is the IS-95 specification.

The third-generation (3G) mobile telephony systems focused on increased data rates to accommodate the increased demand for data content, including audio and video. The EDGE modulation was a significant step towards 3G deployment. An alternative modulation technique specified in the 3G Partnership Project (3GPP) is the Universal Mobile Telecommunications System (UMTS), which uses a wideband version of the CDMA coding technique, WCDMA. The channels are 5 MHz wide, with a symbol rate of 3.84 MSps. These enhanced data rate signals have a high PAPR, and a peak-to-minimum ratio that is infinite, as the signal vectors cross zero, as indicated in Figure 1.8. These characteristics make energy-efficient power amplification a challenge, and several 'new' PA architectures have been developed in recent years to cope with these signals (see Section 1.2).

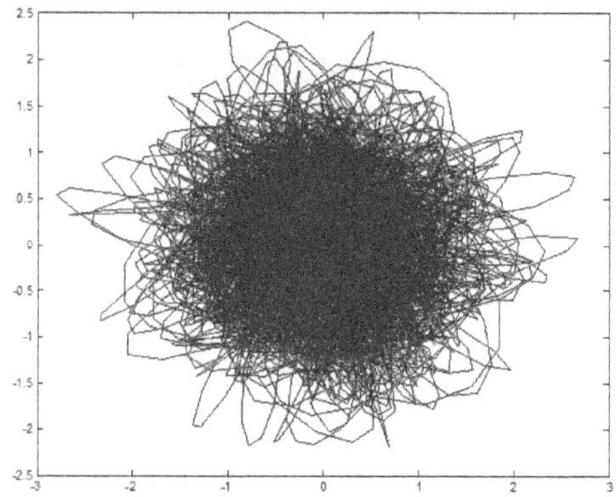

Figure 1.8 The vector diagram in the I-Q plane for unfiltered WCDMA modulation; here 64 physical channels are used.

Quadrature amplitude modulation (QAM) is an efficient method of coding digital data into a relatively narrow bandwidth. The I and Q digital data streams are amplitude-modulated individually. If the I and Q components are each modulated with two amplitude states, then we get four symbols in the IQ-plane; this is known as 4-QAM. With additional amplitude states, we can generate modulated data with many symbol states; 4-QAM, 16-QAM, and 64-QAM constellations are shown in Figure 1.9. As there are no conditions imposed on the symbol transitions, the vector diagram can include zero-crossings, as indicated in Figure 1.8. As mentioned above, QAM is used in the 3G WCDMA communications method, the wide bandwidth in this case coming from the spreading code.

Fourth-generation (4G) systems are currently in development, and in the process of deployment. The 3GPP has defined standards under the umbrella of LTE: Long-term evolution[1]. The LTE standards provide for robust transmission of very high data rate signals, such as video streaming and interactive web access, etc.,

1 a.k.a. Long Term Employment, by those in the mobile radio communications industry ☺.

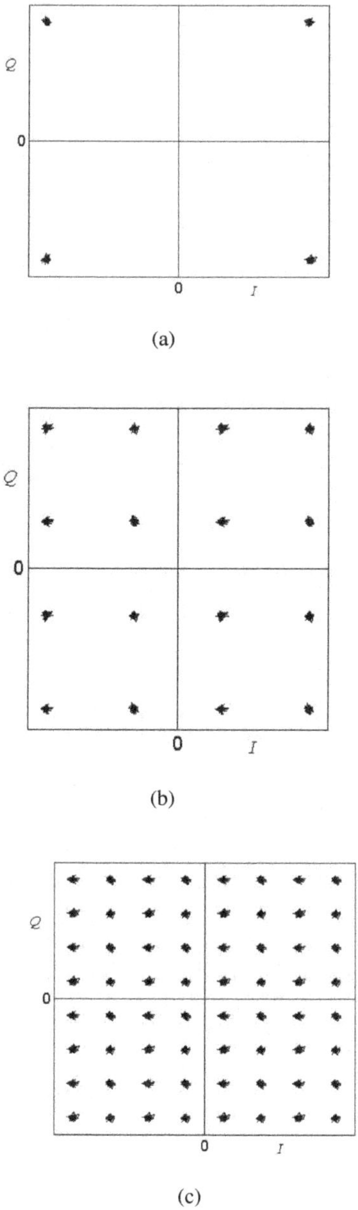

Figure 1.9 Constellation diagrams in the IQ-plane for (a) 4-QAM, (b) 16-QAM, and (c) 64-QAM.

which are susceptible to multipath interference effects in high data rate single carrier systems. Using long symbol times can overcome the interference due to the variable transmission times in a multipath environment, where the multiple paths may arise from movement of traffic between the base-station and handset, even though these are stationary. The overall data rate can be increased by using many subcarriers to carry the information. In traditional multi-carrier radio communications systems the multiple carriers are filtered into their respective frequency bands to prevent interference. In LTE, a different frequency division multiplex scheme is employed. The modulated subcarriers are unfiltered and are separated in frequency by the symbol rate, $1/T_S$. By this means, the peak of any carrier occurs at the nulls or zero-crossings of all the other carriers, as illustrated in Figure 1.10, so effectively the subcarriers are orthogonal, hence orthogonal frequency division multiplex (OFDM). In the LTE standard, the subcarrier spacing is 15 kHz, and the LTE signal can be expanded by the addition of more subcarriers: total bandwidths of 1.25 to 20 MHz can be used. The QAM signal on the subcarrier is randomized by a complex filter, essentially scrambling the constellation. This has the effect of reducing the PAPR compared with the straight OFDM signal.

The OFDM approach is used by several other radio communications standards, including WiMAX (802.16), Wireless LAN (802.11a/g/n), and terrestrial digital video broadcast (DVB-T).

1.1.2 Filters for Digital Communication Systems

Filters are used to control the bandwidth of the transmitted signal; the objective is to reduce the bandwidth without compromising the integrity of the digital data, in other words, to improve the spectral efficiency (symbols/Hz) of the communications system. It is important to recognize the difference between bit rate and symbol rate: the signal bandwidth required for the digital communications signal is dependent on the symbol rate, not the bit rate.

In the RF domain, analog filters can be used to control out-of-band emissions, but the channel widths are generally too narrow for analog techniques. Filters are also used at RF for *impedance matching* of the power transistor to its optimum load impedance. The filtering for control of the signal bandwidth is done in the digital domain, where the concept of a *matched filter* takes on a different meaning.

The digital modulations that we have met so far are characterized by their symbol constellations. The transitions between these states are assumed to be arbitrarily fast. In practice, the digital modulations are still analog signals, and fast transitions between the states will require a wide occupied bandwidth. Filters are used to reduce this bandwidth, by slowing down the transition times. The filters

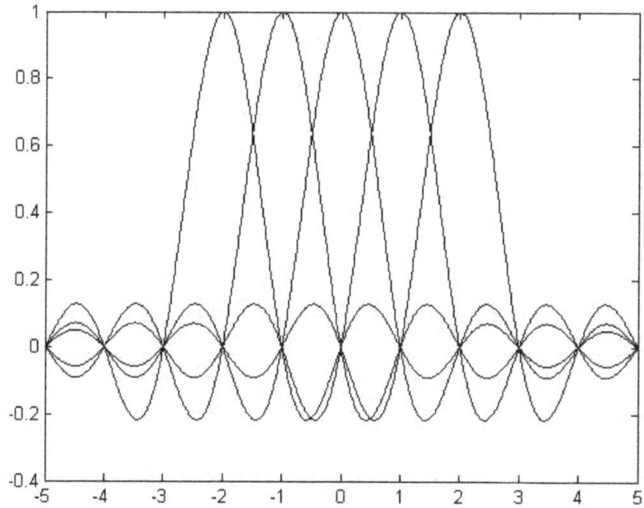

Figure 1.10 The principle of orthogonal subcarriers in OFDM. The subcarrier spacing $\Delta f = 1/T_S$ ensures the maximum of any carrier is located at the zeros of the other carriers.

also act to reduce interference and improve sensitivity at the receiver. Some types of filter will cause the signal trajectory to overshoot the constellation points; this trajectory requires more power from the PA, and so some compromise between bandwidth and power is necessary.

A commonly found filter in digital radio communications is the Nyquist filter. A typical example is the raised cosine filter, whose time response is a *sinc* (sin(x)/x) function, chosen so that its response goes through zero at the symbol clock frequency. This ensures that adjacent symbols do not interfere with one another, because at a given symbol time, all other symbols' response is zero: we are sampling at the Nyquist rate. This filter is typically implemented in two halves, one in the transmitter and the other in the receiver, so that overall we get Nyquist rate sampling:

$$\sqrt{Nyquist_{TX}} * \sqrt{Nyquist_{RX}} = \sqrt{Nyquist} \tag{1.2}$$

This is a known as a matched filter. The filter is implemented in the base-station and the handset as a *root raised cosine* (RRC) filter.

The RRC filter is also characterized by a factor α, which is a bandwidth control parameter.

$$occupied\ bandwidth = symbol\ rate * (1 + \alpha) \tag{1.3}$$

A perfect (brickwall) filter would have $\alpha = 0$. The value of α has other effects on the signal, including reducing the overshoot of the trajectory and hence reducing the PAPR of the signal fed to the PA. The larger the value of α, the smaller the overshoot and the PAPR, but the wider the occupied bandwidth; a compromise must be reached: for WCDMA a value of $\alpha = 0.22$ is used. This requires about 5 dB of excess power from the PA to complete all transitions, including the overshoot.

The GMSK and EDGE signals use Gaussian filters. The Gaussian filter has a Gaussian shape in both frequency and time domains, and therefore, unlike the RRC filters, is unable to zero out the interference between symbols. The Gaussian filters are of short time response so the interference is limited to adjacent symbols. A bandwidth factor is used in the definition of the Gaussian filter to control the response and interference.

1.1.3 Measures for Digital Communication Signals

Many of the measures of the radio signal are performed in the analog RF domain. These include power and frequency measurements such as occupied bandwidth, spectral spreading, and intermodulation products, that is, power spectral density measurements and power distribution functions. These measures can be made using a spectrum analyzer. Modern spectrum analyzers often include signal analysis software, enabling the constellations and vector diagrams of the trajectories between symbol states to be viewed, and measures such as error vector magnitude to be calculated. These metrics are outlined below.

The bandwidth available for a signal channel is well defined by the federal or international regulations; for example, a GSM signal has a 200 kHz channel bandwidth, and WCDMA is 5 MHz. The levels of allowable distortions are also specified. The distortion can be in-band, or out-of-band, that is, in adjacent allowed signal channels, where it will act as interference.

The measure for the out-of-band distortion is the integrated power in the adjacent channel, called adjacent channel power ratio (ACPR) or adjacent channel leakage ratio (ACLR). The distortion power in the next channel over is called the (first) alternate channel power ratio, or Alt-1CPR, and so on. The ACPR can be measured on a spectrum analyzer, and is measured in dB (decibels) relative to the signal channel power. The power is generally specified as that power integrated in a 30 kHz bandwidth of the modulated signal, though GSM has additional specifications for higher-order distortion products. Many modern instruments have this measurement capability built in, for the different signals and regulatory specifications. At most, the user will need to specify the number of transmitted channels, number of adjacent/alternate channels, and the channel bandwidth and separations. A screenshot of a spectrum analyzer showing two WCDMA carriers separated by an empty channel, and the associated out-of-band distortion signals, is shown in Figure 1.11.

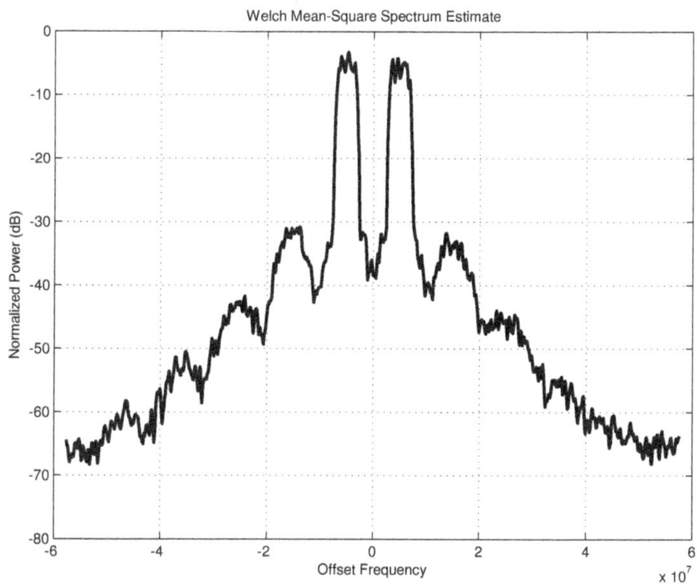

Figure 1.11 Spectrum of a two-carrier WCDMA signal in '101' format, indicating the adjacent and alternate Channel distortion powers. For this signal configuration, the first alternate channels can also be identified as the 3rd-order intermodulation products.

The specified levels of out-of-band distortion for a WCDMA signal are below -45 dBc in the adjacent channels, and below -50 dBc in the alternate channels. At first sight, these might seem quite reasonable maximum values. Converting to absolute power ratio and expressing the distortion as a percentage, the ACPR specification becomes not greater than 0.003% distortion, a figure that makes audiophiles blanch. The customer may also specify 10-dB better performance than this, to accommodate performance variations in several of the components or sub-systems in the PA chain ('margin-stacking'). The specifications for out-of-band distortion and noise for a GSM signal are shown in Figure 1.12. These specifications are significantly more rigorous than the WCDMA or LTE specifications, and are a challenge to meet in wide bandwidth multi-carrier GSM systems (MC-GSM).

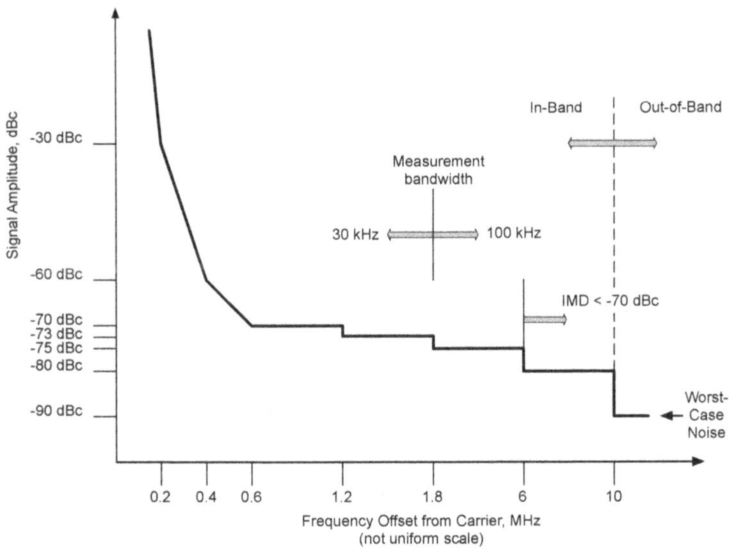

Figure 1.12 Distortion limits specified for a GSM signal.

The in-band distortion of the modulated signal is specified as error vector magnitude (EVM). This is shown schematically in Figure 1.13. The error vector is the difference between the defined constellation point and the actual signal. The EVM is the RMS value of the error vectors over the whole constellation, and is usually expressed as a percentage. Wideband CDMA is relatively tolerant to in-band distortion, as a result of both the spread-spectrum coding and the use of QPSK modulation, in which the constellation points are widely spaced. In contrast, the higher-order QAM signals have constellation points that are more closely spaced, and the EVM specifications are much tighter. Many modern spectrum analyzers have vector signal analysis software built in, and can display the constellations, vector diagrams, and EVM directly.

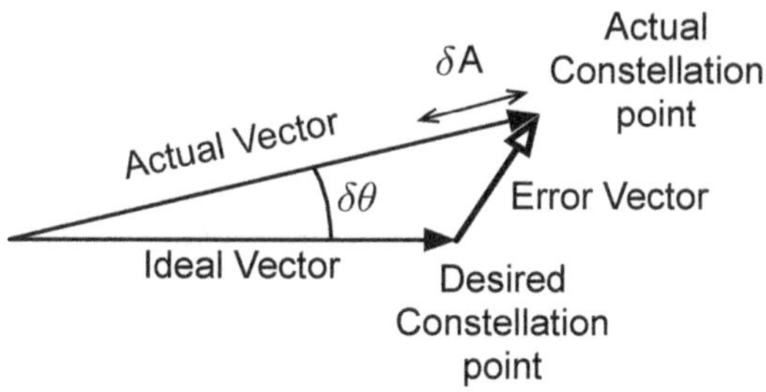

Figure 1.13 Definition of the error vector.

The constellations and vector diagrams of the trajectories of the modulated signals that we have seen show that the power in the signal envelope is not constant. From this we can estimate a power distribution function (PDF): the relative amount of time that the signal is at a given power level. While this is useful information, measures of more immediate interest to the RF PA and DPD designers are the complementary cumulative distribution function (CCDF), and the peak-to-average power ratio, PAPR. The cumulative distribution function (CDF) is the integral of the PDF: the value of the CDF at a given power is the probability the envelope signal has a value at or below that power level. The CDF therefore starts at zero at the origin of the PDF, and has a value of one at the peak power. The CDF is therefore biased towards the lower signal levels, and approaches unity asymptotically. The CCDF is the complement to the CDF:

$$CCDF = 1 - CDF \tag{1.4}$$

and is biased towards the highest signal envelope values, that is, towards the peak of the amplifier power performance, where we normally operate the PA. The CCDF is often expressed as a percentage, and it is common to see the power level specified in terms of dB of power backed-off (below) some PA power metric such as the saturated power, or 1-dB compression point. In Figure 1.14, the probability is expressed as a function of power in dB; here, the peak of the signal is defined at CCDF = 0.01%, that is, 99.99% of the signal is at this power level or below, so this is a statistical measure of the peak power.

The *crest factor* of the modulated signal is defined as the peak signal magnitude divided by the square root of the average power in the signal, that is, the RMS value of the signal power. For these stochastic signals, the peak magnitude is generally specified in terms of the power corresponding to a given value of CCDF,

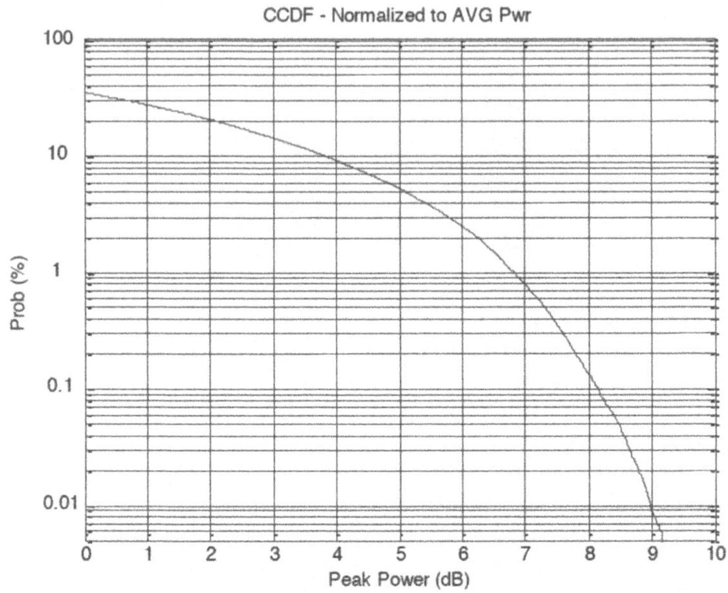

Figure 1.14 The complementary cumulative distribution function (CCDF) as a function of the signal envelope power. In this figure the peak power is defined at CCDF = 0.0001, or 0.01%.

as outlined above. The PAPR is the crest factor, squared: peak power divided by the average power.

1.2 RF POWER AMPLIFIERS: LINEARITY AND EFFICIENCY

The RF power amplifier's job is to boost the modulated signal level so that the signal can reach the receiver at a suitable level to allow detection. Ideally, it should perform this task without adding any extra distortion products and using as little energy as possible: linearity and efficiency are key specifications. A base-station PA will typically be specified at a power level of around 60 to 100 watts. For a single-carrier GSM PA, this represents the 1-dB compression point of the amplifier's output, whereas for a multi-carrier signal, or a 3G or 4G modulated signal with a significant PAPR, this power represents the average power output of the PA: the peak value will be several dB higher, which means the PA will have a peak power capability of several hundred watts. The RF PA has probably the highest current demand of any component in the base-station or handset, and efficient use of that current in generating the transmitted signal is essential, to minimize operating costs or maximize the battery life.

It is worth noting that despite the description above of the digital modulations used for wireless communications, the signal that the RF PA sees is an analog one. The PA is not presented with a series of ones and zeros, but with a continuous, high-frequency waveform whose envelope trajectory will cover the modulation signal.

The 'classical' PA modes of operation are Class A, Class B, Class AB, and Class C. In these modes, the harmonic signals generated by the nonlinear behavior of the power transistor are short-circuited, and so the output voltage is sinusoidal at the fundamental frequency. In the analysis of these modes, the harmonics are assumed to be shorted, and the short circuits are designed into the PA in practice through even harmonic traps, and so forth. Analysis is generally conducted using sinusoidal signals, representing the RF carrier. Modulation is presumed to be a small perturbation around the carrier. The detailed analyses of these modes shall not be repeated here, but can be found in references in the Bibliography (Section 1.5).

The Class A bias, shown in Figure 1.15, can be made to be very linear, especially for smaller signal amplitudes. But the DC-to-RF conversion or energy efficiency is a theoretical maximum of 0.5, or 50% at peak power, falling at lower signal amplitudes. Class A is used for power amplifiers at high millimeter-wave frequencies, but is considered to be too inefficient for PAs in cellular wireless communications applications.

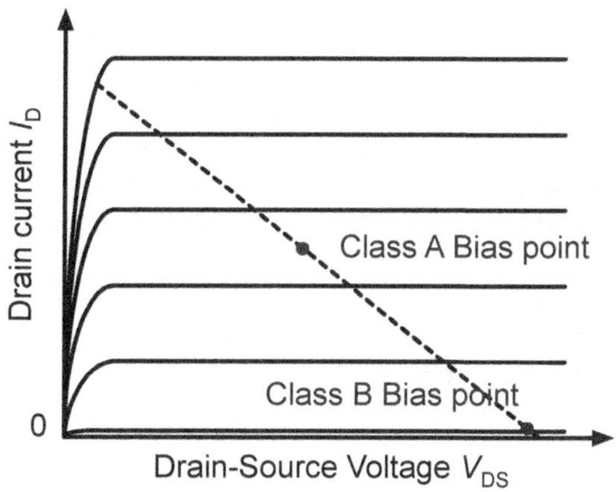

Figure 1.15 Schematic representation of the output IV characteristics of a power transistor, showing the load line and Class A and B bias points.

In Class B, the power transistor is biased just at the turn-on or threshold voltage, so the transistor conducts for only a half-cycle of the RF waveform: the current waveform is a half-wave rectified sinusoid. This waveform generates a significant amount of even harmonic distortion, and some self-bias. Even so, the maximum theoretical energy efficiency rises to $\pi/4$ or 78.6%, compared with Class A, as a result of the reduced DC power dissipation. The even harmonic voltage output can be removed by using a second-harmonic trap in the output matching network. This is often accomplished by using a quarter-wave line for the DC bias input: at the fundamental, the decoupling capacitor at the end of the line is transformed into an open circuit, hence providing no load at the fundamental; at the second harmonic, the line is a half wavelength, and therefore presents a short-circuit load. This condition holds for all even harmonics. The Class B bias point is also shown in Figure 1.15, with the Class A and B conditions sharing the same optimum load value for maximum power output.

The Class A and Class B bias conditions are the only fixed points in the classification of power amplifier modes, representing the 360° and 180° conduction angle conditions, respectively, with all harmonic output voltages shorted. All other amplifier classes or modes represent a range of bias and harmonic conditions.

In practice, it can be difficult to arrange precise Class B biasing conditions, particularly for FET devices, as the threshold voltage variation can be up to ± 100 mV, causing practical variation in the quiescent conditions. A more practical and widely used bias condition is Class AB, where the transistor is biased slightly into the "on" condition. This is usually specified as a constant quiescent drain current, in FETs, with the DC gate bias adjusted to provide the design standing current. This proves to be a useful bias condition, as, irrespective of the precise value of DC gate voltage, the transistor performance at the specified quiescent current is very repeatable. In Class AB, the output current conduction angle lies between 180° and 360°. The optimal load is generally very close to the Class B value, and the achievable energy efficiency can be over 70%, theoretically.

In the Class C bias condition, the power transistor is biased below threshold, so under quiescent conditions, the transistor is switched off. The transistor conducts only when the input signal is large enough to bias the transistor into the on-state. The conduction angle is therefore less than 180°; the output current comprises sharp peaks, a truncated version of the Class B half-wave rectified sine wave. This mode can be very energy efficient, but the gain is low, and the harmonic output and intermodulation distortion can be considerable. Class C tube (or *valve*) PAs were widely used for FM transmission, offering high efficiency and low running costs, and the distortion was not a problem for the constant envelope signal.

For a given power transistor, the optimal load resistance for maximum power output, R_{opt}, can be estimated for Class A and B conditions by drawing a line from the output bias voltage to the maximum current, as illustrated in Figure 1.15. While this value may be used as an estimate, a more accurate load impedance is usually found from load-pull measurements; a schematic source- and load-pull measurements system is shown in Figure 1.16. The source tuner is adjusted to provide a conjugate match at the desired RF power drive level, and the load tuner is adjusted to produce the maximum power or efficiency. In practice, the load impedance plane is generally mapped over many impedance states. This enables performance contours to be created in this plane, indicating impedance contours of equal power, efficiency, gain, and other performance parameters of interest.

Straightforward Class A, AB, and B amplifiers are seldom used nowadays as the final RF power amplifier stage for cellular wireless applications. This is because the energy efficiency of such amplifiers with the high PAPR modulated signals commonly found in modern wireless communications systems is very poor. This is illustrated in Figure 1.17, where the efficiency of the PA at the *average power level* with a signal of 8-dB PAPR ranges from under 10% for Class A to 27% for Class B. This is far below the efficiency currently demanded for commercial base-station PAs. Further, the highest efficiencies with these classic PA modes are reached at only the highest power level, as indicated in Figure 1.17. In consequence, the design of RF PAs has proceeded to seek higher-efficiency architectures.

1.2.1 Harmonic Loading

There are a number of high-efficiency modes of operation of transistors, using harmonic loads to control the output voltage and current waveforms so that their overlap is reduced, to minimize the RF power dissipated in the transistor compared with the power delivered to the load, and hence increase the overall efficiency. The goal is to reduce the time where the current and voltage are simultaneously non-zero. The power amplifiers can be classified on the basis of the harmonic content in the output. The voltage and current waveforms for several common PA classes are shown in Figure 1.18. It is important to realize that these waveforms need to be controlled at the output node of the transistor; the harmonic contributions can be filtered from the fundamental signal later, in the impedance-transforming matching network. The RF properties of the various power amplifier classes with harmonic loads are compared with the classical modes in Table 1.1.

The Class D amplifier is usually described as a form of switching amplifier, where the PA is driven by a pulsed waveform and the output current is switched rapidly between on and off states. It is not to be confused with the audio frequency

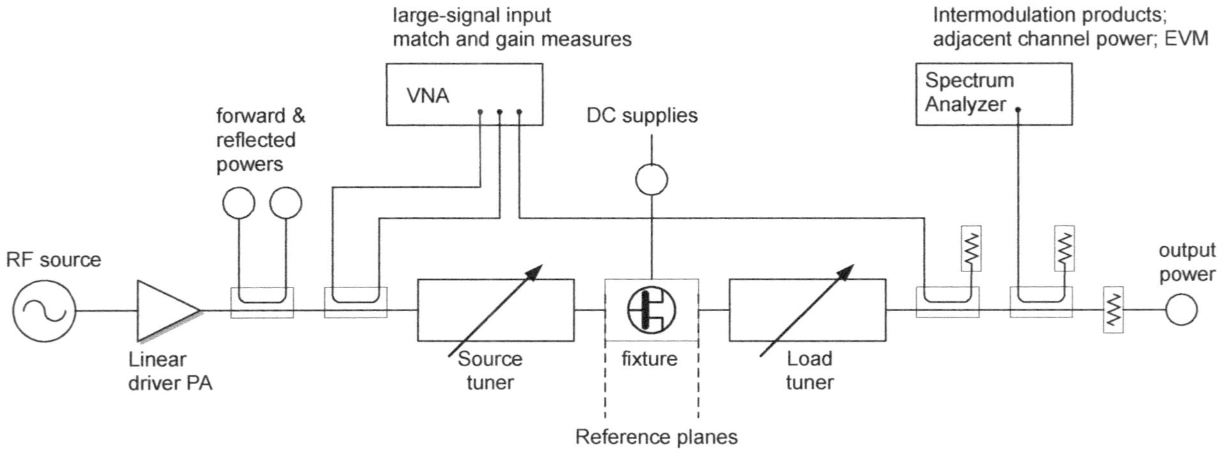

Figure 1.16 A block diagram representation of a source- and load-pull system used for characterizing power transistors to determine the optimum load and source impedances.

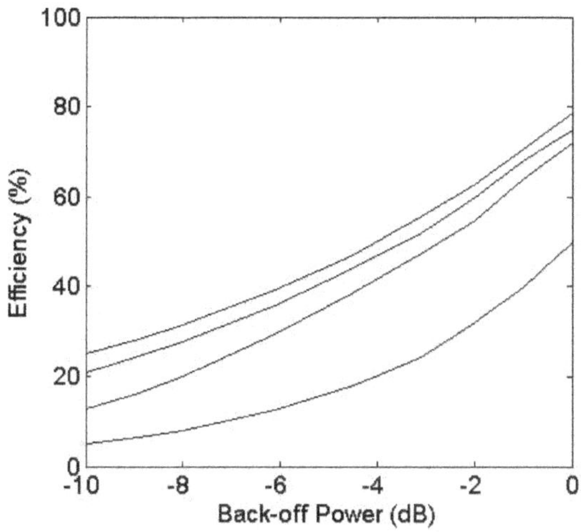

Figure 1.17 The theoretical DC-to-RF conversion efficiencies for a transistor biased in Class A, Class B, and deep Class AB (quiescent currents of 10% and 20% of maximum value), plotted against the RF output power. If the peak value of the modulating signal is arranged to reach no higher than the maximum allowed power, and the signal has a peak-to-average power ratio of 8 dB, the efficiency at the average power is seen to be small.

Class D PAs, which are often found in automobile stereo systems: these are amplifiers driven by a pulse width modulated signal and then lowpass filtered to yield the multi-octave audio signal. These amplifiers are also very efficient. We shall describe the RF Class D operation briefly, next. The Class E, and, to some extent, Class F PAs are also frequently described as switching amplifiers. The original implementation of the Class E amplifier was a switching transistor at a relatively low frequency of 1 MHz. The common implementations of Classes E and F at RF and microwave frequencies accommodate the fact that the power transistors at these frequencies do not behave as ideal switches, but the loadline presented can be designed to present a switch-like trajectory for the transistor output current over the RF cycle. These modes will also be outlined next.

The ideal switching amplifier switches the output current from on to off, under control of the input voltage. The ideal switch has zero on resistance, so both current and voltage waveforms at the load are square. There is no overlap between the voltage and current waveforms, so this amplifier converts the DC to RF with 100% efficiency. A lot of the RF energy is in the harmonics to create the square waveform, so the maximum efficiency at the fundamental is 81%, when the pulse waveform is

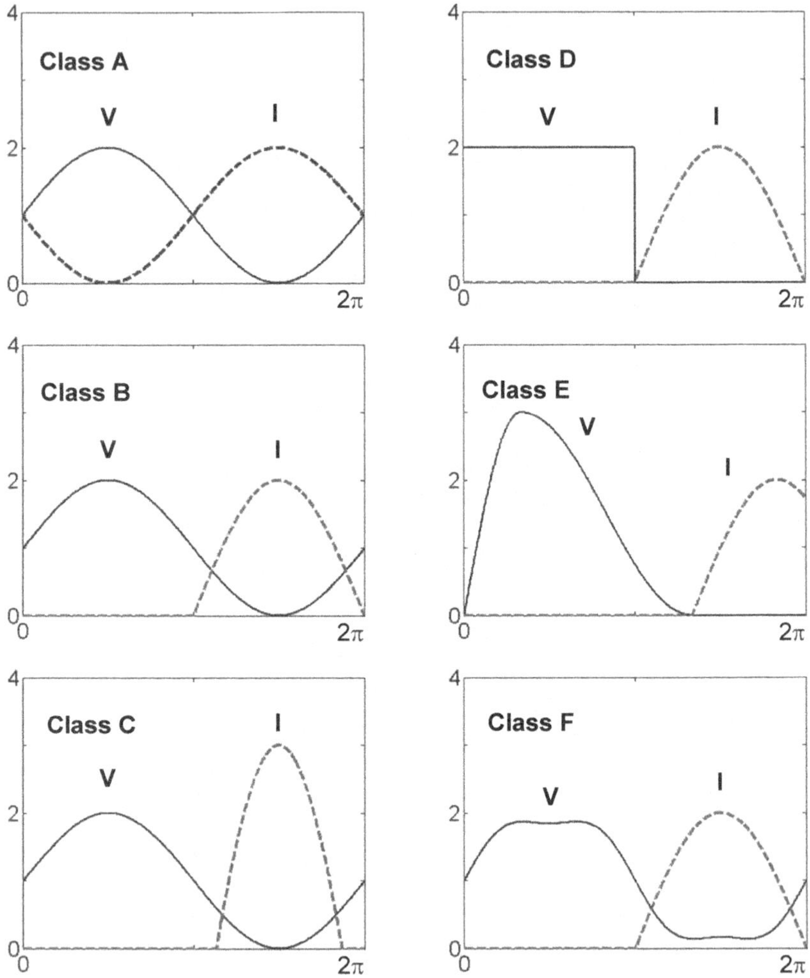

Figure 1.18 The current and voltage waveforms for several common generic classes of power amplifier, after Raab [1]

Table 1.1
The Properties of Various RF Power Amplifier Classes, with Class A as the Reference

Class	A	AB	B	C	D	E	F
Power (dB-Class A)	0	+0.2	0	-3–6	+1	< 0	> 0.5
Input drive required	0	0 to 6	6	> 6	> 6	~ 3	6
Maximum efficiency	50	50–78	78.5	100	100	100	< 100
Efficiency 6-dB OBO	12	25	35	45	75	80	75
Linearity	excellent	good	good	poor	poor	poor	poor
Harmonics (I/V)	1/1	2,3/1	2/1	lots/1	2/odd	lots/lots	2/3
Peak voltage	$2V_{dd}$	$2V_{dd}$	$2V_{dd}$	$2V_{dd}$	$2V_{dd}$	$> 3V_{dd}$	$2V_{dd}$

exactly square, with a 50% duty cycle. A resonator can be used to filter out the harmonics at the load.

The Class D amplifier is a refinement of this switching approach, using two transistors in a push-pull arrangement. In practical implementations, a current-switched approach is typically used, shown in Figure 1.19. This is a convenient arrangement for LDMOS transistors, which have the source terminal of the transistor connected directly to ground, through the package. The resonator is placed across the load, connecting the outputs of each transistor; this provides a short-circuit termination for all current components except the fundamental. From Figure 1.19, we also see that there is no overlap between the current and voltage waveforms at the transistor output; further, the resonator ensures that there is no power dissipated in any harmonics, so the theoretical DC-to-RF conversion efficiency is 100%.

In practice, the Class D switching mode has a number of limitations. For genuine switching operation, the transistors need to have a very short transit time ($1/Transitionfrequency, f_T$), typically not the case for high-power transistors, though newer technologies such a gallium nitride (GaN) FETs and high-voltage heterojunction bipolar transistors (HV-HBTs) can show high f_Ts, and could be considered as candidates for RF switching amplifier applications. The large capacitances of the power transistors will also increase the switching time. In principle, the current-mode Class D (CMCD) switches at the voltage zero-crossing times, so the energy storage associated with the device output capacitance is eliminated; in practice, the signal modulation will result in nonzero voltage switching times, contributing to the switching delay through $C_{out}V^2$ losses. Further, the practical implementation of the circuit, using nonideal baluns for the input and output balanced signals, and the harmonic loads, will also reduce the efficiency. Practical examples

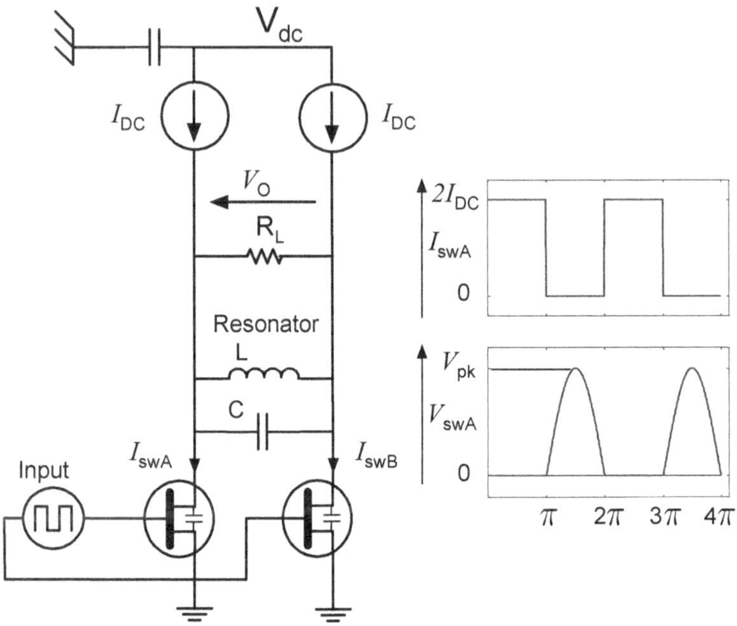

Figure 1.19 The schematic for a current-mode Class D (CMCD) RF power amplifier, showing the output resonator. The current and voltage waveforms at the transistor drain are also presented.

of CMCD for a handset PA [2] and a medium power PA [3] have demonstrated efficiencies greater than 70%.

The Class E and F amplifiers are often called switching amplifiers, but in many cases these amplifiers are presented with the appropriate harmonic terminations so that the instantaneous loadline seen by the transistor approximates to a realistic switching trajectory. This is illustrated in Figure 1.20. When these amplifiers are used in a more traditional RF context, driven by a sinusoidal carrier rather than a digital switching signal, for example, it seems easier to think about these PAs as harmonically-loaded or -tuned amplifiers. A schematic circuit for demonstrating these amplifiers is shown in Figure 1.21. Several circuit variables are open for adjustment: the fundamental load, the loads at second and third harmonic, and the transistor bias. The instantaneous output voltage for such an amplifier can be written:

$$v_{out}(t) = V_{dc} - V_1 \left[\cos(\omega t) + h_2.\cos(2\omega t) + h_3.\cos(3\omega t)\right] \qquad (1.5)$$

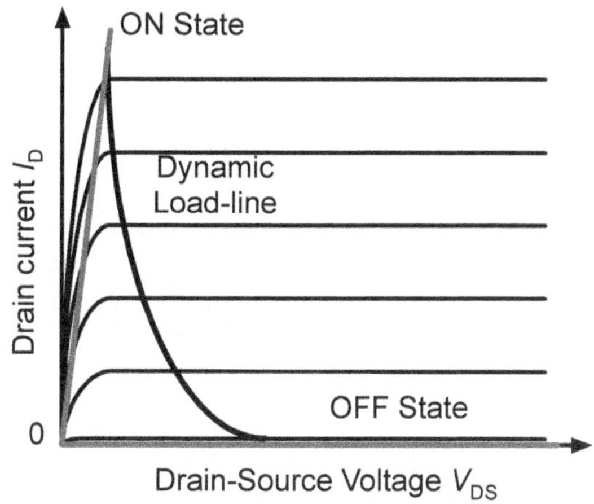

Figure 1.20 A generic dynamic loadline created by harmonic loads that approximates to a switching trajectory in a real transistor.

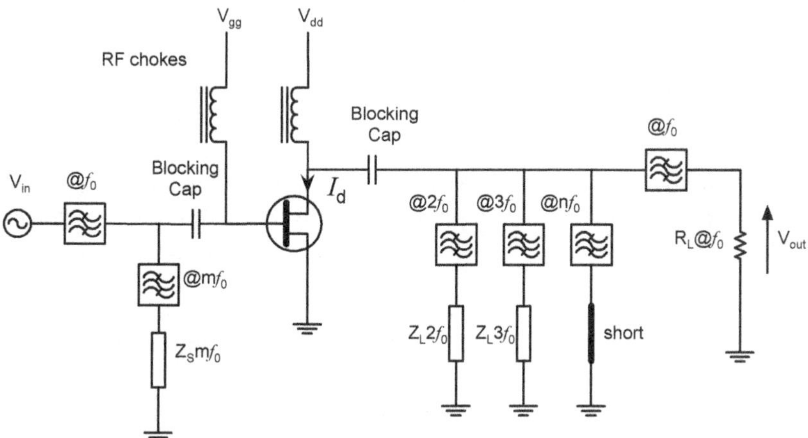

Figure 1.21 Schematic circuit of a general harmonically-loaded PA, using up to three harmonics. All higher harmonics at the load are shorted; all harmonic content at the source is matched.

where $h_2 = V_2/V_1$ and $h_3 = V_3/V_1$ reflect the harmonic content. The broad classes of the harmonically tuned PAs can then be expressed as in Table 1.2.

Table 1.2

The Various Combinations of Second and Third Harmonic Tuning and the Relationship with the Power Amplifier Classes E and F.

$h_2 = 0$	$h_3 = 0$	Bias is A, AB, B, C	Classic modes, tuned PA
$h_2 = 0$	$h_3 \neq 0$	Bias is AB, B	3rd harmonic tuned: Class F
$h_2 \neq 0$	$h_3 = 0$	Bias is A, AB, B	2nd harmonic tuned: Inverse Class F
$h_2 \neq 0$	$h_3 \neq 0$	Bias is AB, B, C	2nd & 3rd harmonic tuned: Class E

In Class F, the third harmonic is tuned to 'flatten' the sinusoidal RF voltage, allowing a higher peak voltage at the fundamental, and hence higher output power. The transistor is usually biased in Class B or deep Class AB, so the RF current contains only second (even) harmonic components. Hence, no power is dissipated at the harmonics. This is all achieved at no change in the DC power supplied to the transistor, and hence there is an increase in the PA efficiency. The 'Inverse Class F' condition swaps the roles of the current and voltage waveforms: the second harmonic load is tuned to create an even-harmonically rich voltage waveform; the transistor is biased nearer to Class A to achieve a symmetrical current waveform, which contains mostly odd harmonics. Again, power is only generated at the fundamental, and the efficiency can be high, despite the Class A bias. Reported efficiencies for Class F and Inverse Class F can be up to 80%. One drawback with the second-harmonic tuning of the voltage is that the peak voltage can be up to three times the drain (or collector) supply voltage: this can lead to voltage breakdown in the device, or be a concern for long-term reliability, unless the supply voltage is reduced accordingly. Such a bias voltage reduction results in lower power output capability.

Both second and third harmonics are tuned for Class E power amplifiers. The harmonic loads need to be carefully tuned so that there is the minimum overlap between the RF voltage and current waveforms. The bias condition is usually close to Class B, though Class C can be used, as this nonlinear amplifier mode will generate a number of harmonics, available for tuning. Again, efficiencies of 70–80% have been demonstrated with second and third harmonic loads for Class E amplifiers.

The amplifier configurations for Class F, Class E, and Class C PAs when loaded by a finite number of harmonics have been analyzed by Raab [4]. While the peak efficiencies for these modes are high, they are attained only at or close to the maximum output power, and the efficiencies fall rapidly with power back-off, in a manner similar to the classical, harmonically-shorted modes. Although this does give some improvement over, say, Class AB, for the high PAPR modulated signals, the benefit is still quite small. To overcome this basic drawback with these amplifier modes, alternative PA architectures have been investigated and developed for wireless communications applications. The principal techniques are outlined next.

1.2.2 The Doherty Amplifier

The Doherty PA architecture was first reported by W. Doherty in 1935 [5]; of course, Mr Doherty used high-power tubes in his original invention. The Doherty PA is the predominant power amplifier architecture in use in wireless infrastructure base-stations today. Its advantage is the ability to maintain reasonably high efficiency even when backed-off from the peak power by several dB. The basic Doherty amplifier structure is shown in Figure 1.22, and resembles the structure of a balanced amplifier. The main difference is that the Peaking amplifier is arranged to switch on only at high power levels. The action of the Doherty PA can be described in terms of active loadpull of the Main amplifier by the Peaking amplifier: *load modulation*.

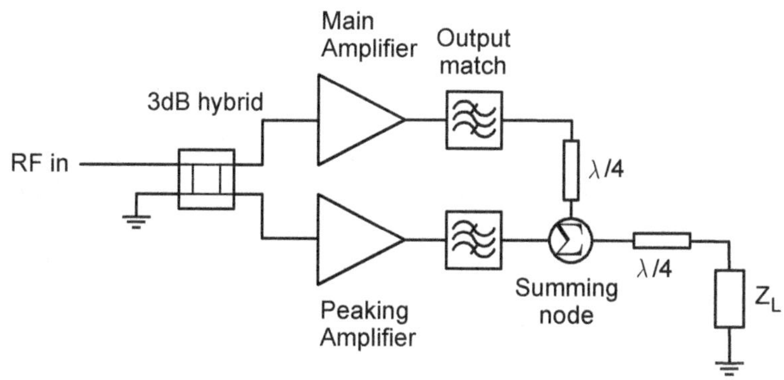

Figure 1.22 Schematic circuit of the Doherty PA.

We shall briefly describe the action of an ideal Doherty amplifier, in which the Main and Peaking amplifiers are sized identically and produce the same maximum power into the same optimum load. We shall consider the Main amplifier to be biased in Class B, for simplicity. In Class B, the peak voltage swing is the drain bias, V_{dd}, and the peak current at the fundamental is $I_{max}/2$, yielding an optimum load resistance for maximum power, R_{opt}, of $2 * V_{dd}/I_{max}$, and an ideal efficiency of $\pi/4$ or 78.5%. This is the classic behavior of the Class B PA operating at its maximum linear power.

The Main amplifier is operating for all signal levels. At lower powers, before the Peaking amplifier is switched on, the Main amplifier is presented with a load of $p * R_{opt}$, and the maximum current that it can drive is therefore $V_{dd}/(p * R_{opt})$, or $I_{max}/(2*p)$. The maximum power that the Main amplifier can deliver into this load is $V_{dd} * I_{max}/(2 * p)$, but the efficiency at maximum power is still $\pi/4$ or 78.5%, as the factor p cancels in expressions for RF and DC powers. For the particular case we are considering, with Main and Peaking amplifiers identically sized, the factor p is 2: the load presented to the Main amplifier is $2 * R_{opt}$.

At the point where the Main amplifier first reaches its maximum efficiency point for the load of the $p * R_{opt}$, the Peaking amplifier kicks in. Its job is to adjust the load seen by the Main amplifier, so that the Main amplifier continues to operate at maximum voltage and current swing, and, hence, maximum efficiency. This can be seen schematically in Figure 1.23. The load modulation effect is shown in simple form in Figure 1.24: the current sources represent the output current sources of the Main and Peaking amplifiers. With the Peaking amplifier turned off, the Main amplifier sees the load of $p * R_{opt}$; as the Peaking amplifier turns on, it drives current into the load, so the effective load seen by the Main amplifier increases. Actually, this is pulling the load in the wrong direction, and the genius of the Doherty amplifier is the addition of the quarter-wave transformer. This is shown in Figure 1.25, with the appropriate values for the load and transformer impedance.

The load impedance is $R_{opt}/2$. With the Peaking amplifier turned off, this load is transformed into $2 * R_{opt}$, as desired, at the output of the Main amplifier. The maximum power produced by the Main amplifier in this condition is one-half of its maximum power capability, or $P_{max}/2$. When both Main and Peaking amplifiers are operating at full power, each amplifier supplies $I_{max}/2$ into the load impedance, so each amplifier sees an effective load of R_{opt}; the quarter-wave transformer has no effect on this value, so the Main amplifier is also presented with R_{opt} at its output. The total power output from the Doherty PA is $2 * P_{max}$, so the Peaking amplifier is operating only over the highest 6-dB range of the delivered power. Over this range, the Doherty PA efficiency remains close to the peak efficiency of the Main amplifier, as the Peaking amplifier is operating in high efficiency Class C mode.

In practical Doherty PAs, the Peaking amplifier is realized by a power transistor operating in Class C bias. The Peaking amplifier is then switched on by the magnitude of the drive signal itself. By sizing the Main and Peaking amplifiers

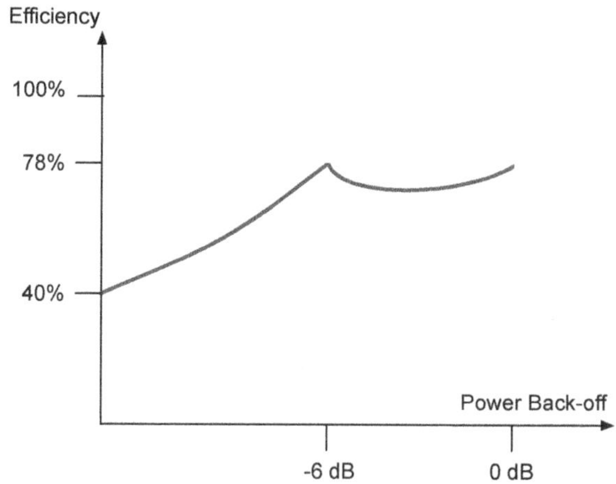

Figure 1.23 Operation of the ideal Doherty PA; the Main amplifier is switched on at all power levels, and delivers peak efficiency at $P_{max}/2$. At this point, the Peaking amplifier switches on, and modulates the load presented to the Main amplifier, so that the Main amplifier continues to operate at maximum efficiency. The peak output power delivered by the Doherty PA is $2 * P_{max}$, with the Main and Peaking amplifiers providing half the power each.

appropriately, and adjusting the value of the Class C bias, the switch-on point can be adjusted to give a peak efficiency range different from the 6 dB described by the simple theory above. Further, in practical PAs, the optimum load resistances for maximum power and efficiency are often different, and so a compromise between the initial and pulled load values is often made to optimize the Doherty PA performance. The efficiency of the Doherty PA can be improved further, by using one of the harmonically-loaded PA structures outlined in Section 1.2.1; Class F has been used successfully in the high-power Doherty PA.

The Doherty architecture, with several variations, is the pre-eminent PA structure deployed in cellular base-stations today. The ability to maintain high efficiency over a range of high output powers means that the Doherty PA can handle the high-PAPR signals typical of modern radio communications, without suffering the efficiency hit associated with operating the conventional PA modes at 6–8 dB backed-off from peak power output.

The bandwidth of typical Doherty PAs is in the range of 5 to 10% fractional bandwidth. This limitation comes from a number of sources. The input and output matches of the power transistors are generally tuned for maximum performance over a limited bandwidth: increasing the RF bandwidth will generally trade off some

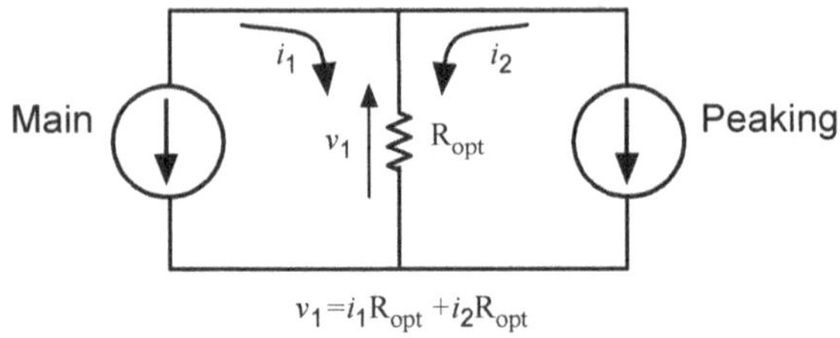

Figure 1.24 The load modulation action: the Main amplifier sees a load of $R_{opt}/2$; as the Peaking amplifier switches on, it also provides current into the load, so the effective load seen by the Main amplifier increases.

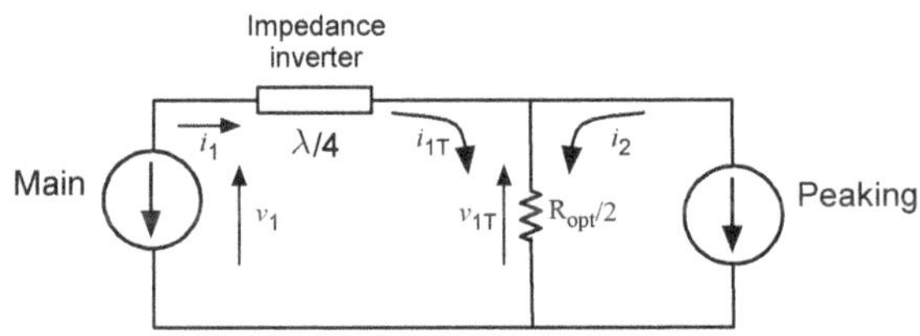

Figure 1.25 The Doherty PA action: the addition of the quarter-wave transformer enables the Peaking amplifier to modulate the load seen by the Main amplifier and keep the Main amplifier operating at maximum efficiency.

gain or power. The quarter-wave transformers also present a frequency limitation. These are naturally narrowband devices and the effects of the phase variation with frequency can compromise the Doherty behavior.

1.2.3 The Polar Amplifier

The polar modulation technique uses changes in the DC bias in response to the magnitude of the modulation signal, and essentially operates the PA at or close to maximum voltage and current swing, and hence peak efficiency, at all times. In *Drain Modulation*, the drain bias is adjusted (in FETs); in contrast to the Doherty PA, the load line presented is kept constant in value, but is moved along the axis of the drain voltage to ensure the transistor is operating at peak efficiency. This is shown in Figure 1.26. There are basically two variants of the drain modulation technique, which are very similar: envelope elimination and restoration (EER), and envelope tracking (ET).

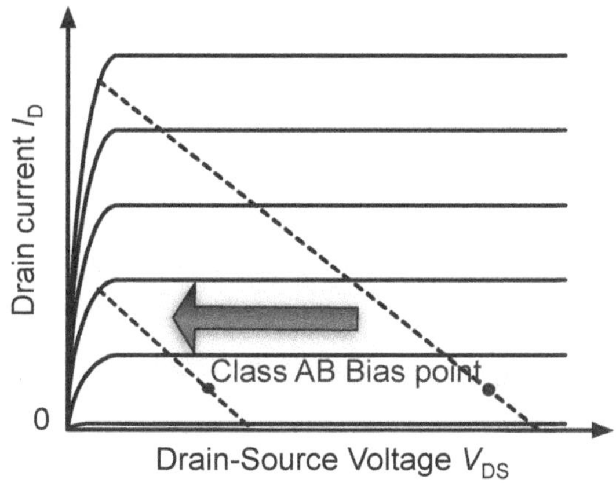

Figure 1.26 The Drain Modulation principle: the drain bias voltage is adjusted dynamically so that the RF voltage swing covers the whole loadline, and the amplifier is operating at maximum efficiency.

The EER technique was first presented by L. R. Kahn in 1952 [6], and is represented in Figure 1.27. In this amplifier, the generic amplitude- and phase-modulated signal is first split into two paths. In one path, the signal is passed through a high-gain limiting amplifier, which effectively removes the amplitude variation from the signal, to give a constant amplitude phase-modulated signal. As mentioned earlier, the constant-amplitude signal is a very friendly signal for the RF PA: the power amplifier can be driven continuously at peak power and maximum efficiency. In the second path, an amplitude detector is used to strip the RF from the signal and leave the amplitude-modulated envelope signal. This envelope signal is then used to control the drain bias of the RF PA, effectively adding the amplitude modulation back into the output. As the PA is operating in a high-efficiency condition at all times, the overall efficiency is high.

This PA architecture is, in principle, capable of accommodating the high-PAPR signals found today in wireless communications systems. The main practical difficulty with this technique is that the power transistor is required to be switched off when the envelope signal is zero. As noted earlier, it is difficult to charge and discharge quickly the large capacitances in the power transistors, resulting in the generation of distortion. One method of overcoming this drawback is to pre-process the envelope data before driving the drain bias modulator, so that the envelope voltage remains above a given minimum value. The pre-processing is usually of a form of changing the trajectory of the signal between constellation points to avoid

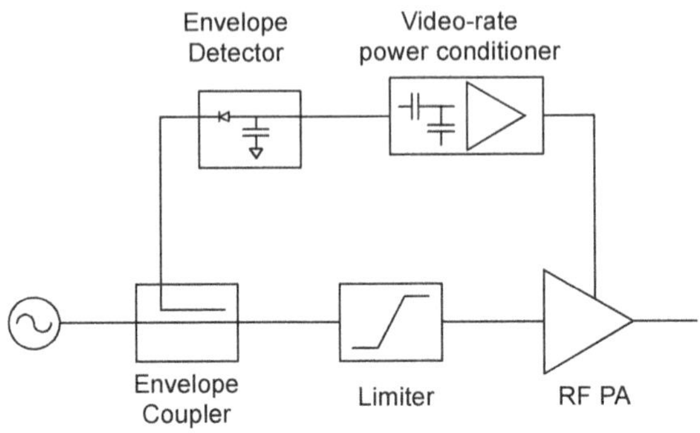

Figure 1.27 Schematic representation of the envelope elimination and restoration amplifier.

crossing zero. This introduces some amplitude and phase distortion into the signal. A simple alternative technique is to apply a bias offset ensuring that the transistor is not switched off, and does not enter the 'knee' region of the I-V characteristics. This will reduce the PA efficiency.

In practice, the EER technique is not used for wireless infrastructure PAs. The alternative drain modulation technique, ET, has been the subject of some interest, particularly for application in higher-efficiency system architectures. The envelope-tracking circuit, shown in Figure 1.28, is similar to EER, except that the limiting amplifier is omitted. This means that the PA sees the full modulated signal, but the drain bias is still modulated by the envelope signal to keep the maximal signal swing and keep the amplifier operating at high efficiency. The PA efficiency as a function of the envelope magnitude can be estimated from efficiency measures taken for power drive-ups over a range of drain bias voltages. An example is shown in Figure 1.29.

In ET PAs, the transistor is operated well into compression to give the highest attainable efficiency that is tracked with the bias voltage. Often these PAs are harmonically-loaded to maximize the efficiency at high power, as outlined in Section 1.2.1, with Class E being a good candidate. One drawback with operating the transistor in saturated mode is that the power gain is reduced: from a system viewpoint, this means that the driver amplifier will have to be made larger to overcome this gain deficit, and this will have a penalty on the system efficiency.

As with EER, it is possible for the envelope signal to fall to zero, switching off the drain bias and hence the transistor. Again, the envelope signal, or even the complete modulated signal, can be pre-processed to prevent zero crossings in the envelope trajectory, at the cost of increased distortion.

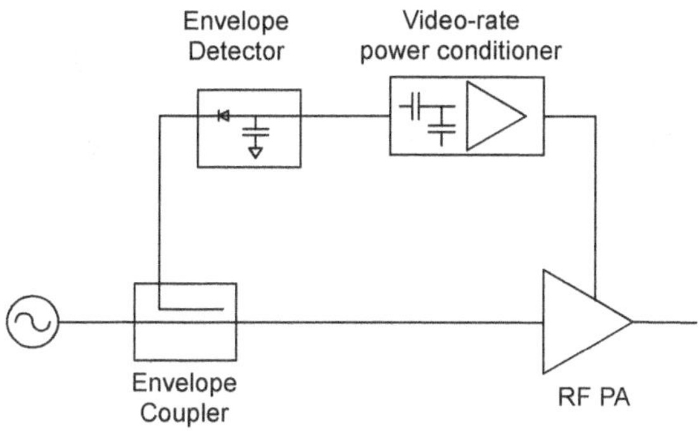

Figure 1.28 Schematic representation of the Envelope Tracking amplifier architecture.

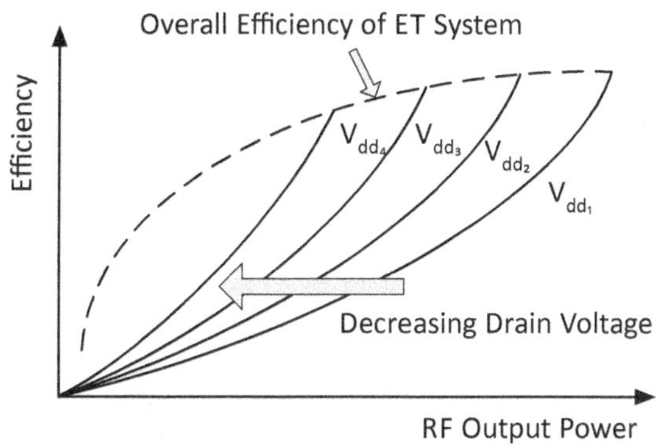

Figure 1.29 Efficiency versus output power for a range of (static) drain biases. The dynamic efficiency locus for ET can be estimated as shown.

The ET technique is 'frequency agnostic.' In other words, the drain modulation is a function of the envelope signal, not the RF signal. In principle, ET can be applied across several frequency applications using the same drain modulation hardware (and signal processing software). This makes it an attractive proposition for multiband PA applications, where the communication signal can appear at widely-spaced carrier frequencies, although the modulation is the same. Examples include TDS-CDMA in China and the DVB-T deployments.

One of the drawbacks with ET systems, currently, is the limited bandwidth available in the drain modulator. This is an outcome of the challenge in the design of the modulator, which must supply the large peak currents demanded by the PA, and accommodate the fast voltage excursions between the constellation points. Typically, DC-DC converters operate in the hundreds of kilohertz regime, not the tens or hundreds of megahertz that characterize 3G and 4G communications signals. Even so, drain modulators with over 20-MHz bandwidth have been demonstrated in the laboratory [7].

1.2.4 Other High-Efficiency PA Techniques

The Outphasing technique, described by Chireix in 1935 [8], combines two PAs driven with signals of different phases, which are controlled so that adding the two PA outputs results in the signal of the correct amplitude. The impedances presented to the PAs are time-varying, and alter their efficiency characteristics. The amplifiers are generally Class B PAs driven into saturation and the overall efficiency can approach that of the Class B amplifier. After some interest in the early 2000s, when high-efficiency performance was demonstrated, outphasing has lost ground to the Doherty PA, which is viewed as a more repeatable and robust design.

Digital amplifiers have been proposed as the answer to the search for high efficiency for many years. The simple digital switching amplifier was outlined in Section 1.2.1. More sophisticated approaches have been presented in recent years, including PAs based upon *Delta-Sigma* D/A converters. While the sophistication of the digital switching and modulation is undeniable, a major problem tends to arise with the reconstruction filter that must be placed after the final PA stage. It would appear that in many designs, a significant amount of RF energy is present out-of-band, and when this is filtered out, the fundamental efficiency is relatively low: that is, not displaying an overwhelming advantage over present analog techniques. Digital PAs may make their mark as enhancements of current analog techniques, to improve the phase control in Chireix, for example.

1.2.5 High-Efficiency PA Summary

The high-efficiency PA architectures described in this section – the harmonic loading, Doherty, Chireix (outphasing), Kahn (EER), and envelope tracking – all run the power transistor at close to peak power, in compression or even saturation to achieve this high efficiency. This is a very nonlinear regime of operation, and even if a low-pass output match is used to prevent the harmonics from escaping the PA, there will still be considerable distortion in-band, as intermodulation, power leakage into the adjacent channels, or constellation errors in the signal channel. The conclusion is, therefore, that some form of linearization technique is essential if the PA is to comply with the regulatory limits outlined in Section 1.1.3, 'Measures for Digital Communication Signals.'

1.3 LINEARIZATION TECHNIQUES FOR RF PAS

Distortion in amplifiers has been a problem since the early days of telephone communications. The feedforward and feedback circuit techniques for reducing distortion in repeater amplifiers were patented by Harold Black in 1928 [9] and 1937 [10], respectively, although the feedback work was initially carried out some time earlier. Feedback has been used very widely in many analog and digital systems since its inception, to improve the stability and distortion performance of the circuit or system. It is probably the more familiar circuit concept.

1.3.1 Feedback

The basic idea of a feedback circuit is to sample the signal at the output, and then apply an attenuated version of this signal at the input, in antiphase with the original signal. This simple notion is shown in Figure 1.30. The forward amplifier has a gain of A, and the feedback path has a gain of β, in practice, a gain of much less than one. A couple of lines of simple algebra returns the familiar expression for the system gain in the presence of (negative) feedback:

$$\frac{v_{out}}{v_{in}} = \frac{A}{1+\beta A} \tag{1.6}$$

This simple technique has found widespread application in low-frequency circuits, where the time delay around the loop is short enough to be discounted. At RF, this time delay through the PA may be comparable to the signal frequency, and the practical ramifications of causality are a concern. Further, the presence of parasitic capacitance and inductance can introduce other unwanted feedback paths into the system, complicating the analysis and design. Naive application of feedback at RF and microwave frequencies can often result in unwanted oscillations. Further, at these frequencies, the available gain and bandwidth from the device are usually limited, which means that since the feedback factor itself $A\beta$ is quite small, any potential benefits are therefore only modest.

Consequently, the application of feedback in RF PAs is usually carried out in the modulation domain, comparing the input signal and the sampled signal

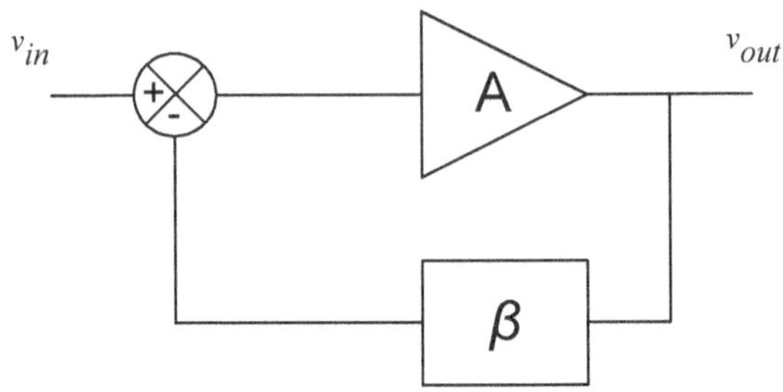

Figure 1.30 Simple schematic showing the feedback principle.

after down-conversion to some intermediate frequency. These feedback methods are generally called *indirect* feedback. The techniques are the precursor to digital pre-distortion.

1.3.2 Polar Feedback

In this technique, the *envelope* amplitude and phase are compared individually to provide the correction. The schematic representation of this technique for an RF PA is shown in Figure 1.31. The amplifier output is sampled, attenuated, and down-converted to the baseband IF frequency for comparison with the input signal. The input and feedback signals are both split, for comparison of the magnitude and phase qualities of the signals. The envelope phases are compared using a wide bandwidth phase-lock loop (PLL) circuit, whose output is a voltage that is proportional to phase difference. This slowly-varying voltage is used to drive a voltage-controlled oscillator (VCO) that then produces the desired instantaneous frequency, complete with the frequency-modulation component. The envelope amplitudes are detected and compared using a wide bandwidth (video) differential amplifier. This amplifier drives a modulation amplifier, which impresses the amplitude modulation onto the VCO output. This modulation amplifier can be used to drive the gate terminal of the PA, to adjust its gain dynamically, or it can also be used to drive the drain bias voltage of the PA, as in polar modulation using EER or ET techniques.

As noted above, the comparison circuit must be of wide bandwidth to capture fully the IM distortions present in the output signal. On the positive side, as the VCO and amplitude modulator are in the feedback loop, their linearity performance is not critical, though we are now also having to correct for their deficiencies.

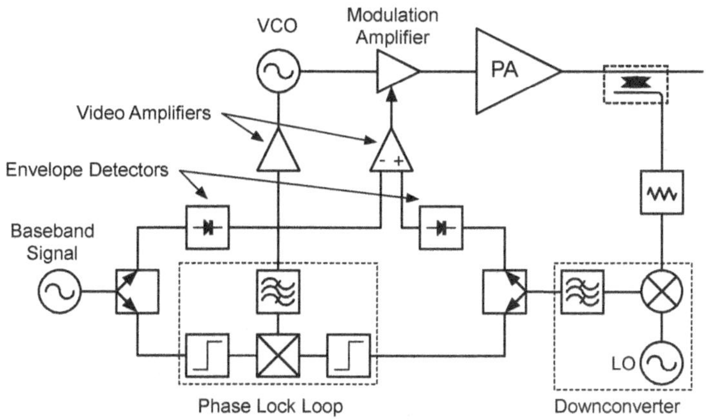

Figure 1.31 Schematic showing the polar feedback principle.

1.3.3 Cartesian Feedback

In this technique, the correction is carried in the *I*-*Q* domain, directly on the in-phase and quadrature components of the modulating signal. The schematic for this technique is shown in Figure 1.32. The PA output is sampled and down-converted using quadrature demodulators to obtain the *I* and *Q* components at the baseband frequency. The components are compared directly with the baseband *I* and *Q* components at the inputs of wideband differential amplifiers. The outputs from these amplifiers are then up-converted using quadrature modulators to form the input to the power amplifier. The same LO is often used for both up- and down-conversion processes.

This is a standard simple architecture that can be applied to any modulation. The key to good performance is that the in-phase and quadrature paths can be very well matched, resulting in good AM-to-PM performance; there is no need for a PLL/VCO system as tracking and phase errors are eliminated. Differential video amplifiers with high bandwidth and excellent noise and distortion performance are readily available, as are vector modulators and demodulators. Typically, the Cartesian loop architecture is an entry point into digital signal processing and pre-distortion in the baseband.

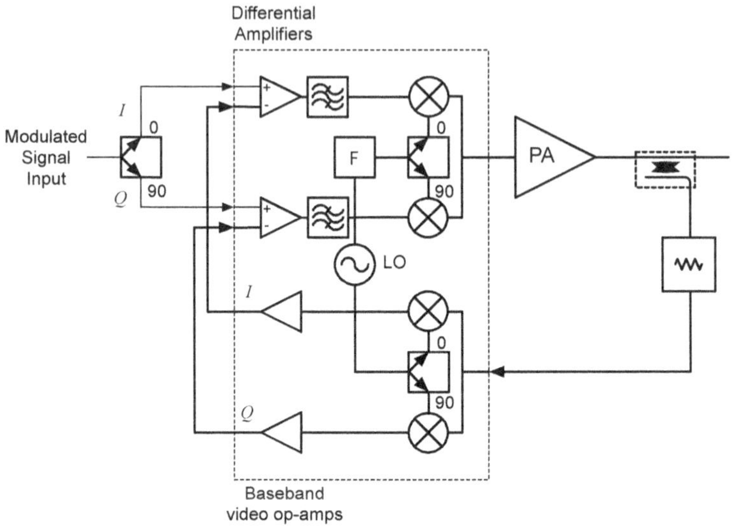

Figure 1.32 Schematic showing the Cartesian feedback principle.

1.3.4 Feedforward

As noted earlier, feedforward has been around for even longer than feedback. The essential difference between the two techniques is that in feedforward the error signal is compared at the output of the amplifier, whereas in feedback the comparison is at the input. The schematic circuit is shown in Figure 1.33.

The feedforward circuit works in the following way. A delayed version of the undistorted input is compared with a sampled and attenuated version of the output of the power amplifier. The result of this comparison is a measure of the gain and phase distortion introduced by the PA. This distortion signal is then passed through an error amplifier and combined with a delayed version of the distorted output of the PA, to yield a distortion-free signal. The delay lines must be carefully aligned to match the group delays of the PA and of the error amplifier. The correction process is direct, and performed at RF. The bandwidth is defined by the bandwidth of the PAs and the accuracy of the error tracking. As a result, well-designed feedforward PAs can accommodate wide bandwidth signals and provide up to 30 dB of correction.

To work well, the feedforward PA requires accurate gain and phase tracking in all elements of the loops. This is often achieved in practice through the use

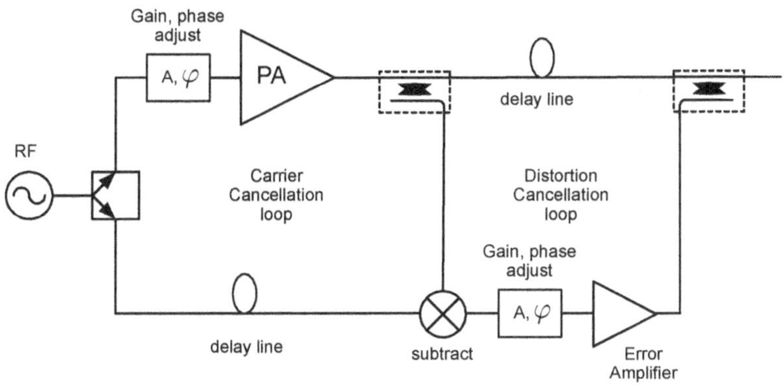

Figure 1.33 Schematic showing the basic feedforward principle.

of a pilot tone, which is placed outside the signal channel, but close enough that gain and phase behavior are well-characterized by the action of this tone through the feedforward PA: the delay loops can be adjusted electronically based on the information from the pilot tone behavior [11].

The feedforward architecture also requires a second power amplifier: the error amplifier. This PA needs to be linear and well-behaved, since any tracking or gain errors that are introduced by the error amplifier are uncompensated. The error amplifier must also be powerful enough to overcome the coupling factor at the output combiner of the overall PA. These requirements generally mean that the error amplifier is sized similarly to the main PA: this introduces both cost and efficiency issues that need to be faced.

1.3.5 Pre-distortion

In contrast to the feedforward and feedback techniques outlined above, in pre-distortion we actually add distortion components to the signal input to the PA, in such a way as to cancel the inherent nonlinearity of the PA so that its output is a linear replica of the original input signal. It is the phrase "in such a way" that presents most of the practical difficulty in this exercise. The pre-distorter can be thought of as a component that is inserted into the amplifier chain immediately in front of the power amplifier, transforming the input signal into this new signal, as indicated in Figure 1.34.

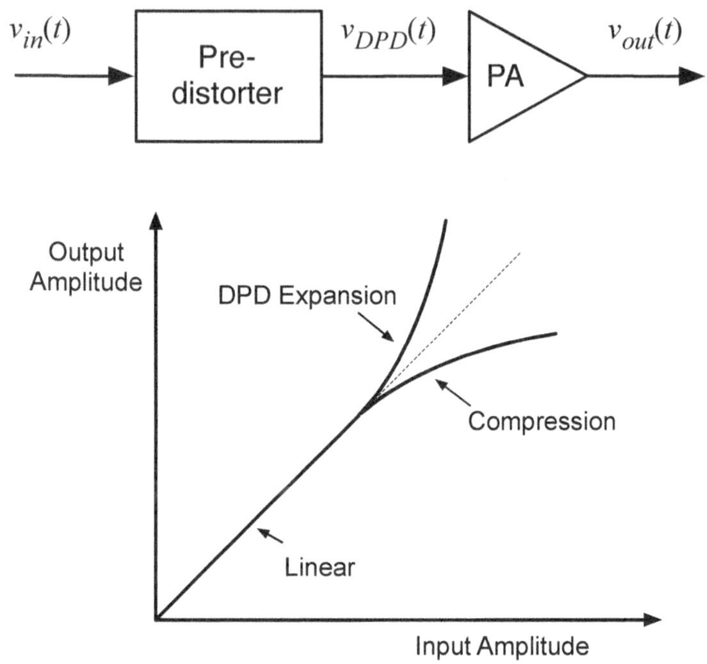

Figure 1.34 A prototype pre-distorter and PA arrangement. The Gain or AM-to-AM responses of the power amplifier and the pre-distorter are shown, along with the resulting linear output, up to the maximum saturated output power available from the PA.

The principle of pre-distortion can be explained simply; here we will consider only nonlinearities in the power amplifier's gain function, though phase nonlinearities can be handled in the same manner. The AM-to-AM or gain function of the amplifier is shown in Figure 1.34, indicating that at high input powers the amplifier enters compression and, eventually, saturation. It is clear from this graph that the gain response is nonlinear. In fact, it should be noted that in a typical RF PA, the transfer function is nonlinear even at low input powers: we still obtain measurable intermodulation products when well backed-off from the saturated power output. To obtain a linear AM-to-AM response, we need to apply a higher amplitude signal at the input of the PA, to compensate for the compressive characteristic. This superlinear function is also shown in Figure 1.34, along with the resulting linear gain response. This pre-distortion action can only apply at power levels below the saturated output power of the PA: pre-distortion does not give us any more output power. In fact, if we apply a higher input power than the point of saturation of the PA, we end up with more distortion because of the sharper characteristic.

We can also view pre-distortion action in the frequency domain. The PA is nonlinear at all power levels; it creates intermodulation (IM) products in the frequency channels of interest. We can think of the pre-distorter as a component that adds these IM products at the input of the PA, at such a signal level and phase that once they have been amplified by the PA, they have the same magnitude but appear in antiphase with the IM distortion products generated by the PA from the original signal. The pre-distorter behaviors are summarized in Figure 1.35, which shows the wanted signal and the pre-distortion signal spectra, indicating also their amplitudes. The output signal from the PA is free from IM distortion: linear.

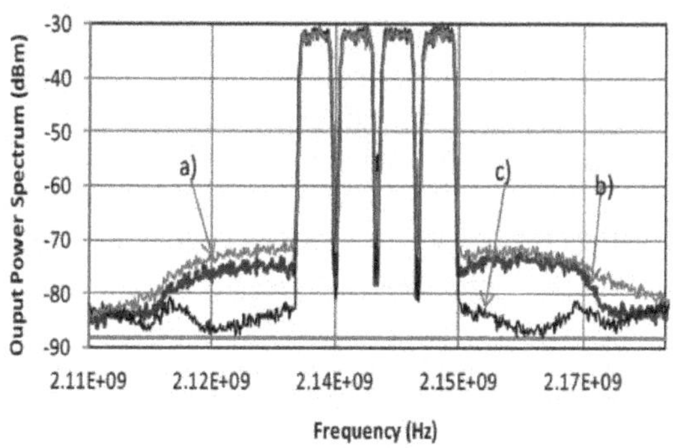

Figure 1.35 Measured output spectra from a high-power Doherty PA, showing: a) the uncorrected signal; b) with memoryless pre-distortion; and c) with memory polynomial predistortion. The signal is 4-carrier WCDMA, occupying 20 MHz of bandwidth, from [12].

The action of the pre-distorter is therefore to increase both the signal bandwidth and its amplitude at the input to the RF power amplifier. These are serious concerns. First, the power amplifier must be designed to accommodate wider bandwidth signals than those specified by the channel allocation of the customer; wider by perhaps five times or more to accommodate the added intermodulation products required for the pre-distortion. Second, the power amplifier must be designed to accept a higher input signal than would be initially determined from the maximum power capability and gain performance of the PA, to accommodate the signal expansion characteristic of the pre-distorter.

The pre-distorter is a complex gain (magnitude and phase) adjuster placed at the input of the RF PA. If the power amplifier has a nonlinear gain function $G(v)$, then the output voltage from the PA for an input voltage v_{in} is

$$v_{out}(t) = G(v_{in}(t)) = A.v_{in}(t) + nonlinearterms \tag{1.7}$$

The pre-distorter function F must be chosen so that the output voltage with the pre-distorter in place is:

$$v_{out}(t) = G(F(v_{in}(t))) = A.v_{in}(t) \tag{1.8}$$

a scaled linear version of the input voltage. In simple terms, one can think of the function F as being the inverse of the nonlinear gain function G, such that $G.F = G.G^{-1}$ within a linear scaling factor. The implementation of the pre-distorter function F can be in the analog or digital domain, as either a fixed function, or one whose parameters are continually adapting to the signal.

Analog pre-distorters have been used in applications as widespread as space-borne communications and in cellphone handsets, where their prime advantage is small size. In space applications, the PA is often a traveling-wave tube (TWT) amplifier, whose nonlinear behavior can be described with a relatively simple function, and a fixed linearization function can be determined reasonably easily. Often, a junction diode can be used to generate the dominant nonlinear products used for the correction signal; this approach has also been applied in early cellphone amplifiers. An interesting analog approach to pre-distortion in the RF domain for cellular applications is outlined in Chapter 9.

Digital pre-distortion (DPD) techniques are nowadays pre-eminent in the cellular communications world. The digital correction takes place at the baseband or the modulation frequency. As noted earlier in Section 1.3.3, the Cartesian feedback architecture is commonly used as the basis of the pre-distorter system. Once the sampled output from the PA has been down-converted and demodulated into *I* and *Q* data streams, this data is converted into the digital domain by a fast, high-resolution analog-to-digital converter (ADC). This data is then compared with the outgoing *I* and *Q* data, which is then adjusted by the pre-distortion algorithm.

The DPD algorithm has parameters that are continually being updated to obtain a better model of the amplifier's nonlinearities. Usually a block of several samples of *IQ* data is collected to perform the comparison with the input for the identification of the PA nonlinearities. This new model is then used for the pre-distortion of the input signal. This adjustment is performed continuously, block by block. The nonlinear model identification and the pre-distortion algorithm will be described in more detail in later chapters.

1.4 BEHAVIORAL MODELING

The essence of the modern DPD technique is the adaptive identification of the dynamic nonlinear behavior of the power amplifier. The PA is generally operating close to saturation, and so is very nonlinear; therefore, the model of the PA needs to capture this nonlinearity. Dynamical effects arising from reactances (usually nonlinear) in the transistors themselves, reactive matching networks, charge storage in the bias networks, and so forth, will also be present. These dynamical effects are often called *memory effects*, and they account for the influence of signals at previous instants on the value of the present signal.

Basically, we are building a nonlinear dynamical model of the RF power amplifier from knowledge of its input and output signals: a 'behavioral model.'

A behavioral model is a model of a system that relates its outputs to its inputs. This relationship may well be entirely mathematical, and may have little or no relationship to the physical phenomena that are taking place inside the system. The use of any such knowledge, known as priors, may lead to more compact model. The natural home for expressing a nonlinear dynamical behavioral model is the time domain: this allows the description of transient as well as steady state phenomena.

Until relatively recently, measured data for RF and microwave systems could be obtained accurately and reliably only in the frequency domain, using periodic signals for test. RF engineers generally work in the frequency domain, hence the popularity of frequency domain-based nonlinear models. As noted above, dynamical effects are naturally described in the time domain. While the frequency-domain data can be transformed into the time domain using Fourier transform techniques, the transient behavior is lost. In any case, the frequency-domain data itself is usually relatively sparse. Such data sets would typically include AM-to-AM and AM-to-PM curves, over a span of RF frequencies, intermodulation signal levels as functions of drive level and tone spacing, harmonic output as function of drive, and so forth. This means that the resulting behavioral models are relatively simple in structure, and obtained by essentially 'curve-fitting'[2] of a family of functions to the given data.

2 There is quite a difference between 'curve-fitting' and 'function approximation:' function approximation can be thought of as taking a known but complicated function and approximating it with a simpler function, possibly over a narrow range of input values; data- or curve-fitting implies that we are choosing a function from a large set of candidate functions (such as polynomials), and adjusting the parameters or coefficients of the function until the data is represented with sufficient accuracy. We can also think of this process as using a simple function to approximate an unknown, but possibly quite complicated function that is described by the data; usually at measurement points that we have not chosen. This will be expanded on in Chapter 2.

One example of this form of behavioral model for RF PAs is due to Saleh [13]. This is a simple model of the AM-to-AM and AM-to-PM characteristics, containing only two adjustable parameters for the gain and two for the phase response:

$$g(r) = \frac{\alpha_A r}{1 + \beta_A r^2} \tag{1.9}$$

$$\Phi(r) = \frac{\alpha_\varphi r}{1 + \beta_\varphi r^2} \tag{1.10}$$

where r is the amplitude of the envelope signal.

These expressions are analytically invertible, which means that it is straightforward to obtain the linearizer functions. This model has been applied successfully to TWT amplifiers used in space communications applications, though it has been less successful with solid-state PAs.

Another popular curve-fitting technique is to use a polynomial function to represent the curve. This approach is based on the Taylor series expansion of an arbitrary function about a given point. The polynomial of degree N can be expressed in the following form:

$$y(t) = a_0 + \sum_{n=1}^{N} a_n u^n(t) \tag{1.11}$$

If we choose the polynomial coefficients to be complex, this function can represent the complex gain (magnitude and phase) of the power amplifier. The polynomial approximation is *linear in parameters*, which means that the coefficients a_k are not part of the nonlinear expression. This allows us to use standard and straightforward least-squares techniques to determine the values of these coefficients or parameters of the polynomial from the measured data.

The degree of the polynomial approximate can also be chosen to minimize the difference between the measured data and the value of the polynomial, in a least squares sense. We need to be careful in choosing the degree of the polynomial for our model. Initially, we may well find that the fitting error will tend to decrease as the degree of the polynomial is increased, but after some point, the error will begin to increase again. We may also observe that the measured data points are reproduced well, but as the polynomial degree is increased, the values of this function between the measured data points describe unphysical oscillatory behavior. These phenomena are characteristic of overfitting, illustrated in Figure 1.36. Note also that the polynomial approximation can fail dramatically outside of the measured data domain: the extrapolation performance of polynomial approximators is, generally speaking, not terribly good, though with care it can be managed.

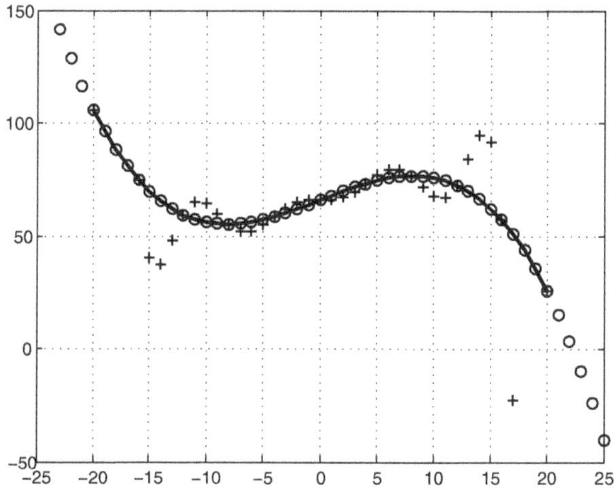

Figure 1.36 Approximating discrete measured data by a continuous polynomial function can give erroneous results. Here, a cubic function is approximated by a a cubic polynomial, and an 11th-degree polynomial over a discrete, small number of points, then replotted over a denser grid of points. For the 11th-degree polynomial, the behavior of the function between the measured data points becomes unrealistic. Poor extrapolation behavior may also be evident, outside the measured data range.

Typically, frequency-domain polynomial models will be used to model the AM-to-AM and AM-to-PM characteristics of the PA, and, if the model is created over a reasonably wide RF bandwidth, the intermodulation behavior can also be predicted by the model. In general, frequency-domain models can describe the RF frequency response phenomena quite well, but are unable to accommodate the memory effects associated with long time constants, for example, bias line reactance and charge storage.

To include the memory effects, we turn to a more complicated, although still polynomial-based, model description: the *Volterra series*. The Volterra series model retains the advantage of polynomial models in that the model structure is linear-in-parameters: that is, the parameters, or coefficients of the nonlinear terms can be found using linear techniques such as least-squares minimization. We shall investigate the structure and properties of Volterra series in more depth and detail in Chapter 6, but we shall introduce them briefly here. Simply, the Volterra series can be thought of as a Taylor series with memory; that is, a Taylor series defined not only at the present instant in time, but including terms at previous instants, up to some specified delay. Just from this simple view, it can be appreciated that the number of parameters or coefficients in the Volterra series will grow rapidly as the polynomial degree is increased and the memory depth is increased. For many years, this in itself was a good enough reason to avoid using the Volterra description: for a modestly-sized model, the number of coefficients that has to be calculated was too large to calculate efficiently or accurately.

But nowadays, this picture is changed. With the introduction of microwave instruments such as vector signals analyzers and microwave oscilloscopes that can capture time-domain waveforms at rates of hundreds of megasamples per second (MS/s), and desktop computers with gigabytes of accessible memory and processors operating a gigahertz clock rates, we have the ability to capture and analyze time-domain waveforms consisting of millions of data points. This changes completely the way in which we can work with the data.

Time series analysis is a rapidly growing method of data analysis, using large data sets, in which the data itself is used to uncover the inherent nonlinearities and dynamics of the system under test. When allied with Volterra series modeling, convenient and compact descriptions of the nonlinear dynamical system can be generated quickly, enabling the online adaptation of DPD systems to the RF PA nonlinear behavior. We shall investigate these new behavioral modeling paradigms in more detail in this book.

1.5 CONCLUDING REMARKS AND BIBLIOGRAPHY

This opening chapter is intended to give a broad introduction to the various technical topics that converge in the design and realization of high-power RF amplifiers for radio communications. The material presented here is only scratching the surface, to present some of the features that are relevant to PA modeling and linearization. There are many texts that deal in detail with these topics, specifically in the context of wireless communications; a few are recommended here as insightful background to these topics, providing a level of detail and description that is beyond the scope of this introduction.

Digital Communications:

- B. Sklar, *Digital Communications: Fundamentals and Applications*, 2nd ed., Upper Saddle River, NJ Prentice-Hall (2001)

This is an excellent introduction to the topic.

- A. Goldsmith, *Wireless Communications*, Cambridge, UK, Cambridge University Press (2005), and
- E. McCune, *Practical Digital Wireless Signals*, Cambridge, UK, Cambridge University Press (2010)

These provide excellent descriptions of the signals and modulations found in modern wireless communications, along with the mathematical background to enable us to make use of these signals.

RF Power Amplifiers:
Three books spring to mind when discussing RF PA design; the book by Cripps provides a detailed review of the PA modes and classes, and of the high-efficiency architectures outlined in this chapter. Colantonio *et al.* offers an interesting insight into the design of harmonically-loaded PAs, and other high-efficiency techniques for RF and microwave applications. The book by Kennington is perhaps a little dated, but still an excellent read.

- S. C. Cripps, *RF Power Amplifiers for Wireless Communications*, 2nd ed., Norwood, MA, Artech House (2006)
- P. Colantonio, F. Giannini, and E. Limiti, *High Efficiency RF and Microwave Solid State Power Amplifiers*, New York, Wiley (2009)
- P. B. Kennington, *High-Linearity RF Amplifier Design*, Norwood, MA, Artech House (2000)‘

Behavioural Modeling
As a review, I can recommend the following book, and the recent review by Pedro and Maas [14]:

- J. Wood and D. E. Root (editors), *Fundamentals of Nonlinear Behavioral Modeling for RF and Microwave Design*, Norwood, MA, Artech House (2005)

– has chapters by the leading nonlinear behavioral modeling practitioners worldwide.

Linearization and Digital Pre-Distortion
...actually, there isn't so much here, particularly for RF power amplifier applications; I hope this book provides a useful reference.

The rest of this book will focus on the behavioral modeling techniques appropriate for high-power RF amplifiers, and how these techniques are applied in modeling and for pre-distortion. We will provide the necessary mathematical background for the modeling, and examples of practical systems for pre-distorters will also be presented.

References

[1] F. H. Raab, P. Asbeck, S. Cripps, P. B. Kennington, Z. B. Popovic, N. Pothecary, J. F. Sevic, and N. O. Sokal, "Power amplifiers and transmitters for RF and microwave," *IEEE Trans. Microw. Theory Tech.*, Vol. 50, No. 3, pp. 814–826, Mar. 2002.

[2] T. P. Hung, A. G. Metzger, P. J. Zampardi, M. Iwamoto, and P. M. Asbeck, "Design of high-efficiency current mode class-D amplifiers for wireless handsets," *IEEE Trans. Microw. Theory Tech.*, Vol. 53, No. 1, pp. 144–151, Jan. 2005.

[3] H. Nemati, "Design, implementation, and evaluation of a current-mode class-D power amplifier," Master's thesis, Chalmers University of Technology, 2006.

[4] F. H. Raab, "Class-E, class-C, and class-F power amplifiers based on a finite number of harmonics," *IEEE Trans. Microw. Theory Tech.*, Vol. 49, No. 8, pp. 1462–68, Aug. 2001.

[5] W. H. Doherty, "A new high efficiency power amplifier for modulated waves," *Proc. IRE*, Vol. 24, No. 9, pp. 1163–82, 1936.

[6] L. R. Kahn, "Single-sideband transmission by envelope elimination and restoration," *Proc. IRE*, Vol. 40, No. 7, pp. 803–806, 1952.

[7] M. Hassan, M. Kwak, V. W. Leung, C. Hsia, J. J. Yan, D. F. Kimball, L. E. Larson, and P. M. Asbeck, "High efficiency envelope tracking power amplifier with very low quiescent power for 20 MHz LTE," in *IEEE Radio Frequency Integrated Circuits RFIC Symposium*, Baltimore, MD, Jun. 2011, pp. 1–4.

[8] H. Chireix, "High power out phasing modulation," *Proc. IRE*, Vol. 23, No. 11, pp. 1370–92, 1935.

[9] H. S. Black, "Translating system," U.S. Patent 1,686,792, 1928.

[10] H. S. Black, "Wave translation system," U.S. Patent 2,102,671, 1937.

[11] R. N. Braithwaite, "Positive feedback pilot system for second loop control in a feedforward power amplifier," *IEEE Trans. Circuits Syst. I*, Vol. 55, No. 10, pp. 3293–3305, Oct. 2008.

[12] M. C. Fares, D. Y.-T. Wu, S. Boumaiza, and J. Wood, "Inward nonlinear characterization of Doherty power amplifiers," in *IEEE MTT-S Int. Microwave Symp. Dig.*, Boston, MA, Jun. 2009, pp. 1–4.

[13] A. A. M. Saleh, "Frequency-independent and frequency-dependent nonlinear models of TWT amplifiers," *IEEE Trans. Commun.*, Vol. 29, No. 11, pp. 1715–20, Nov. 1981.

[14] J. C. Pedro and S. A. Maas, "Comparative overview of microwave and wireless power-amplifier behavioral modeling approaches," *IEEE Trans. Microw. Theory Tech.*, Vol. 53, No. 4, pp. 1150–63, Apr. 2005.

Chapter 2

Models, Modeling, and Simulation

"Laws of Simulation and Modeling[1]"

- A simulation is only as accurate as the models it is based on.
- A model is (mostly) useless unless it is embedded in a simulator.
- Models are, by definition, inaccurate; it's just a matter of degree.
- Models generally trade off complexity or simulation time for accuracy.

Although not laws like gravity or income tax, these observations hold true.

1 attributed to Colin McAndrew, *Freescale Semiconductor*

2.1 A PERSPECTIVE

What is a model? What do we mean by modeling? What is the difference between data and a model? What is the difference between a model and a simulation? or modeling and simulation?

These are all pertinent questions, and it is paramount to be clear on what we mean by these terms; though I am reluctant to propose a definite taxonomy on this subject, as the 'rules' and borders between the terms and their usage can become fuzzy. Nevertheless, let's do our best.

The *data* are the basic items of information from which we will construct the model. The data will usually come from a set of measurements[2] and provide independent and dependent variable sets, for example, input voltage and output current. The *model* is the relationship between the dependent and independent data; and *modeling* is a set of procedures that we follow to ensure that our model describes this relationship between the measured qualities within some specified level of accuracy. The *simulator* is (usually) a computer program that solves a specified set of equations using dependent information provided by the model. For example, a circuit simulator essentially solves Kirchhoff's Current and Voltage Laws by using models of the circuit elements that, in the case of a resistor, will return a value of the branch current through the resistor at a given impressed voltage.

2.2 THE MODEL

What is a model? For any given system, a model describes the relationships between its input and output signals, or the independent and dependent signal values. In our case, system can mean a transistor, a power amplifier, or an RF transmitter including the linearization scheme. This rather abstract description can be summarized in Figure 2.1. In this figure, we also show disturbance signals, which may be noise, or adjacent channel interferers, for example. The noise or disturbance may be a measurable input, from which we can estimate its effect on the measured output, perhaps in the absence of the true input. An unmeasurable disturbance can only be observed through their influence on the output signal.

The model is generally a mathematical abstraction of the input-output relationship. It may be based on knowledge of the physics of the system, and such physical description may in many circumstances be good enough to describe the

2 or from simulation, though where do we get the model from to use in this simulation? We shall investigate this further, later in this chapter.

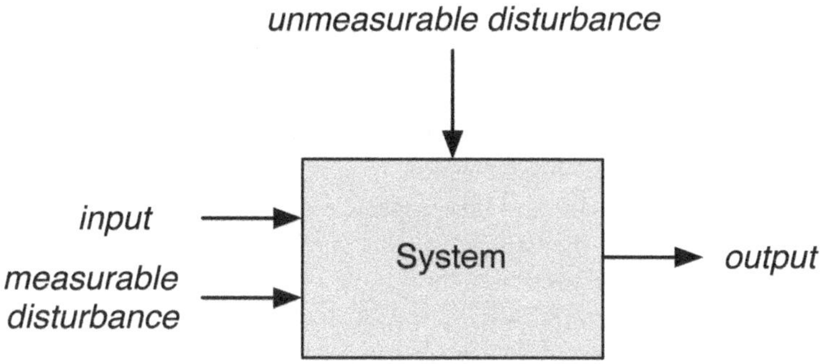

Figure 2.1 A simple system indicating the input and output signals, including disturbance signals. A model represents the relationship between these inputs and outputs.

input to output function. In other cases, the mathematical description may be expressed more in terms of the convenience of parameterizing the functions that we choose to represent the input-output relationship.

In any case, our creation and use of the model should be guided by its usefulness in the simulator, describing the behavior of the device or system. The model will not be an exact replica of the physical reality, but its usefulness lies in how much insight it provides us in solving the larger problem of the circuit or system design. Some models may only serve to guide the designer in the right direction, by informing which are the better compromises to make in a design. Other models may provide more accurate results, at the expense of describing the physical features that were ignored in some particular circumstances.

In the limit, we can use the data itself for the model. This provides an exact relationship between the independent and dependent values, within the experimental error of the measurements themselves. If the data comes from a simulation of a lower-level circuit or subsystem that we want to model, it will be free from noise, but only as accurate as the models used in the simulation.

Usually, this is not a satisfactory approach, as the 'data-model' can become extremely large if we want to cover a wide range of application space for our model, because it has no ability to generalize beyond the measurements themselves [1]. How this model describes the behavior of the measured system between the data points is left to the simulator: how does the simulator interpolate between the data? For example, the frequency response of a low-loss resonant circuit crosses the real axis near the open-circuit condition; if the measured data straddles this point, how the simulator interpolates between them can determine the resonant frequency and the quality factor (Q) of the circuit. Knowing these values accurately is quite important for the design of matching networks for high-power transistors. Different simulators may behave differently in this respect, and even different interpolation modes in the same simulator, returning different results for the same excitation. Extrapolation is usually out of the question, though again the simulator may impress its own particular interpretation in this respect. This sort of unpredictable behavior is probably not something we want from our model.

One should also be aware of the above concerns even if all we want to do is visualize the data graphically. The visualization of multi-dimensional data may require the graphics tool to interpolate between the measured data points. For example, load-pull data comprising power sweeps measured at several load impedances is often contoured: that is, we prefer to view this data at, say, the 1-dB compression points of each of the drive-up power sweeps, and join these points to form a continuous curve on the impedance plane. This requires the visualization tool to perform several mathematical processes on the data, including interpolation of the measured power sweeps to find the compression point on each one, and then some surface fitting algorithm to estimate the path of the equi-power contour. Is this mathematical treatment a model, or are we still dealing with the data?

Another common example of such a data-model includes S-parameters of complicated passive circuit layouts, such as filters or matching networks in power amplifiers. One key quality here is that the circuit components that we describe in this way are passive, linear components, and therefore independent of the current or voltage impressed at their terminals. Having measured the S-parameters of the component over a range of frequencies, we can then use these S-parameter models in a larger circuit and use the simulator to investigate its frequency response. This larger circuit may contain nonlinear elements, but the S-parameter models will still be valid in a large-signal simulation as they are linear models.

Generally, though, we ask more of a model.

To build a working definition for a model, one requirement could or should be that it reduces the data set to a more *compact* representation. For example, we have just remarked that a set of measured S-parameters is really just data, not a model, even though we can use the S-parameters in a simulation, by importing them through a datafile reader component in the simulator. We can transform the S-parameter data into a different space or parameter set, such as Y- or Z-parameters. This transformation neither reduces the size of the data set, nor provides any more information, so in itself does not create a new model.

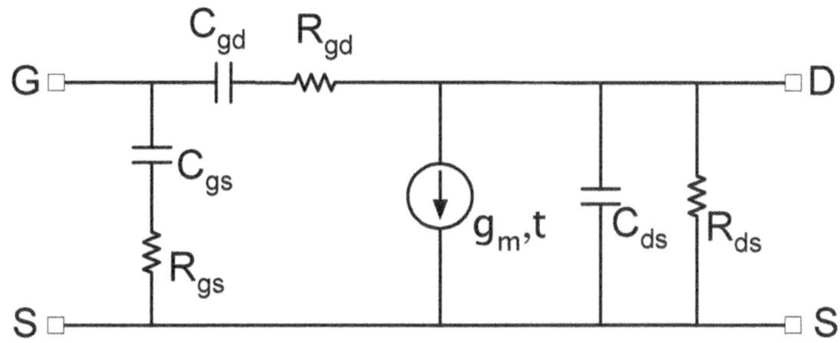

Figure 2.2 Small-signal equivalent circuit representation of the intrinsic FET.

But, if we take the Y- or Z-parameters, and convert the real and imaginary parts of the immitances into resistive and reactive circuit components, we have extracted the frequency information and created a smaller description of the data: an equivalent circuit representation. A good example of this is the small-signal modeling of microwave transistors: to get a good representation of the small-signal behavior of the transistor, we measure the S-parameters over a range of applied bias voltages and frequencies. This can easily be a very large data set. After converting the measured data to Y-parameters, we can then eliminate the frequency dimension by replacing the Y-parameters by equivalent circuit parameters at each bias point. For a FET, the resulting bias-dependent small-signal equivalent circuit model is shown in Figure 2.2, for the intrinsic device, where the circuit elements have their usual definitions [2]. The value of each component in this circuit depends on the applied bias voltages, but the frequency dependence of the device is described by the circuit's frequency response. A good check on the validity of the model, that is, this circuit representation, is to compare its frequency response predicted by simulation against the measured data. If there are differences, this may be because the circuit representation that we have chosen is incorrect, and there is a more appropriate circuit choice, although this may be more complex than our initial model; or the differences may be due to inaccurate measurements.

The circuit of Figure 2.2 is a bias-dependent linear model of the intrinsic FET device. The complete small-signal model for the practical transistor will include additional components that describe the physical structure of the device: the metallization resistances, the metal-to-semiconductor contact resistances, so-called access resistances due to the semiconductor material between the contacts and the intrinsic device in the channel region beneath the gate contact, inter-conductor

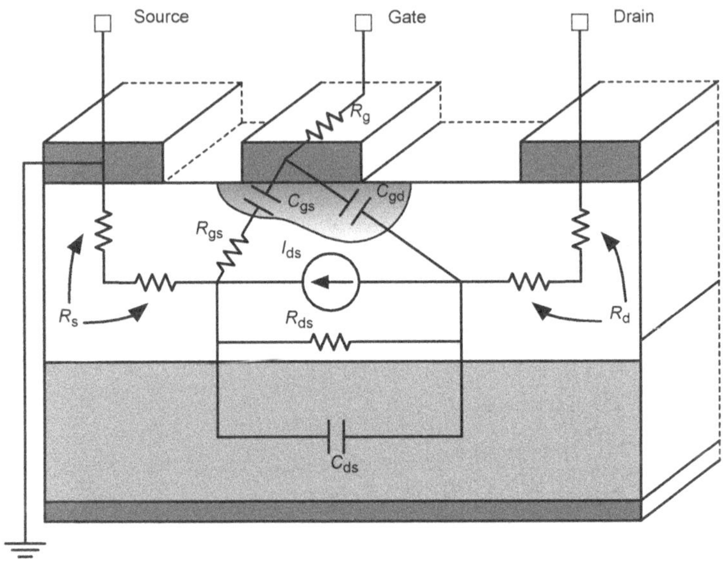

Figure 2.3 Schematic cross-section of a microwave GaAs FET showing the physical origins of the extrinsic components of the equivalent circuit model.

capacitances, metal inductances, and so forth. A schematic cross-section of a microwave GaAs FET indicating these extrinsic elements is shown in Figure 2.3.

So far we have introduced the concept of a circuit model, where the elements of the circuit and their values are extracted from the measured data, so that the simulation of the circuit mimics the measured data. This is a data-based circuit model, often called an equivalent circuit model. We could also derive a circuit model from other considerations, for example, by combining the results from several different measurements, from simulations of simpler or more complex circuits, or from results of simulations in other physical domains. We can also have models from data or simulation that do not have a direct circuit representation. These are often called behavioral models, as they represent the relationship between the inputs and outputs of the device, circuit, or system that we are interested in modeling. Behavioral models are usually mathematical descriptions, and may or may not have any physical relationship with the object we are modeling.

These model descriptions are somewhat vague, and often models contain elements of circuits, physics, and behavioral approaches, so a clean distinction is not always possible. Here below we shall present briefly some general 'classifications' of models, with examples to illustrate how the models are constructed and used.

2.2.1 Data- or Measurement-Based Models

This seems at first something of a truism: surely all models are to a greater or lesser extent based upon some measured data? This is certainly true, as even the physical simulations that we shall describe later require some physical or material parameters that are generally derived from experimental data.

What we mean in this context is that the model is derived fairly directly from a specific set of measurements, an example being the equivalent circuit model for the intrinsic FET derived from (small-signal) S-parameter measurements, as described earlier. This approach is also known as a table-based model: the bias-dependent linear FET model can be constructed as a circuit model whose element values are read from a table that is indexed by the bias voltages. By this means, the circuit simulator can interpolate between the discrete tabulated values to arrive at the correct element values for the model.

We can also develop this idea further, by performing more sophisticated transformations on the measured data to obtain the parameters and parameter values for our model. An excellent example of this is the 'Root' table-based large-signal FET model that is derived from the same bias-dependent S-parameter data set as the bias-dependent linear model [3].

We cannot create a large-signal model from the bias-dependent linear model simply by applying the large-signal RF voltages at its terminals. The bias-dependent capacitances in the model are the sensitivities of the charges in the device with respect to the applied voltages. Using these voltage-dependent capacitances directly in a large-signal context results in a failure to conserve charge in the model. The capacitances must be integrated over the voltage bias space to yield the charge as a function of the terminal voltages, and this charge can then be used in the large signal model. In one dimension, the procedure follows as

$$C(V) = \frac{dQ}{dV} \tag{2.1}$$

and hence

$$Q = \int dQ = \int C(V)dV \tag{2.2}$$

In the FET, we integrate over the two dimensions of the bias voltage space, V_{gs} and V_{ds}. For example, the charge on the gate node is obtained by integrating the gate-source and gate-drain capacitances [2]:

$$
\begin{aligned}
Q_g\left(V_{gs}, V_{ds}\right) = & \int_{V_{gs0}}^{V_{gs}} \left[C_{gs}\left(v_{gs}, V_{ds0}\right) + C_{gd}\left(v_{gs}, V_{ds0}\right)\right] dv_{gs} \\
& - \int_{V_{ds0}}^{V_{ds}} C_{gd}\left(V_{gs}, v_{ds}\right) dv_{ds} + Q_g\left(V_{gs0}, V_{ds0}\right)
\end{aligned} \tag{2.3}
$$

where V_{gs} and V_{ds} are the instantaneous large signal voltages, and V_{gs0} and V_{ds0} are the starting voltages for the integration; these could be the DC bias voltages. In the simulator, the starting point and consequent constant of integration are of no concern in RF simulation, where the dynamic current is given by dQ/dt.

Why is it important to have a charge-conservative model, aside from the fact that obeying the fundamental laws of physics seems a reasonable thing to do? The large-signal circuit models for microwave transistors, developed in the 1990s, were more or less extensions of the small-signal models and used voltage-dependent capacitors, even though a conservative charge-based model had already been demonstrated in the 1980s [4]. The simple capacitance-based approach was generally good enough for the relatively low-power applications that were commonplace in microwave circuit design at the time: Class A amplifiers driven with sinusoidal signals. Modern RF power amplifiers must be designed to accommodate very dynamic digitally-modulated signals, and generally operate in deep Class AB or in more sophisticated PA architectures. Linearity and efficiency of operation are key design requirements, and the prediction of these qualities in circuit simulations is highly desirable. It has been shown that charge-conservative models can predict the low-level nonlinear behavior with these dynamical signals more accurately than simple capacitance-based models [5]. This is because the non-conservative models will accumulate some extra, nonphysical charge over the simulation of a cycle of the RF frequency; over several cycles, this charge will build up. The simulator simply solves the circuit equations. It does not matter where the charge comes from; the simulator will converge to a point where the error in the matrix of equations that are solved is minimized. This means that with a non-conservative model, the simulation will probably converge to the wrong solution, if indeed it converges at all. Using a charge-conservative model avoids this problem: the nonphysical charge is not present, and the simulator can solve the correct problem. As it seems so straightforward to include a physical, conservative charge description into the large signal model, it seems surprising that anyone would do otherwise.

The circuit implementation of this charge-conservative model is shown in Figure 2.4. The gate and drain current and charge sources are all functions of V_{gs} and V_{ds}. These sources can also be thought of as 'large-signal Y-parameters,' where

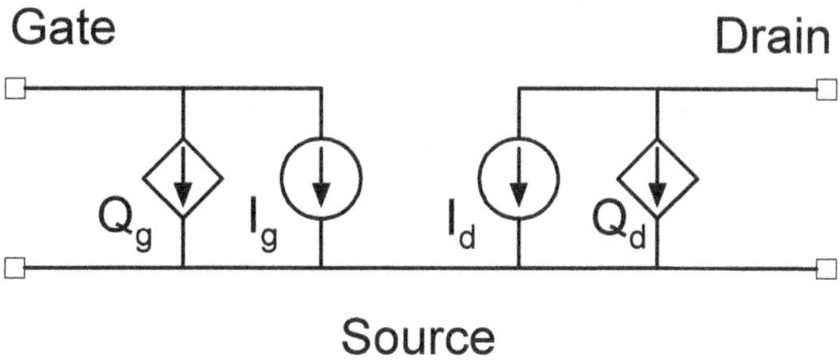

Figure 2.4 Large-signal circuit representation of the intrinsic charge-conservative FET model.

the real parts of Y_{11} and Y_{12} are collected into the gate current source, and the imaginary parts (divided by ω, the radial frequency) are collected into the charge source. The values of the current and charge sources are stored in a table indexed by the gate-source and drain-source voltages; and again, the simulator can interpolate between the discrete tabulated values to obtain the correct charge. The total current into the gate node is given by

$$i_g = I_g\left(V_{gs}, V_{ds}\right) + \frac{dQ\left(V_{gs}, V_{ds}\right)}{dt} \tag{2.4}$$

Another example of a data- or table-based model is the X-parameters™ model from Agilent Technologies, Inc. [6]. This model formulation relates the large-signal A- and B-waves at the ports of the device-under-test (DUT), for each harmonic of the test signal under consideration. For instance, the B-wave that is output at Port 2 at the jth harmonic is given by the nonlinear functional relationship $\mathbf{F_{2}}_j$

$$B_{2j} = \mathbf{F}_{2j}\left(A_{11}, A_{12}, ...A_{1n}, A_{21}, A_{22}, ...A_{2n}, A_{p1}, A_{p2}, ...A_{pn}\right) \tag{2.5}$$

where n is the number of ports of the DUT, and p is the total number of harmonics. The model relates the output B to inputs A at all other ports and harmonics, as a function of the drive signal at the fundamental at port 1, given by A_{11}. In the implementation, each input is related linearly to each output at every port and harmonic, through a complex parameter, and the nonlinearity is achieved through the sum of the various components and the drive level. These parameters are extracted from the measured data and stored in a table. For a 2-port DUT, and considering five harmonics, then each B component at each harmonic is described by a total of ten complex coefficients. This can yield a large table fairly quickly, as the user considers more power levels, fundamental frequencies, and so forth.

The X-parameters model is described in more detail in Chapter 10, in the context of frequency-domain behavioral models. Similar model formulations have been developed and published by other workers [7, 8].

2.2.2 Physically Based Models

Another common approach to creating a transistor or system model is to use the basic physics of operation of the device to relate the terminal voltages and currents. A very successful example of this approach is the circuit model for the bipolar junction transistor (BJT). The physics of operation of this device is based on the current flow across two *p-n* junctions that are spatially very close together, so that the flow across the first junction determines the flow across the second. Hence the name: the fundamental behavior is that the current flow through the output resistor is controlled by the voltage at the input terminals: a *trans*fer res*istor*, or transistor. This physical behavior can be explained by the simple block structure of the BJT shown in Figure 2.5.

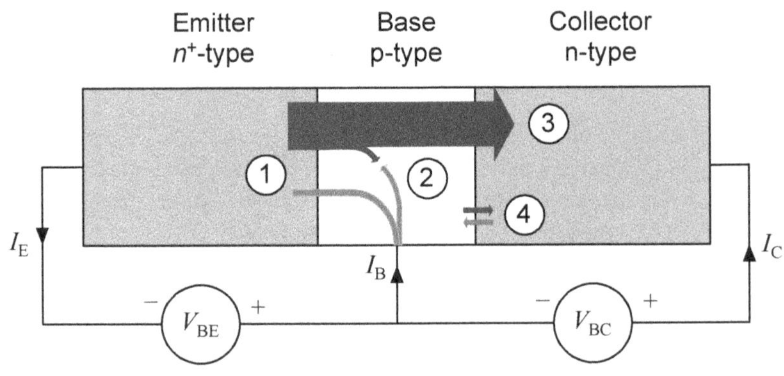

Figure 2.5 Simple block or 'bar-of-soap' structure used for describing the operation of the bipolar junction transistor.

The transistor depicted in Figure 2.5 is an *n-p-n* device, where the majority charge carriers are electrons. This structure is seen to consist of two *p-n* junctions placed back-to-back, and very close together. Reading the figure from the left, the first junction, between the *emitter* and *base* regions of the structure, is forward biased, so that a large current flows across the junction. From our knowledge of the physics of charge carrier transport across a *p-n* junction [9], this current flow is proportional to the exponential of the voltage across the junction, V_{BE}

$$I_E = I_{E0}.\left\{\exp\left(\frac{eV_{BE}}{k_B T}\right) - 1\right\} \tag{2.6}$$

where the proportionality constant I_{E0} is known as a *saturation current*; e is the unit electronic charge, k_B is Boltzmann's constant, and T is (absolute) temperature.

The current flow across this junction comprises electrons from the emitter, and holes from the base. This is the current ① in the figure. In silicon BJTs, the ratio of the electron to hole flow is proportional to the doping of these regions; therefore, it is usual to have the emitter much more heavily doped than the base, so that the current flow across this junction is essentially due to the electron flow. This ratio of electron to hole current is called the emitter efficiency.

The second *p-n* junction is reverse biased. For an isolated *p-n* junction, this would mean that the current flow across the junction would be very small. In fact, the current is due to the *minority* carriers that reach the junction and are swept across by the applied voltage: the minority holes on the *n*-side, and minority electrons on the *p*-side. The minority carriers are present in the semiconductor as a result of the thermal energy in the material; their concentrations are several orders of magnitude smaller than the *majority* carriers that are created by the doping. Because the numbers of these carriers are small, in equilibrium, the reverse current across a *p-n* junction is very small, as indicated by the current ④.

The transistor action arises from the close proximity of the two junctions. The electrons that cross the emitter-base *p-n* junction, under the control of the base-emitter voltage, appear in the base region where they are now minority carriers. The base region is very narrow, so that these carriers have almost no opportunity to return to their equilibrium density (by combining with majority holes in the base,②), but instead they diffuse across the base and appear at the base-collector junction where they are swept across, and become the collector terminal current, ③.

The current in the output or collector circuit is determined by the current flow across the base-emitter junction, which is determined by the input voltage across the base-emitter junction, V_{BE}:

$$\begin{aligned} I_C &= \alpha_F I_E \\ &\propto V_{BE} \end{aligned} \tag{2.7}$$

The proportionality constant α_F is the forward current transfer ratio. As our bar-of-soap structure is fairly symmetrical, we could apply the bias voltages in the

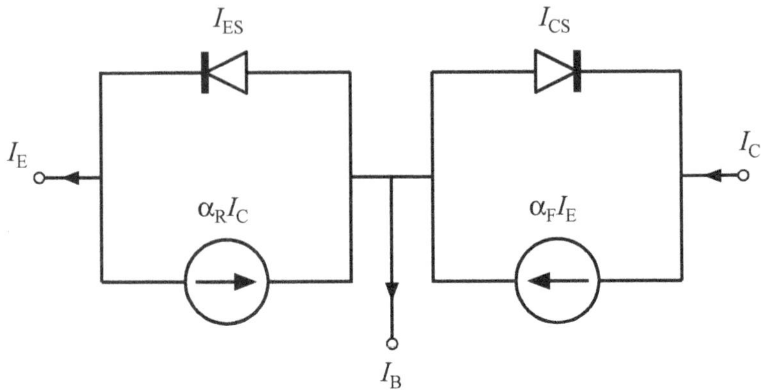

Figure 2.6 The Ebers-Moll circuit model of the bipolar transistor.

opposite sense and get a reverse operation of the BJT. Because we would designate structure to optimize for operation in the forward sense, this reverse transistor would not be as good.

From (2.6) and (2.7), we can construct a simple circuit model that describes the operation of the BJT and is based on the physics of operation of the semiconductor structure. This circuit model is shown in Figure 2.6, and includes the forward and reverse operating modes. This model is the Ebers-Moll model of the BJT [9].

The Ebers-Moll model is usually augmented by capacitances that describe the depletion capacitances of the two *p-n* junctions, and the diffusion capacitance of the mobile electrons in the base region. In most circuit simulators, the two controlled current sources are combined into a single element; this circuit model is the Gummel-Poon model [10], and is the basis for many derivative BJT models.

At microwave and RF frequencies, bipolar transistors are built from the *III-V* semiconductors. These materials have some properties that are advantageous for high frequency operation, such as the electron mobility and velocity at high electric fields, enabling rapid transport of the electrons through the device, and hence high switching speeds and transition frequency.

For bipolar transistor applications, the *III-V* semiconductors and their alloys have other interesting properties. For example, the alloy indium-gallium phosphide, InGaP, has a mixture of indium and gallium atoms at the *III* location, and phosphorus at the *V* location in the crystal. This material can be made with the same crystal structure dimensions as gallium arsenide, so that if InGaP is placed (grown) on top of GaAs, the crystal structure is effectively seamless, but with an abrupt junction between the two materials. For a bipolar transistor application, the InGaP region is the emitter, doped to be n-type, and the GaAs is the p-type base, with the n-type collector beneath. The interesting feature about this structure is that the electron energy bands in the InGaP are offset from those in the GaAs, and they have a wider gap between the electron and hole energy bands. These physical features affect the carrier transport across the emitter-base junction in the following way. The electron flow is enhanced by the energy band offset, by a factor that is exponentially related to the offset energy:

$$I_E = I_{E0}.\left\{\exp\left(\frac{eV_{BE}}{k_B T}\right) - 1\right\}.\exp\left(\frac{\Delta E}{k_B T}\right) \tag{2.8}$$

Similarly, the hole flow is diminished by the energy band offset for holes.

The exponential factor is a huge multiplying factor on the electron flow, much more significant than can be obtained by the doping ratio between emitter and base. This allows us to dope the base region very heavily, reducing its access resistance without affecting the emitter efficiency. This allows us to reduce the loss at microwave frequencies. The high electron velocity in the GaAs collector is also an advantage for microwave operation, minimizing the output loss. These device structures are known as *heterojunction* bipolar transistors (HBTs).

While the Gummel-Poon model is an effective description of silicon BJT behavior, it lacks some features and equations that describe the physical transport behavior in the *III-V* HBTs. Recent physically-based model developments for HBTs include models developed by the team at UCSD [11], Rudolph and co-workers [12], and the commercial implementation of the AgilentHBT™ model [13] from Agilent Technologies, in their ADS™ simulator.

On the other hand, it is much more difficult to create a physically-based model for a FET; such models are usually far from simple, though significant progress has been made in recent years. The first description of the FET by Shockley [14] was based on the electron transport in the undepleted channel under the gate of the junction field effect transistor. This model was only valid in the so-called linear region, below the knee of the output current characteristic. The saturation region behavior was accounted for by assuming an infinitesimally-thin sheet of charge at the drain end of the channel, through which the current could flow in an otherwise closed or pinched-off channel described by the model. While this seems rather unsatisfactory, this description held sway for many years until the explanation of the saturation region by Leichti in 1975 [15], as being caused by the electrons in the channel reaching their saturated velocity at the high electric fields that were present in the structure at the drain end of the channel. The saturated velocity can be attained with the classical pinch-off of the channel taking place. Practical evidence of the velocity saturation limiting the current is that the drain current is a more nearly-linear function of the gate voltage, rather than the square-law predicted by the classical analysis.

At low frequencies (sub-RF), the Berkeley 'BSIM' models [16] for MOS FETs have been very widely used in analog and digital IC design. This model is based on threshold voltage description of the interface charge density, the fundamentals of which are well known [9]. The BSIM4 model accommodates many physical phenomena that are present impractical transistor structures [17], including nonuniform doping in the channel, drain-induced barrier lowering, and the short channel effects and channel-length modulation effects observed in submicron devices. The electron transport is described by field-dependent carrier mobility and velocity saturation to account for the saturation region of the $I-V$ characteristics, and breakdown effects resulting from impact ionization and gate tunneling effects can be included in the model description. The richness of this physical detail is reflected in the large number of parameters that are needed in the model to describe its electrical behavior. Many of these parameters can be determined or obtained from the structural and material properties of the transistor, and others determined through measurement.

The BSIM4 capacitance model is based on the charges in device. Generally, the following relationship is used to describe the charge in a MOSFET:

$$Q_G + Q_{INV} + Q_B = 0 \tag{2.9}$$

where Q_G is the charge on the gate electrode, Q_{INV} is the charge in the inversion layer under the gate, and Q_B is the charge in the depleted semiconductor beneath the gate. The inversion layer charge is usually partitioned into source and drain charges associated with the respective nodes. Different charge partition schemes can be used, controlled by a user-defined parameter in the model. The charges are calculate from the physically-based equations as mentioned above, and the small-signal capacitances are derived from the charges in the usual way:

$$\begin{aligned} C_{ij} &= \frac{dQ_i}{dV_{ij}} \quad i \neq j \\ C_{ij} &= -\frac{dQ_i}{dV_{ij}} \quad i = j \end{aligned} \tag{2.10}$$

where i, j are the gate, drain, source, and bulk terminals of the MOSFET.

Unlike measurement-based models (see Section 2.2.1), the charges above are calculated from the applied voltages, and so the charge is conserved in the structure, even though the partitioning may appear arbitrary. The capacitances are derived from these charges. Conversely, in the measurement-based approach, the capacitances are the measured quantities, and the charges must be derived from them: in the Root and large-signal model outlined earlier, care is taken to ensure that these derived charges are conservative.

The BSIM4 model can be extended to high frequencies by the inclusion of extrinsic elements in the model, to account for access resistances such as the gate resistance, and the capacitances and inductances that become more important at RF and microwave frequencies.

More recently, a surface potential approach to the calculation of the inversion charge has been pursued. This is because as the device dimensions and applied voltages become smaller and smaller with technology scaling, the threshold voltage becomes more difficult to define unambiguously. The Surface Potential model of Gildenblat and coworkers [18] and the MOS Model 11 from NXP (formerly Philips Semiconductor N.V.) are good examples of this approach: these models have been combined to create the *PSP* MOSFET model [19].

The PSP model is surface-potential based, and substrate referenced, and features an analytic formulation that applies from accumulation through inversion regions, providing a description of the charges and currents that are continuous and differentiable to high order. The key physical effects are included, as in the BSIM models, as well as material and structural features pertinent to modern small-gate technology, for example: gate tunnelling current, gate-induced drain and source leakage, influence of pocket implants, polysilicon gate depletion, quantum-mechanical effects, gate and substrate currents, and bias-dependent overlap capacitances. The model also includes non-quasi-static effects so that the electrical behavior at high frequencies or switching rates can be predicted accurately. The charges are calculated from the local potentials and are thus conservative, and small-signal capacitances are derived as in (2.10), above.

2.2.3 Equivalent Circuit Models

The equivalent circuit representation of S-parameter data has been outlined already in Section 2.2. This approach to implementing a model into the circuit simulator seems very natural: the data set is reduced and represented by fundamental circuit elements such as resistors, capacitors, inductors, and controlled sources in active devices. This is fine as long as one realizes that the circuit elements are representations of the n-port network parameters that have been used to characterize the device. Where we can run into difficulty is when these circuit components are presumed to represent specific physical features in the device. For example, the circuit component defined by

$$C_{11} = imag\frac{Y_{11}}{\omega} \tag{2.11}$$

is very easy to associate in one's mind with the gate-source capacitance of the FET; this is not helped by having this component so labeled in many small-signal circuit models (Figure 2.2 is an excellent example of this). In fact, with a little bit of algebra, this association can be shown to be close. The problem arises when device design and process choices are made on the basis of simulation results showing what happens when this model parameter is varied. Why is this a problem? Because in general the values of the model parameters are inter-related: changing one physical variable will likely result in changes in several of the model parameters. This is especially true for nonlinear models, which, as we have seen, can contain tens to hundreds of inter-dependent parameters.

Equivalent circuit models can also be created from this limited physical viewpoint: focusing on one aspect of the device structure. This has been the case for several large-signal FET models, where circuit components are chosen to model specific features, often ignoring the physical interaction that occurs in the real transistor. An often-seen example is the representation of the behavior at the gate node of the FET in terms of the two circuit branches, from gate to source and from gate to drain. Each branch is modeled by:

- a junction diode, to represent the rectifying metal contact, even though in modern pHEMT devices the gate $I - V$ characteristic can be somewhat different from the classical diode equation; in parallel with
- a nonlinear capacitor, which is sometimes a function of two voltages and therefore not charge conservative;
- a series resistor, to account for the resistance of the semiconductor in the channel, even though the charge transport in the channel is described by the controlled drain-source current source.

While such an approach is OK, but limited, for the metal-semiconductor contact at the gate, it fails to represent the behavior of the charge at the gate node in an FET.

Care should be taken when devising an equivalent circuit representation of a structure, that the physics of the device are included at the outset. This then becomes more of a physically-based model approach.

2.2.4 Compact Models

Apart from the table-based models, essentially all of the models described above fall into the general category of compact models. That is, the model, implemented either as a circuit or as a set of equations to solve in the simulator, is a more compact representation than the data that it represents. Compact models in general will be created from a set of measurements, or material parameters in the model equations, and will have the ability to predict the device behavior under electrical conditions that were not used in its creation. This ability to generalize is the value of a model.

An industry association called the Compact Modeling Council (CMC) exists to determine which models shall be implemented in the circuit simulation tools as standard device models. They ensure that the implementation is consistent across the simulator platforms, and also that the models obey certain structural and electrical requirements to ensure robustness in operation.

2.2.5 Behavioral Models

A behavioral model is a simplified model that describes the essential relationship between the inputs and outputs of a device or system. The model is simplified, so that only the features of interest in the given application need be included. This input-output relationship may be entirely mathematical, and make no reference to the underlying physics of the system, though inclusion of prior knowledge may make the model simpler. The behavioral model must be quick to run in the simulator, but provide sufficient accuracy in the domain of interest. In other words, the objective of a behavioral model is to trade off the speed of simulation against accuracy and features in the model. This is the fourth "Law of Simulation and Modeling," and is illustrated in Figure 2.7.

In the context of pre-distortion, the main behavior of interest is the nonlinear RF behavior of the PA at a certain drive level. Other features, desirable in a compact model representation, such as load sensitivity, frequency response, and DC behavior, are not relevant in this context, and therefore need not be included in the behavioral model. What we are interested in are the speed of simulation and sufficient accuracy of the results for this particular application. A compact model may be at least as accurate, but because it is dealing with several other inputs and outputs, it takes much longer in simulation.

The features of the larger system or device model that we may be willing to trade off against speed of simulation can include: DC functionality; temperature dependence; noise model; electromagnetic interactions; frequency response; capability of operation in different simulator types, such as transient, harmonic balance,

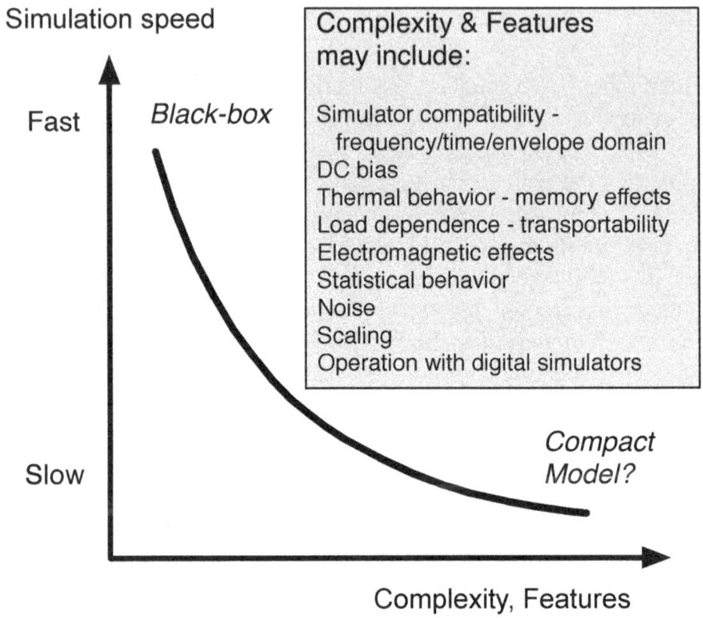

Figure 2.7 A behavioral model trades off speed against complexity. Some of the model features that we may be willing to give up are shown here.

or circuit envelope; scaling capability; load sensitivity or transportability; class of signals to be used; or any combination of these, in return for fast and accurate simulation of the behavior of interest. In many cases, including such effects for all the components in the system would result in an enormous netlist, and the simulation time, if the simulator converged at all, would be huge. Behavioral models enable these simulations to be run, and the system to be analyzed and designed.

The behavioral model can also act as a bridge between different design environments. For example, linking the detailed RF circuit design of the power amplifier to the larger RF transmitter system, or to the digital domain that is used for the pre-distorter design. An example of such a bridge in a design hierarchy is shown in Figure 2.8. This bridge may also be between different simulation environments: the RF design being done in a circuit simulator in the frequency domain, and the pre-distorter design being done in a time-stepping digital signal processing simulator. The behavioral model enables experts in these two domains to communicate with each other.

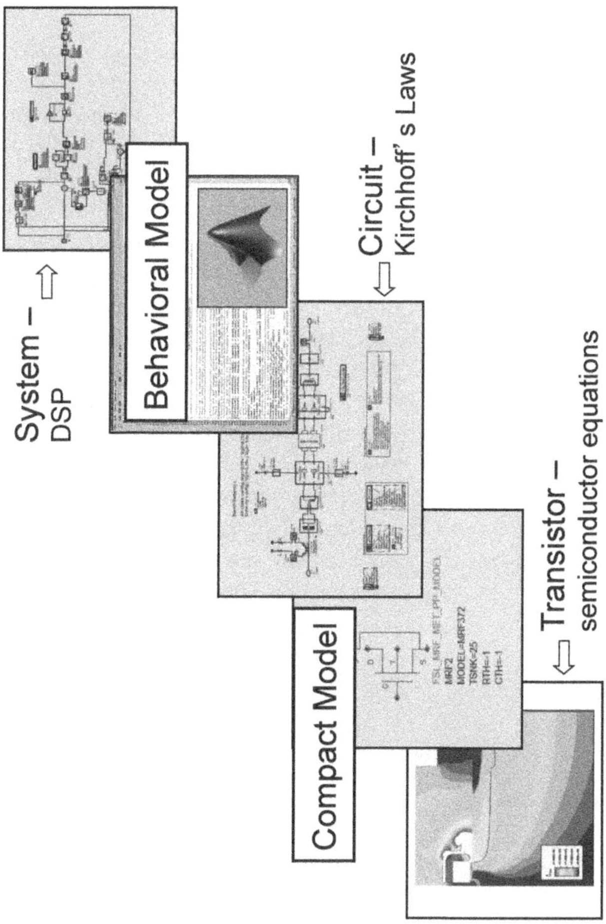

Figure 2.8 A hierarchy of simulation and models. The models provide effective bridges between the different simulation domains.

Interestingly, several compact FET models can be described as behavioral models: using mathematical descriptions to relate the input and output currents and voltages, without engaging in any physics. For example, the drain current characteristic is often modeled using a hyperbolic tangent function, with a simple expression shown below:

$$I_D = \beta \left(V_{GS} - V_T \right)^{\gamma} . \tanh \left(\alpha V_{DS} \right) . \left(1 + \lambda V_{DS} \right) \tag{2.12}$$

where V_T is the threshold voltage, and $\alpha, \beta, \gamma, \lambda$ are fitting coefficients. We take the measured $I_D - V_{DS}$ data and fit the above function to it. While we can make some observations about the effects of the various parameters on the nature of the function, for example, λ controls the slope of the current in the saturation region, that is, the output conductance, none of these fitting parameters has a meaning in terms of the physics of operation of the device. So this large-signal FET model is a behavioral model (and a compact model because it has few parameters?).

A general 'modeling landscape' is presented in Figure 2.9, indicating the range of model approaches, from physical models that describe the internal details of the semiconductor device, through to behavioral models that relate the input to output signals. In Figure 2.9, we also show that some of the modeling approaches are unilateral, in the sense that the input and output are as we think of them as labels on the PA, whereas other approaches consider that all ports of the system can be subject to input signals and deliver output signals in response. The input-output models are simpler to identify and implement, but the full, multi-port models offer an understanding of the loading and reflection effects that are important in RF and microwave systems.

2.3 MODELING

Modeling is the process that we follow in translating the measured data or device physics into a model, compact or behavioral, that is implemented in the simulator. This activity is the creation of a new model architecture or structure. The generation of parameters for an existing model implementation is called model extraction or parameter extraction. One goal of model extraction is to make this activity as procedural as possible, so that the quality of the resulting model is uniformly high.

An example model extraction process is shown in the flowchart in Figure 2.10. This flow describes the extraction of a power transistor model, which is a dynamic electrothermal model with conservative charge functions [20]. Power transistors are large devices, and the distributed effects of the metallizations are accounted for by

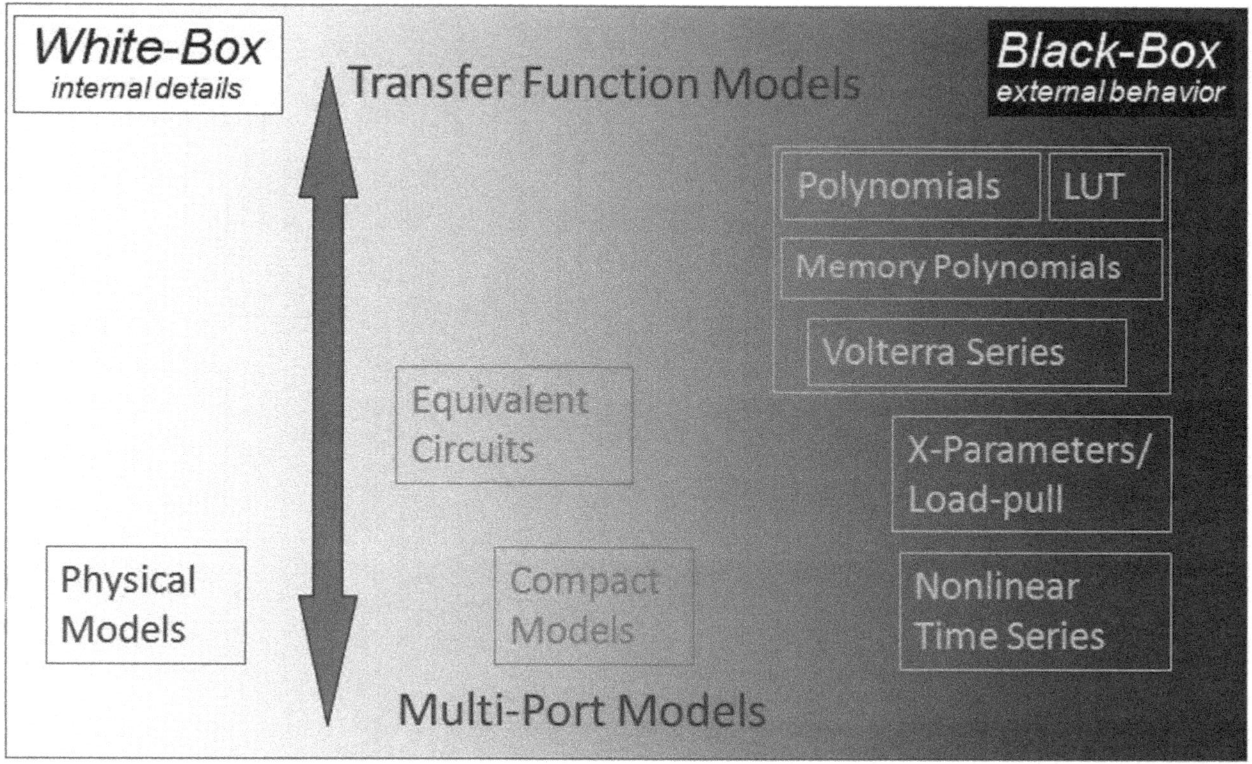

Figure 2.9 A modeling landscape, showing the range of modeling approaches from physical through behavioral.

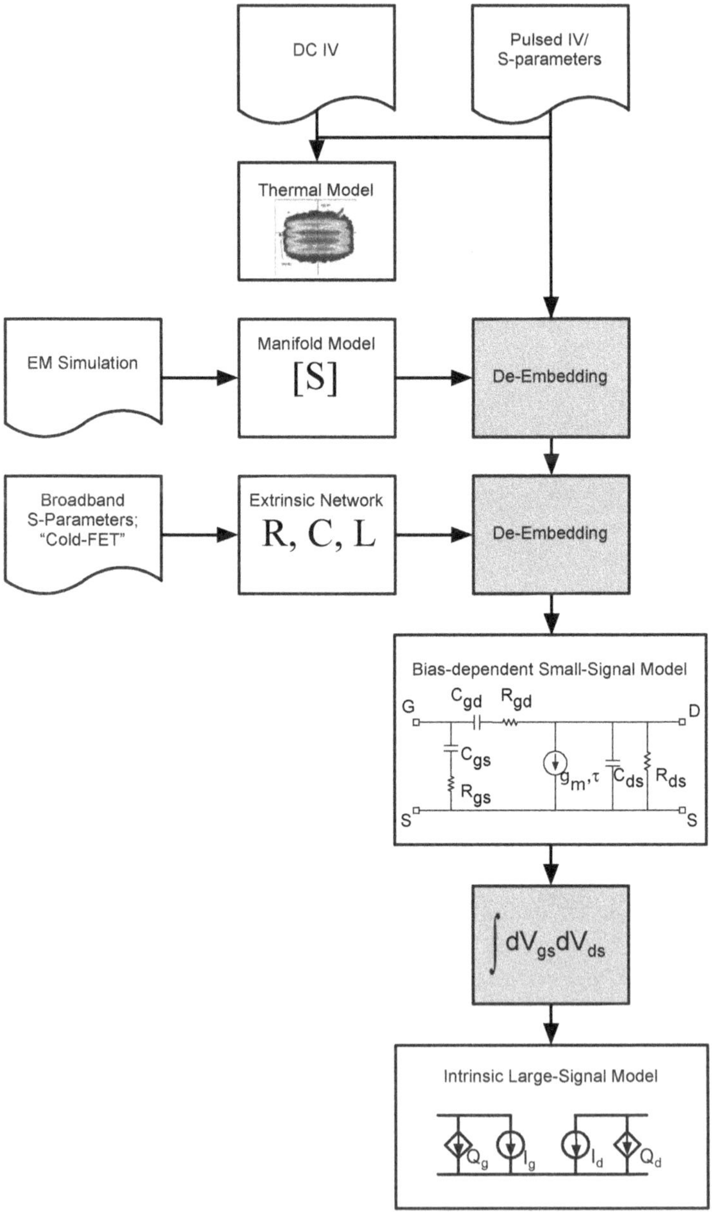

Figure 2.10 The model extraction flow for a power FET modeling procedure.

electromagnetic simulation of the gate and drain manifolds. Each of these effects plays a role in the definition of the model.

2.3.1 Function Fitting and Function Approximation

One of the key activities in modeling is finding the parameters of a function to represent the data. As we have mentioned, a function is a more compact way of describing the data. If we know that the data has a specific physical functional description, then by finding the parameters of this function we shall have a physically-based model for our data. An example is the fitting of the diode parameters, the saturation currents and ideality factors, in the Ebers-Moll BJT model. This activity is really what we mean by function fitting.

On the other hand, we may choose a more behavioral description of the data, where the physical relationship between the input and output data is not known, but we can approximate the relationship with a generic or specially-devised function. An example of the former is the use of a polynomial approximator for an unknown function:

$$y(x) = a_0 + a_1 x + a_2 x^2 + ... + a_n x^n = \sum_{i=1}^{n} a_i x^i \tag{2.13}$$

or a more generic nonlinear basis function approximation $f(x)$:

$$y(x) = \sum_{i=1}^{n} a_i f_i(x) \tag{2.14}$$

where the choice of the parameters a_i determine the accuracy of the fit of the function. An example of a specially-devised function is the drain current expression of (2.12). In these cases, we are choosing a simpler function to represent the data than the actual (physical) relationship, so we are approximating the true function with another one: this is function approximation. In practice, we find that both terms are frequently used interchangeably to describe the activity of fitting a known functional form to our data.

How well our function fits the data is governed by the error or cost function that is used to measure the difference between the data and the model. The parameters of linear models, and *linear-in-parameters* nonlinear models such as those described by (2.13) and (2.14) can be determined by well-established methods such as Gaussian elimination or least-squares fitting, which is described in more detail for linear modeling in Chapter 3, on Linear Systems and Identification.

One of the most popular functional forms for fitting data is the polynomial approximation of (2.13). This function approximation technique is built into many data and math software packages, such as Microsoft® Excel® and Mathworks™ MATLAB®, to name but two. All the user is required to do is choose the degree of the polynomial, and the software does the rest: the least-squares fitting, and even provides the error measure of the regression. While this all appears very easy, there are some pitfalls that the user needs to be aware of in the search for the best model or approximation.

A polynomial approximation will use all of the data that we present to it: it is a global approximation. This means that the function fit is the best we can make on average over the whole data set. We can exclude some data points from the approximation, for example by limiting the data range to exclude low-level noise, but it is more difficult to weight a polynomial approximation so that the fit is better in a given region of the data. This usually requires some filtering or resampling of the data set prior to fitting. If weighting the data is important, it may be better to use a different approximation technique, such as cluster-based modeling.

The polynomial fit is usually achieved using a least-squares technique, which minimizes the distance between the data and the function; this is the error measure. A better fit, that is, a smaller error measure, is generally obtained by using a higher degree of polynomial. But one should choose the degree with care: a low degree may give too large an error and not be predictive; too high a degree can result in the function passing through every data point, yielding a low error measure, but failing to predict the general trend of the data; for example, the function may be oscillatory between the data points to achieve this fit, and is therefore not predictive outside the actual data points. In this case, the interpolation properties are poor. Polynomial models tend to extrapolate poorly, whatever the degree, though again it is much worse for high-degree polynomials. In other words, the approximation is not guaranteed to be close outside the data set. Nevertheless, with care, polynomial approximations can be used successfully to create models with relatively few parameters that can predict the device or system behavior accurately within specified limits. Extrapolation functions can be built onto the polynomial approximator; often, these serve only to guide the simulator back to the region where the data is known to be valid.

Rational polynomial functions, also known as Pade approximations, are the ratios of polynomials:

$$y(x) = \frac{a_0 + a_1 x + a_2 x^2 + ... + a_n x^n}{b_0 + b_1 x + b_2 x^2 + ... + b_m x^m} \tag{2.15}$$

where m and n are non-negative integers, and b_0 is usually set to one. The numerator and denominator polynomials define the zeros and poles of the model function. Rational function approximations offer several advantages over the simpler polynomials: they can fit a wider range of shapes, including high curvatures with lower degrees in the two polynomials, resulting in fewer model parameters that need to be found. This means that they are generally smoother and less oscillatory, giving better interpolation properties. The extrapolation of the rational function expression can also be adjusted to model the observed (or desired) behavior of the data set,

$$y|_{\lim x \to \infty} \approx \frac{a_n}{b_m} x^{n-m} \tag{2.16}$$

although this expression must be used with care: if $n < m$, then the asymptotic behavior is $y \to 0$; if $n > m$, then the function tends towards infinity at large x.

Least-squares techniques can be used to find the parameters of the rational function approximation. This is generally done by recasting the expression, for each data pair (x_i, y_i):

$$y_i = y(x_i) = a_0 + a_1 x_i + a_2 x_i^2 + ... + a_n x_i^n - b_0 y_i + b_1 x_i y_i + b_2 x_i^2 y_i + ... + b_m x_i^m y_i \tag{2.17}$$

One caution when using rational function approximation is that poles will be introduced into the model, even though there may be no poles defined by the data or the theoretical analysis of the system.

Spline functions are often used to interpolate between data points. Splines are polynomial functions, but with very limited support, only over the number of points required to find the polynomial parameters. This means that they are local approximators. Often, these parameters are determined during the simulation itself, fitting the spline over the local data points from a table-based model. The spline in this case is usually very limited in degree, with cubic splines being popular. As the input signal changes, the spline is re-evaluated to determine the output. Splines are implemented in such a way as to ensure the continuity of the function and first derivative, and so provide a smooth representation of the data. Despite the low degree, this local approximation generally gives a very smooth appearance.

Piecewise polynomials are also local approximations. Low-degree polynomials are fitted to small regions of data, and then coupled together to form the model. As with splines, care is taken to ensure smoothness at the boundaries between the local polynomials. Unlike splines, though, these functions are fitted to the data to create a fixed model that is then used in the simulator.

While smooth and accurate fits to the data can be achieved with splines and piecewise polynomials, which often yield smaller fitting error measures than global polynomial approximations, their derivative behavior is often a concern in nonlinear modeling. For example, a cubic spline-interpolated table model of a power transistor may provide an accurate representation of the $I_d - V_{ds}$ data. When this device (model) is placed in an amplifier circuit that is driven into compression, the nonlinear behavior may also be reasonably well modeled, as the large-signal nonlinear behavior is traced by the spline functions over a range of amplitude. But at low signal levels, the amplitude of the signals may be too small to trace over several data points, and so only the cubic, or third-order behavior is captured. Higher-order distortions are not well-modeled.

A nonlinear function fitting technique that has received a lot of attention in recent years is the use of artificial neural hetworks (ANNs) [21–23]. This is a more complex technique of model generation than the methods outlined above, in that the parameters of the basis functions are included in the nonlinear structure. Finding the values of the parameters for a given data set requires a nonlinear optimization exercise. Much of the difficulty in applying this technique of data or function approximation has been reduced with the availability of easy-to-use software packages such as Mathworks™ MATLAB® *Neural Network Toolbox* and *NeuroModeler*® [22]. Other attractive features of ANNs are related to their mathematical properties, such as the *universal approximation theorem*, which states that a single-layer ANN can approximate any nonlinear function within an arbitrary degree of accuracy [24].

A neural network consists of a number of nonlinear basis functions, called *neurons*, often identical in form, whose inputs are a weighted sum of the input variables to the function. The outputs from the nonlinear basis functions are combined to yield the overall output of the model. The nonlinear basis function is usually chosen to be a *sigmoid* function, the hyperbolic tangent being the typical implementation. If $\varphi(\cdot)$ is this nonlinear basis function, then the overall model output y for a vector of inputs $\vec{x} = [x_1, x_2, ..., x_K]$ is

$$y = \sum_{j=1}^{N} a_j \varphi \left(\sum_{i=1}^{K} w_{ij} x_i + b_j \right) + b_{out} \tag{2.18}$$

where the sum over i is the sum of the weighted inputs to neuron j, plus a *bias* b_j, and the sum over j is the sum of the weighted outputs from the neurons, again with the possibility of an added bias value, b_{out}. The function fitting is an optimization over the weights w_{ij} and a_j and the biases b_j and b_{out}. The error measure that is minimized in the optimization is the distance between the function and the data, usually the L_2 norm.

Equation (2.18) can be expressed graphically, as in Figure 2.11, showing that the neural network can be thought of as a series of layers: the input layer, the processing or *hidden* layer of neurons, and the output layer. An ANN can have more than one hidden layer; having two hidden layers is found to improve the generalization properties of the ANN [25]. As the nonlinear basis function of each neuron is continuous and infinitely differentiable, so is the neural network output, giving a smooth function with the possibility that in nonlinear modeling, harmonic and intermodulation distortions are modeled accurately.

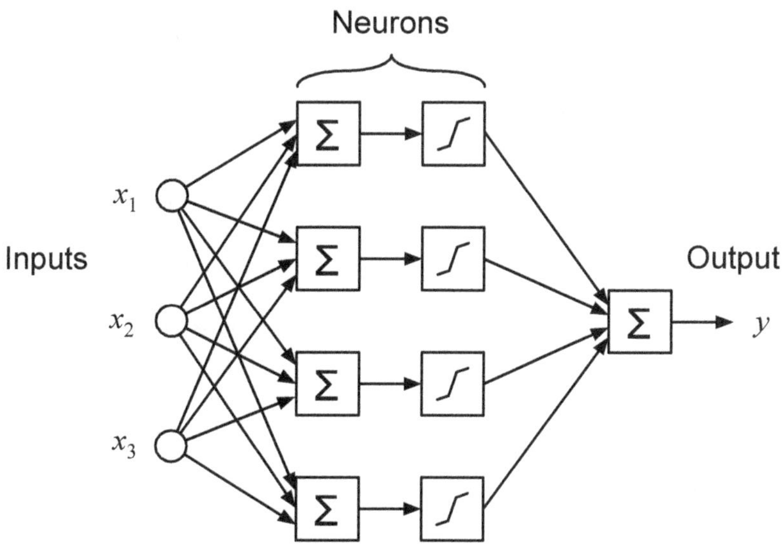

Figure 2.11 Structure of an artificial neural network, showing an input layer of K=3 inputs, a hidden layer of N=4 neurons, and an output layer comprising one linear summing node.

Neural network function approximations are particularly useful when the function or data that we want to fit has many inputs. As we shall see later, in Chapter 6, we construct a nonlinear dynamical model for a power amplifier whose inputs include the present and past values of the input voltage signal. With a polynomial approach, this would result in many cross-product terms between the present and past input signals, requiring many parameters to be found. In an ANN, the extra inputs are accommodated by using the present and past or delayed voltages as the input vector, and any increased complexity is handled by adding more neurons to the hidden layer, if required; the optimization process to solve the function fitting remains the same.

2.3.2 Model Validation

Validation and verification are two essential, non-negotiable steps in making a model.

> **Validation** answers the question "Are we doing the right thing?"
> **Verification** answers the question "Are we doing the thing right?"

Verification essentially tells us whether we have implemented our model correctly, and that the functions do what we expect them to do. Tools that can be used in the verification of the model include design reviews, code walk-throughs, and functional or *white-box* testing. White-box testing is where we verify the operation of our model by applying known input-output data to our function, and testing whether we get the right answer. In the verification process, we should also check for illegal inputs to the function, and how these are handled in the implementation of the model; "gracefully" should be our goal.

Validation is where the simulation data produced by the model is compared with an independent set of data that was not used in the construction of the model. This is *black-box* testing. Validation provides confidence in the predictive qualities of the model. It is not enough for a model just to be able to predict the data from which it was constructed. An example of model validation would be the comparison of measured and simulated large-signal load-pull data for a transistor model that was constructed using only DC and S-parameter data.

Note that validation is not the same as tuning. An *ad hoc* model fit obtained by adjusting the parameters of the model to agree with measured data may result in a model that is not predictive under other test circumstances. Validation should be an exercise carried out without model adjustment.

For behavioral models, validation can be more difficult. As one of the objectives of behavioral modeling is to reduce the scope of the model to the essential features of interest in exchange for simulation speed, it is often the case that this reduction in scope is realized by limiting the class of signals that the model can perform with. For example, a behavioral model of a PA that will be used in predicting its linear behavior when subject to pre-distortion of the signal may well be optimized for a given class of digitally-modulated signal, such as WCDMA. The WCDMA signal is continuous in time. This model may then not perform well with a time-division signal such as GSM, where the signal is switched off momentarily between the transmission slots. The time-switching of the signal allows the PA to cool slightly between the transmission times, giving rise to a thermal memory effect that may not have been captured in the original model that was optimized for WCDMA.

The validation of the behavioral model should be done with a consistent signal class. If the model is designed to be used with WCDMA signals, for example, then we can construct the model using one set of WCDMA samples taken over a specific time period, and validate against the same signal taken over a different set of time samples. This technique is known as cross-validation, and is often used in the optimization of the parameters of an ANN.

2.4 SIMULATION

Simulation is the process of solving a set of equations that describe some physical behavior, that are applied to a specific structure. The structure will include models of systems and objects whose behavior we want to understand, and the connections between these objects. Typically in the simulation and modeling of power transistors and power amplifier systems we will encounter the simulation environments outlined below.

2.4.1 Circuit Simulation

Here we solve Kirchhoff's equations for a given network of components, to find the branch currents and node voltages for given signal stimuli. The network will contain models of the components that describe the current-voltage relationship at their terminals. In the simplest case, the model will reflect Ohm's law, and in the most complex, we could have multi-port behavioral models that describe multivariate nonlinear dynamical relationships between the input and output variables. The circuit simulators use a modified form of nodal analysis that can accommodate both voltage- and current-controlled components in the (modified) admittance matrix. This arrangement enables the matrix to be partitioned for efficient solution. An excellent review of modern circuit simulation techniques and how convergence of the solution is achieved is the book by Kundert [26].

Microwave and RF design is traditionally carried out in the frequency domain. The small-signal S-parameter simulation is the sensitivity of the component response around the DC operating point, calculated in the frequency domain. For linear components, the S-parameters are independent of the DC conditions; for the transistors, the S-parameters are bias-dependent. The S-parameter simulation is often used to establish matching conditions for linear amplifier, and so forth. The large-signal behavior in the frequency domain is calculated using harmonic analysis, solving the admittance matrices at each harmonic or intermodulation frequency of interest. This simulation technique is known as *harmonic balance*. This simulation tool enables us to determine the level of harmonic and intermodulation distortion that is produced by the transistor or amplifier.

Harmonic balance is well-suited to dealing with a small number of sinusoidal input signals. As we have seen, the signals used in modern RF communications systems are digitally-modulated signals, with a bandwidth that can extend to several MHz. The circuit envelope [27] or envelope transient [28] simulation technique was devised to address the simulation of modulated signals in analog simulators. The envelope technique is a combination of harmonic balance and transient simulation methods. The harmonic balance engine is used to solve the large-signal frequency domain problem at RF; the transient simulator is used for the modulation. The simulators are coupled so that the harmonic balance solution at RF is found at each modulation-rate timestep. This avoids the problem of having to run a transient simulation at the RF period rate. Even so, the simulations require many time samples of a typical digitally-modulated signal to be able to converge to an accurate estimate of ACP, and so on. Using compact models for the power transistor devices can result in very long simulation times, hence the motivation for finding fast and accurate behavioral models that can be applied in the circuit envelope simulation environment.

2.4.2 Electromagnetic Simulation

Electromagnetic (EM) simulation solves Maxwell's equations for the electric and magnetic fields in a given physical structure. Maxwell's equations are vector differential equations expressed in three dimensions (3D); the electric and magnetic fields are integrated over the structure where the terminal potentials and currents are calculated. There are full 3D EM simulators using finite-difference and finite-element discretization of the differential equations. Such tools are appropriate for simulation of 3D structures such as transistor packages, interactions between arrays of bondwires, PA pallets, and so forth, as illustrated in Figure 2.12. By using symmetry and the properties of EM wave propagation, simulation tools have been developed for solving planar EM problems as are found on printed circuit boards, integrated circuit metallizations, and so forth. Such tools are known as '2.5D' or planar EM solvers, and for these problems are much faster and often more accurate than the 3D simulators. In the design and simulation of a power amplifier, a combination of these two approaches is required.

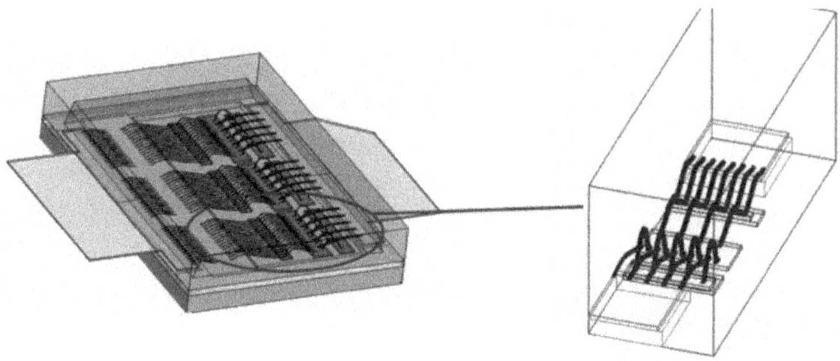

Figure 2.12 Illustration of package and bond-wire array structures for simulation in a 3D electromagnetic simulation software. The structure is the physical dimensions of the package and the electrical properties of the materials.

The EM simulation of the structures used in PAs is usually performed in the frequency domain. The output from the EM simulation is often in the form of an S-parameters file, but can also be an equivalent circuit using inductors, capacitors and resistors, or an RLGC equivalent transmission line matrix, or as a mathematical model describing the poles and zeros of the structure. In other words, the output of the EM simulation is a model. The S-parameter file or model can be used directly in the circuit simulator.

The S-parameters are usually calculated over a wide frequency range to accommodate the harmonics of the RF carrier frequency, and a DC point is added to enable convergence of the circuit simulator at DC to establish the quiescent operating conditions.

2.4.3 Device Physical Simulation

The physical simulation of a device structure solves the semiconductor equations that govern the charge transport and concentrations in the semiconductor material, for the given device structure, subject to the effects of internal and externally-applied potentials. The structure of the transistor includes sizes and shapes of the various semiconductor, insulator, and metal layers that make up the device, and their material properties, for example:

- gate length, gate oxide thickness, epitaxial layer thickness, or implant depth;
- dielectric constants, resistivities, semiconductor doping densities.

The semiconductor equations are Poisson's equation, the charge continuity equation, and, in many modern microwave devices, Schrödinger's equation to describe quantization effects in the conducting channel of the FET [29]. The solution of these equations in the given structure yields the terminal current-voltage relationship of the device. While the physical simulation is too demanding to solve in the context of a circuit simulator, we can use it to generate the DC and RF terminal characteristics from which we can extract a compact model.

Commercial semiconductor physical simulation tools can solve for sophisticated device structures, as 2D cross-sectional solutions or even in full 3D. Many of these simulation tools also offer semiconductor process simulation, so a model of the manufactured structure is generated as the framework for solving the semiconductor device equations. This enables a device manufacturer to develop a close linkage between the manufacturing process and the transistor performance. An example of a 2D cross-sectional simulator interface is shown in Figure 2.13, showing the process automation GUI and device visualization GUI.

The solution of the semiconductor equations even in two dimensions is a challenging task for the device simulator. One reason for this is that the Poisson and Continuity equations are coupled by the local potential at every point in the semiconductor, and need to be solved self-consistently. One of the challenges is that the mesh size for the numerical solution is determined by the local charge density through the Debye Length, λ_D, which is the characteristic minimum distance in a semiconductor over which the charge can change:

$$\lambda_D = \sqrt{\frac{\varepsilon_S \varepsilon_0 k_B T}{e^2 n}} \tag{2.19}$$

where $\varepsilon_S \varepsilon_0$ is the dielectric permittivity of the semiconductor, and n is the local charge density. In the regions of the transistor where the charge density is very high, such as in the channel of a FET in saturation, the mesh size must be very small to be able to describe changes in the charge density over that distance. An example of the mesh in an LDMOS channel is shown in Figure 2.14, showing that the mesh density has to track the anticipated charge density to obtain an accurate solution to the semiconductor equations. This requires a large number of calculations, and can be computationally very expensive.

An alternative to the direct coupling of the semiconductor equations in a 2D simulator is to use two coupled one-dimensional (1D) equations to model the dynamic charge behavior in the device. This approach works well for surface-mediated devices such as FETs, where the charge transport along the channel can be solved as a 1D continuity equation problem, and the charge density as a function of the depth into the device is found from a 1D solution of the Poisson equation (or

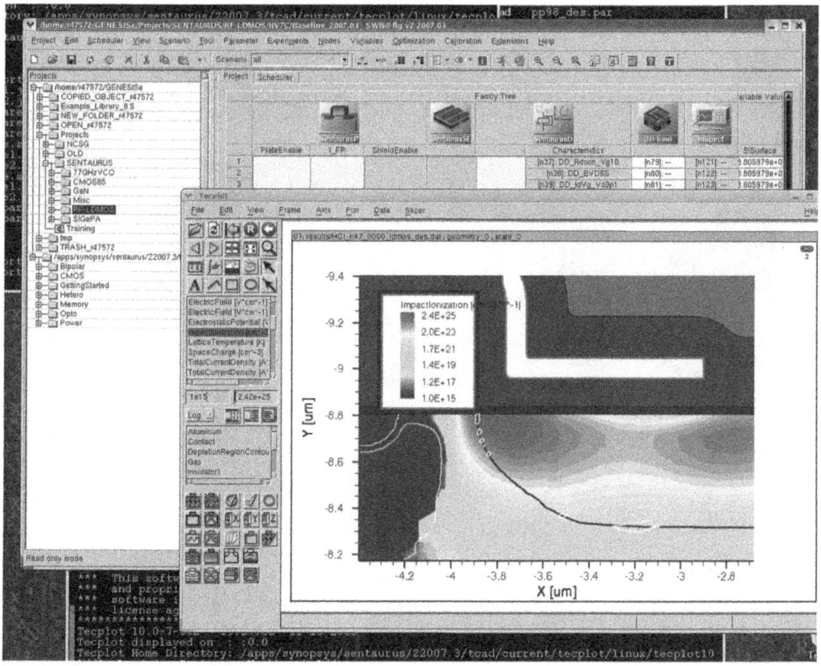

Figure 2.13 Screenshots of TCAD modeling software interface. This is the Synopsys® Sentaurus simulator. (Used with permission of Synopsys, Inc.)

Poission/Schrödinger equations for quantized channels). The coupling is by using Gauss' Law integrating around a unit cell of the channel to get the charge enclosed. Charge continuity is enforced by moving this charge from one cell to the next along the channel. Accurate results in terms of the prediction of the mobile charge density in the channel and the terminal DC and RF currents and voltages has been obtained in GaAs pHEMT and LDMOS power FET devices, over a range of manufacturing process conditions [30, 31]. The solution techniques in one dimension have better stability than 2D solvers, and the two 1D equations can be solved extremely quickly. This simulation technique is known as *Quasi-Two-Dimensional* (Q2D) simulation.

2.4.4 Thermal Simulation

The thermal simulation solves the Diffusion equation. Again, this is a three-dimensional solution accounting for the physical shape and material properties of

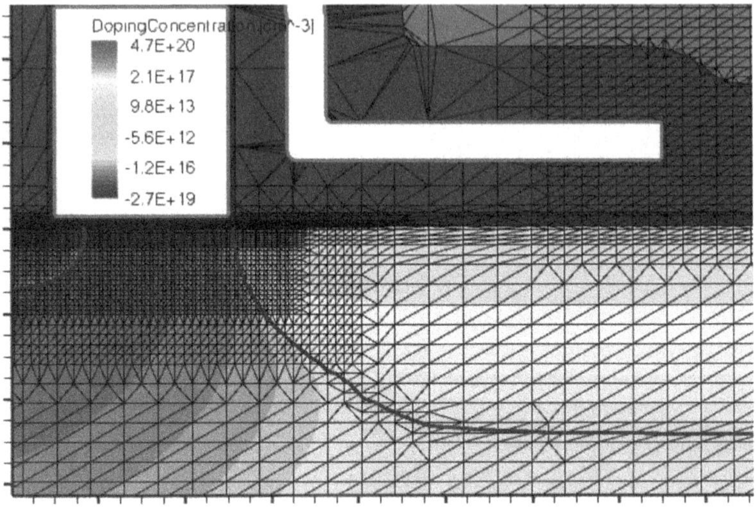

Figure 2.14 Screenshot of mesh density in TCAD modeling software. This is the Synopsys® Sentaurus simulator. (Used with permission of Synopsys, Inc.)

the semiconductor die, the package, and other components that present a path for the flow of thermal energy. The thermal simulation is often done using a steady-state input power to determine the thermal resistance of the packaged transistor, to inform the PA and system designers about heatsink requirements.

Thermal simulation at the device level is generally performed to estimate the thermal resistance and capacitance presented to the active device. The temperature rise in operation can then be estimated, and fed back to the physical simulator or to the compact model. The compact model should include a dynamic electrothermal component to the charge and current state equations [20]. It can then be used in the circuit simulator in a realistic thermal environment to predict the transistor or PA performance in practical conditions of operation. Such a model enables the inclusion of thermal memory effects, described in more detail in Chapter 5, in the circuit simulation, thereby enabling a better description of the dynamical nonlinear behavior of the RF power amplifier.

2.4.5 System Simulation

System simulators are generally time-stepping simulators, mimicking the clock cycle of a digital system or the sampling of a digitally-modulated signal in a mixed-signal simulation. The simulation is driven by the solution of mathematical descriptions of the signal processing functions that take place at the system level. In a mixed-signal context, such functions include filtering, modulation and demodulation, coding and decoding of the digital information, up- and down-sampling (filtering, again), and so forth. The mathematical operations are performed in simulators such as Mathworks Simulink® and MATLAB, which provide Toolboxes containing specialized functions, as well as an environment in which custom functions can be developed and implemented. Other EDA vendors are also providing system-level simulators that can link with their circuit simulation tools, examples being Agilent's SystemVue™ and the Visual System Simulator™ from Applied Wave Research®.

Behavioral models provide the bridge from the physical world and circuit simulation to the system-level simulator.

2.5 CONCLUDING REMARKS

In this chapter we have introduced the ideas of models and simulation, presented the differences between them, and shown that both are necessary to describe the behavior of a physical system, be it a transistor or a transceiver. Models are descriptions of the components of the system, providing output qualities in response to specific stimuli; the inputs and outputs can represent voltages and currents, or complex signals, for example. The model must be embedded in a simulator to be of any value. The simulator provides the numerical framework for solving the fundamental equations describing the physical phenomena. In this way, we can use the simulator to help us design our system, using models of the component parts of that system. The model and the simulator form a symbiotic partnership. We have presented several types of model and simulation environments, all of which are used at some point in the design of an RF power amplifier or linearized transmitter.

References

[1] N. Gershenfeld, *The Nature of Mathematical Modeling*. Cambridge, UK: Cambridge University Press, 1999.

[2] P. Aaen, J. Plà, and J. Wood, *Modeling and Characterization of RF and Microwave Power FETs*. Cambridge, UK: Cambridge University Press, 2007.

[3] D. E. Root, S. Fan, and J. Meyer, "Technology independent large signal non quasi-static FET models by direct construction from automatically characterized device data," *Proc. 21st European Microwave Conf.*, Stuttgart, Germany, Sep. 1991.

[4] D. E. Ward and R. W. Dutton, "A charge-oriented model for MOS transistor capacitances," *IEEE J. Solid State Circuits*, Vol. 13, No. 5, pp. 703–707, Oct. 1978.

[5] J. Staudinger, M. C. de Baca, and R. Vaitkus, "An examination of several large-signal capacitance models to predict GaAs HEMT linear power amplifier performance," *Proc. IEEE Radio and Wireless Conf. (RAWCON)*, Colorado Springs, CO, Aug. 1998, pp. 343–346.

[6] J. Verspecht and D. E. Root, "Polyharmonic distortion modeling," *IEEE Microwave Mag.*, Vol. 7, No. 3, pp. 44–57, June 2006.

[7] M. Myslinski, F. Verbeyst, M. V. Bossche, and D. Schreurs, "S-functions extracted from narrow-band modulated large-signal network analyzer measurements," 74th *ARFTG Microwave Measurements Symposium*, Broomfield, CO, Dec. 2009, pp. 1–8.

[8] S. Woodington, T. Williams, H. Qi, D. Williams, L. Pattison, A. Patterson, J. Lees, J. Benedikt, and P. J. Tasker, "A novel measurement-based method enabling rapid extraction of a RF waveform look-up table-based behavioral model," *IEEE MTT-S Int. Microwave Symp. Dig.*, Atlanta GA, June 2008, pp. 1453–1456.

[9] B. G. Streetman, *Solid State Electronic Devices*. Englewood Cliffs: Prentice-Hall International, Inc, 1990.

[10] H. K. Gummel and H. C. Poon, "An integral charge-control model of the bipolar transistor," *Bell System Tech. J.*, Vol. 49, pp. 827–852, May 1970.

[11] L. H. Camnitz, S. Kofol, T. Low, and S. R. Bahl, "An accurate, large-signal, high-frequency model for GaAs HBTs," *IEEE Gallium Arsenide Integrated Circuit GaAs IC Symposium*, Orlando, FL, Nov. 1996, pp. 303–306.

[12] M. Rudolph, *Introduction to Modeling HBTs*. Norwood, MA: Artech House, 2006.

[13] M. Iwamoto, D. E. Root, J. B. Scott, A. Cognata, P. M. Asbeck, B. Hughes, and D. C. D'Avanzo, "Large-signal HBT model with improved collector transit time formulation for GaAs and InP technologies," *IEEE MTT-S Int. Microwave Symp. Dig.*, Philadelphia, PA, June 2003, pp. 635–638.

[14] W. Shockley, "A unipolar field-effect transistor," *Proc. IRE*, Vol. 40, pp. 1365–1376, 1952.

[15] C. A. Leichti, "Microwave Field Effect Transistors–1976," *IEEE Trans. Microw. Theory Tech.*, Vol. 24, No. 6, pp. 279–300, JunE 1976.

[16] W. Liu, *MOSFET Models for SPICE Simulation, including BSIM3v3 and BSIM4*. New York, NY: Wiley-IEEE Press, 2001.

[17] T. Ytterdahl, Y. Cheng, and T. A. Fjeldly, *Device Modeling for Analog and RF CMOS Circuit Design*. Chichester, UK: John Wiley & Sons, 2003.

[18] G. Gildenblat, H. Wang, T.-L. Chen, X. Gu, and X. Cai, "SP: An advanced surface-potential-based compact MOSFET model," *IEEE J. Solid State Circuits*, Vol. 39, No. 9, pp. 1394–1406, Sep. 2004.

[19] H. Wang, X. Li, W. Wu, G. Gildenblat, R. van Langevelde, G. D. J. Smit, A. J. Scholten, and D. B. M. Klaassen, "Unified non-quasi-static MOSFET model for large-signal and small-signal simulations," *IEEE Custom Integrated Circuits Conference*, Santa Clara, CA, Sep. 2005, pp. 823–826.

[20] J. Wood, P. H. Aaen, D. Bridges, D. Lamey, M. Guyonnet, D. S. Chan, and N. Monsauret, "A nonlinear electro-thermal scalable model for high-power RF LDMOS transistors," *IEEE Trans. Microw. Theory Tech.*, Vol. 57, No. 2, pp. 282–292, June 2009.

[21] S. Haykin, *Neural Networks: a Comprehensive Foundation*, 2nd ed. Upper Saddle River, NJ: Prentice Hall, 1995.

[22] Q.-J. Zhang and K. C. Gupta, *Neural Networks for RF and Microwave Design.* Norwood, MA: Artech House, 2000.

[23] H. Kabir, L. Zhang, M. Yu, P. H. Aaen, J. Wood, and Q.-J. Zhang, "Smart modeling of microwave devices," *IEEE Microwave Mag.*, Vol. 11, No. 3, pp. 105–118, May 2010.

[24] G. Cybenko, "Approximation by superposition of sigmoidal functions," *Math. Control, Signals Systems*, No. 2, pp. 303–314, 1989.

[25] D. L. Chester, "Why two hidden layers are better then one," *Proc. Int. Joint Neural Networks Conf.*, 1990, pp. 265–268.

[26] K. S. Kundert, *The Designer's Guide to SPICE & SPECTRE.* New York, NY: Springer, 1995.

[27] D. Sharrit, "New method of analysis for communication systems," *IEEE MTT-S Int. Microwave Symp. Workshop, 'Nonlinear CAD'*, San Francisco, CA, June 1996.

[28] E. Ngoya and R. Larcheveque, "Envelop transient analysis: a new method for the transient and steady-state analysis of microwave communication circuits and systems," *IEEE MTT-S Int. Microwave Symp. Dig.*, San Francisco, CA, June 1996, pp. 1365–1368.

[29] J. Wood and C. G. Morton, "An analysis of the effective position of the two-dimensional electron gas in the channel of MODFET epitaxial layer structures," *IEEE Trans. Electron Devices*, Vol. 45, No. 7, pp. 1622–1624, Jul. 1998.

[30] C. M. Snowden and R. R. Pantoja, "Quasi-two-dimensional MESFET simulations for CAD," *IEEE Trans. Electron Devices*, Vol. 36, No. 9, pp. 1564–1574, Sep. 1989.

[31] J. P. Everett, M. J. Kearney, E. M. Johnson, H. Rueda, P. H. Aaen, J. Wood, and C. M. Snowden, "A quasi-two-dimensional model for high-power RF LDMOS transistors," *IEEE Trans. Electron Devices*, Vol. 58, No. 9, pp. 3081–3088, Sep. 2011.

Chapter 3

Linear Systems and Identification

A system is linear if the output is proportional to its input. There are some more formal requirements for linear behavior as we shall outline below. Essentially any linear system, when driven hard enough, will become nonlinear. In other words, the output will no longer be proportional to the input, and may contain some other signal components that were not in the original input signal, for example, noise and distortion. Power amplifiers are good examples of such nonlinear behavior.

Why study linear systems properties when we are trying to develop nonlinear models of PAs? The answer is that the linear theory offers a framework where the signal and system properties can be understood, and a foundation of techniques that can be applied, either directly or with some modification, to nonlinear systems. One such technique is the process of *linearization* of a problem about some point in its nonlinear operating conditions. In the context of modeling an RF power amplifier, we could linearize about its average output power for a given communications signal to create a model of its transfer characteristics.

In this chapter, we shall briefly review some of the basics of linear systems, covering topics such as superposition, decomposition, transfer functions, and convolution, in both the time domain and the frequency domain. We shall also provide a link between the continuous-time expressions found in the theoretical derivations and the time-sampled data that we obtain from measurement. We shall outline some approaches to solving linear system problems using the procedures of system identification, and, finally, indicate how these system identification approaches can be adapted for nonlinear systems.

3.1 A REVIEW OF LINEAR SYSTEM PROPERTIES

The output of a linear system is proportional to its input:

$$v_{out} = Av_{in} \tag{3.1}$$

This must be the case for any inputs, or any combination of inputs, so, given two input signals v_{1in} and v_{2in}, and a scalar quantity k,

$$v_{1out} = Av_{1in} \tag{3.2}$$

$$v_{2out} = Av_{2in} \tag{3.3}$$

and

$$k\left(v1_{out} + v2_{out}\right) = kA.\left(v1_{in} + v2_{in}\right) \tag{3.4}$$

This is the principle of superposition.

The output of a linear system for time $t > 0$ is the sum of two factors: the initial conditions at $t = 0$ and the result of the input signal from $t = 0$ to t. The zero-input response is the system output resulting from only the initial condition: the input signal is zero. The zero-state response is the system response arising from the input signal from $t = 0$ to t, with the initial conditions set to zero – the zero state. The ability to separate the output into these two components is the decomposition property. This can be illustrated with a simple example of a series RC network, as shown in Figure 3.1.

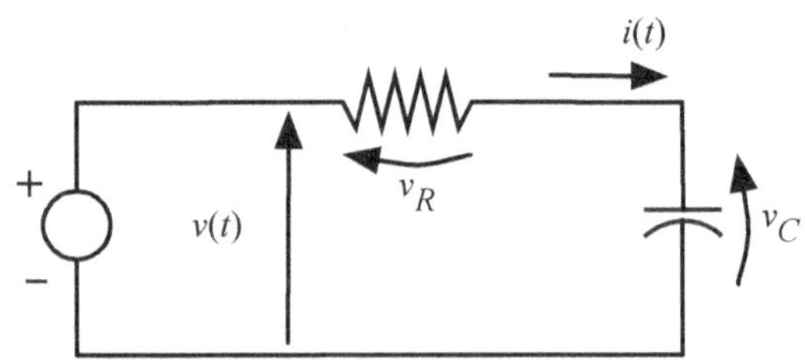

Figure 3.1 RC network with voltages and currents shown.

The total voltage across the RC network is given by

$$v(t) = i(t)R + \frac{1}{C}\int_{-\infty}^{t} i(t)dt \tag{3.5}$$

which we can rewrite as

$$v(t) = i(t)R + \frac{1}{C}\int_{-\infty}^{0} i(t)dt + \frac{1}{C}\int_{0}^{t} i(t)dt \tag{3.6}$$

where the middle term on the right-hand side is the initial voltage condition on the capacitor, $v_C(t)$, and the remaining terms are the zero-state outputs:

$$v(t) = v_C(0) + i(t)R + \frac{1}{C}\int_{0}^{t} i(t)dt \tag{3.7}$$

This expression can be generalized to initial conditions at any time, t_0:

$$v(t) = v_C(t_0) + i(t)R + \frac{1}{C}\int_{t_0}^{t} i(t)dt \tag{3.8}$$

A linear system is one that possesses these properties of superposition and decomposition. *A system is nonlinear if it is not linear*[1].

3.1.1 Instantaneous and Dynamic Systems

An instantaneous system is one where the output at some time *t* depends only on the input at the same time *t*. In electrical circuits this is true for resistors with no temperature coefficient, and ideal amplifiers with no group delay. In more practical linear circuits, we have capacitors and inductors present, and the outputs will depend on integrals or derivatives of the voltage or current associated with a given component, as we saw earlier for the RC network. These circuits are *dynamic* circuits, or circuits with *memory*, covered in more detail in Chapter 4.

For a system with an input $u(t)$, and output $y(t)$, we can write the general dynamical linear relationship

$$\frac{d^n y}{dt^n} + a_{n-1}\frac{d^{n-1}y}{dt^{n-1}} + ... + a_1\frac{dy}{dt} + a_0 y = b_m\frac{d^m u}{dt^m} + b_{m-1}\frac{d^{m-1}u}{dt^{m-1}} + ... + b_1\frac{du}{dt} + b_0 u \tag{3.9}$$

where, without any loss of generality, we have divided through by a_n.

1 This statement comes from B.P. Lathi, *Signals and Systems*, 1987, page 7.

- If the a_i, b_j are constant, the system is said to be *linear time-invariant* (LTI).
- If the a_i, b_j are functions of time, the system is *linear time-varying* (LTV).
- If the a_i, b_j are functions of anything else, the system is nonlinear, and these functional dependencies also become inputs to the system.

An example of a nonlinear system is a simple modification of the RC network in Figure 3.1. If the linear capacitor is replaced by a p-n junction capacitor, whose capacitance is dependent on the voltage across it, then the voltage across the RC network depends not only on the current but also on the voltage across the capacitor. This is a nonlinear circuit that is surprisingly complex to solve.

3.1.2 Time Domain or Frequency Domain?

Linear circuit problems are often solved in the frequency domain, using the Laplace transform. From the differential equation (3.9) we obtain:

$$\begin{aligned}&\left(s^n + a_{n-1}s^{n-1} + ... + a_1 s + a_0\right) Y(s)\\&= \left(b_m s^m + b_{m-1}s^{m-1} + ... + b_1 s + b_0\right) U(s)\end{aligned} \tag{3.10}$$

where $U(s)$ is the Laplace transform of the input signal $u(t)$, and the output signal $y(t)$ can be found from the inverse Laplace transform of the output in the frequency domain $Y(s)$. Writing $D(s)$ for the polynomial in a, and $N(s)$ for the polynomial in b, we get the following expression for $Y(s)$

$$Y(s) = \frac{N(s)}{D(s)}U(s) = H(s).U(s) \tag{3.11}$$

where $H(s)$ is the transfer function, a frequency-domain property.

The solution for $Y(s)$ is found by factoring the denominator polynomial, and splitting $H(s)$ into separate poles:

$$Y(s) = \frac{N(s)}{(s-\lambda_1)(s-\lambda_2)...(s-\lambda_n)}U(s) \tag{3.12}$$

$$Y(s) = \left\{\frac{k_1}{s-\lambda_1} + \frac{k_2}{s-\lambda_2} + ... + \frac{k_n}{s-\lambda_n}\right\} U(s) \tag{3.13}$$

for which the time-domain solution for $y(t)$ is

$$y(t) = \int_0^t \left\{k_1 e^{\lambda_1(t-\tau)} + k_2 e^{\lambda_2(t-\tau)} + ... + k_n e^{\lambda_n(t-\tau)}\right\} u(\tau)d\tau \tag{3.14}$$

+ the zero-input response.

This time-domain result contains both the transient and steady state responses to the input excitation $u(t)$. The transient components are found in the exponentials in λ, which will die out to a constant value over time, for a stable system, leaving a steady gain response, $\sum_n k_n$, multiplied by the drive signal $u(t)$. If the drive is sinusoidal, the output will also be sinusoidal, of proportional amplitude once the transients have died away: from (3.1),

$$v_{out} = Av_{in} \tag{3.15}$$

we get

$$v_{out} sin(\omega t) = Av_{in} sin(\omega t) \tag{3.16}$$

With linear passive circuits, we often work in the frequency domain at RF and microwave frequencies. The measurement instrumentation is based on periodic signal excitation and display: network and spectrum analyzers. The steady-state amplitude and phase response over frequency tell us what we need to know for linear passive networks such as filter, and even for linear amplifiers operating far from the large-signal regime.

3.1.3 Convolution Integral and Impulse Response

The zero-state response above can be written in a more familiar form:

$$y(t) = \int_0^t \left\{k_1 e^{\lambda_1(t-\tau)} + k_2 e^{\lambda_2(t-\tau)} + ... + k_n e^{\lambda_n(t-\tau)}\right\} u(\tau)d\tau \tag{3.17}$$

$$y(t) = \int_0^t h(t-\tau)u(\tau)d\tau \tag{3.18}$$

which is the convolution integral, where

$$h(t) = k_1 e^{\lambda_1 t} + k_2 e^{\lambda_2 t} + ... + k_n e^{\lambda_n t} \tag{3.19}$$

defines the transfer function response in the time domain.

The limits [0,t] indicate that the functions $h(t)$ and $u(t)$ are both causal. The general form for the convolutional integral has limits $[-\infty, \infty]$ allowing for non-causal signals; this also permits changes of variable that otherwise would require detailed proofs, which we will not go into here.

The convolution integral is a linear function. It is the solution for the output of a linear system to a given driving function, $u(t)$. If the driving signal $u(t)$ is the impulse function, then the convolution integral defines the impulse response of the circuit or system.

$$y(t) = \int_0^t h(t-\tau)\delta(\tau)d\tau = h(t) \tag{3.20}$$

Using the Laplace transform, convolution can also be expressed in the frequency domain:

$$y(t) = \int_0^t h(t-\tau)u(\tau)d\tau \tag{3.21}$$

$$Y(s) = H(s)U(s) \tag{3.22}$$

which is the product of the signal and the transfer function. If the signal is the impulse function $\delta(t)$, then the frequency domain expression yields the frequency response of the system:

$$Y(s) = H(s) \tag{3.23}$$

or, in the steady state,

$$Y(j\omega) = H(j\omega) \tag{3.24}$$

Convolution can be described graphically, as shown in Figures 3.2 and 3.3. The frequency-domain figure is probably the more natural to interpret: it can be thought of as a filtering function of the frequency response.

3.1.4 State Space Representation

An alternative formalism for expressing the relationships between the input and output signals in a linear system is to use *state space* representation:

$$\begin{aligned} \dot{x}(t) &= Ax(t) + Bu(t) \\ y(t) &= Cx(t) + Du(t) \end{aligned} \tag{3.25}$$

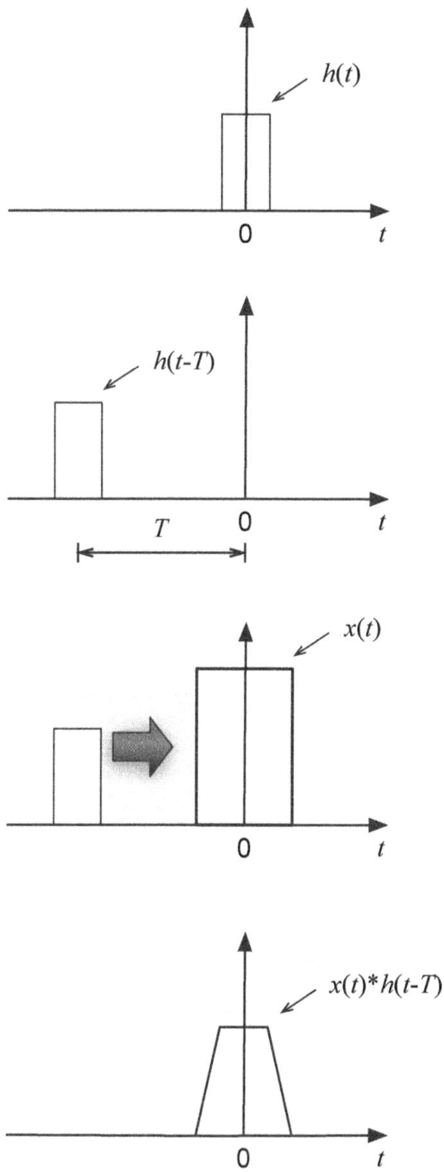

Figure 3.2 Representation of convolution in the time domain. The kernel $h(t)$ is first shifted by τ; it is then swept in time, across the signal $x(t)$, and the convolution integral plot (3.18) is calculated, yielding the convolution of $h(t-\tau) * x(t)$ in the final graphic.

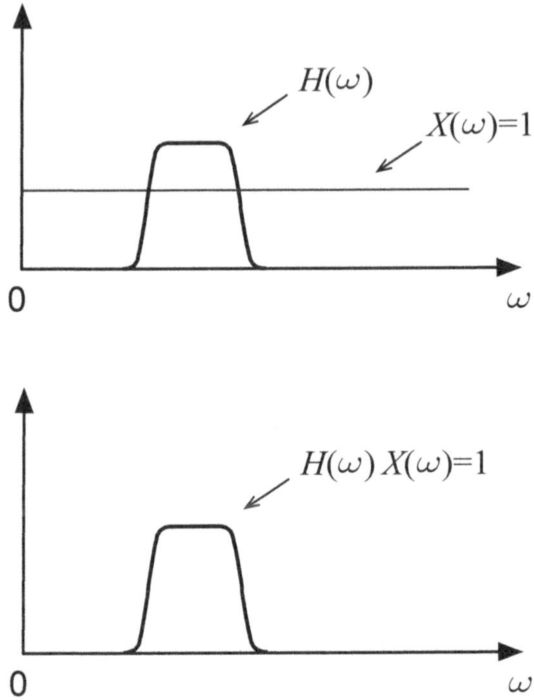

Figure 3.3 Representation of convolution in the frequency domain: filtering.

where x is a state variable of the linear system, and may be a vector describing several internal states, and $\dot{x}$ is its time derivative. The output $y(t)$ may be a function of the internal states only, with the influence of the drive $u(t)$ coming through the states, or the output may have also a direct relationship to the input signal through the parameter *D*.

In the simple RC network of Figure 3.1, the state variable is the charge on the capacitor, q. The current is then $\dot{q}$, and the voltage in the system is:

$$u(t) = v_R + v_C = \dot{q}R + {}^{q}\!/_{C} \tag{3.26}$$

and if we choose the output to be the voltage across the capacitor, v_C, we obtain

$$\begin{aligned} \dot{q} &= -\tfrac{q}{RC} + \tfrac{u}{R} \\ y &= \tfrac{q}{C} \end{aligned} \tag{3.27}$$

although this is a somewhat trivial example. An RLC series circuit is more instructive, with the inductor flux being the second state variable, leading to a vector in x.

The state space approach can be extended to nonlinear systems, by replacing the linear coefficients $\{A, B, C, D\}$ by the nonlinear functions:

$$\begin{aligned} \dot{x}(t) &= f(x, u) \\ y(t) &= g(x, u) \end{aligned} \tag{3.28}$$

3.2 DISCRETE TIME REPRESENTATION

The measured data that we shall use to characterize a system, linear or nonlinear, will generally be in a time-sampled form. This is true whether we generate the data from simulation, where the data will be calculated at the discrete time steps of the simulation, or from measurement, as most instruments nowadays will take the analog measured signal and digitize as early as possible in the receiver chain to obtain benefits of improved noise performance, resolute bandwidth, and so forth. We shall in general work directly with this time-sampled data, and so we can recast the continuous-time equations presented earlier in a suitable discrete data form.

We take the continuous time description of a linear dynamical system from (3.9),

$$a_n\frac{d^n y}{dt^n}+a_{n-1}\frac{d^{n-1}y}{dt^{n-1}}+...+a_1\frac{dy}{dt}+a_0 y = b_m\frac{d^m u}{dt^m}+b_{m-1}\frac{d^{m-1}u}{dt^{m-1}}+...+b_1\frac{du}{dt}+b_0 u \tag{3.29}$$

and write in terms of a linear difference equation

$$y(t)+a_1 y(t-1)+a_2 y(t-2)+...+a_n y(t-n) = b_0 u(t)+b_1 u(t-1)+...+b_m u(t-m) \tag{3.30}$$

where $a_0 = 1$ for convenience. This expression (3.30) is written in terms of discrete time samples, the delays are in unit steps of the sampling rate, the m & n are *memory depths* in the input and response signals, respectively, and the a_i & b_j are the parameters to be found. In this expression we have included the current input, $b_0 u(t)$, though this is sometimes omitted for a general prediction. This equation can be recast in terms of finding a solution for $y(t)$ given the past observations, and the current input $u(t)$:

$$y(t) = -a_1 y(t-1) - a_2 y(t-2) ... - a_n y(t-n) + b_0 u(t) + b_1 u(t-1) ... + b_m u(t-m) \tag{3.31}$$

which can be simplified by vectorizing

$$y(t) = \phi^T(t)\theta \tag{3.32}$$

where

$$\phi(t) = [-y(t-1) \ -y(t-2) \ ... \ -y(t-n) \ u(t) \ u(t-1) \ ... \ u(t-m)]^T$$
$$\theta = [a_1 \ a_2 \ ... \ a_n \ b_0 \ b_1 \ ... \ b_m] \tag{3.33}$$

$\phi(t)$ is known as the regression vector, and θ is the set of model parameters.

The continuous time and discrete time descriptions are physically equivalent. This comes from the numerical expression for a derivative; the first derivative, using backward difference, is

$$\frac{dy}{dx} = \frac{y(x) - y(x - \Delta x)}{\Delta x} \tag{3.34}$$

and the second derivative is, using the central difference formula:

$$\frac{d^2y}{dx^2} = \frac{y(x + \Delta x) - 2y(x) + y(x - \Delta x)}{\Delta x^2} \tag{3.35}$$

Recasting for only past time samples, this can be approximated by

$$\frac{d^2y}{dx^2} = \frac{y(x) - 2y(x - \Delta x) + y(x - 2\Delta x)}{\Delta x^2} \tag{3.36}$$

In practical terms, this means that the dynamical behavior can be captured by a sampled data system; provided the sample rate is high enough, the time derivatives will be well approximated.

3.3 GENERAL TRANSFER FUNCTION MODELS

In Section 3.1.3, we developed the continuous time convolution integral relating the output signal to the input signal and the system impulse response:

$$y(t) = \int_0^t g(t-\tau)u(\tau)d\tau \tag{3.37}$$

For the discrete time system, this becomes

$$y(t) = \sum_{k=1}^{\infty} g(k)u(t-k) \tag{3.38}$$

where we have assumed a unit sample period and t is the sampling instant.

Following Ljung (see *Bibliography*), we introduce the forward shift operator q, and the backward shift operator q^{-1},

$$\begin{aligned} qu(t) &= u(t+1) \\ q^{-1}u(t) &= u(t-1) \end{aligned} \tag{3.39}$$

which allows us to write the following shorthand notation:

$$\begin{aligned} y(t) &= \sum_{k=1}^{\infty} g(k)u(t-k) = \sum_{k=1}^{\infty} g(k)q^{-k}u(t) = \left[\sum_{k=1}^{\infty} g(k)q^{-k}\right]u(t) \\ y(t) &= G(q)u(t) \end{aligned} \tag{3.40}$$

where we have introduced $G(q)$ as the term in square brackets; $G(q)$ is the *transfer operator*. The system G is said to be stable if the impulse response function is bounded in the following way:

$$\sum_{k=1}^{\infty} |g(k)| < \infty \tag{3.41}$$

3.3.1 Noise and Disturbance Signals

A practical system is also subject to additional inputs due to noise and unwanted disturbance signals. This is illustrated in the schematic in Figure 3.4, where we

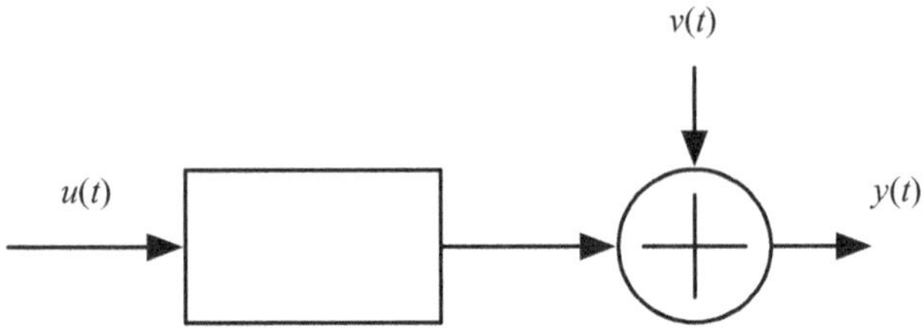

Figure 3.4 Simple system with a noise signal input.

have placed the unwanted signal at the output of the system. The output signal from the system in Figure 3.4 is given by

$$y(t) = \sum_{k=1}^{\infty} g(k)u(t-k) + v(t) \tag{3.42}$$

For the sampled data system, we can write

$$v(t) = \sum_{k=1}^{\infty} h(k)e(t-k) \tag{3.43}$$

where the signal $e(t)$ is the unwanted disturbance. If we consider this as a noise signal, then $\{e(t)\}$ is white noise, a sequence of independent random variables, and $h(k)$ is the noise transfer function of the system. The noise signal is characterized by a probability density function, giving a statistical mean and variance. For white noise, the mean is zero; the variance is some value σ:

$$\begin{aligned} \langle v(t) \rangle &= \sum_{k=1}^{\infty} h(k) \langle e(t-k) \rangle = 0 \\ \langle v(t)v(t-\tau) \rangle &= \sum_{k=1}^{\infty} \sum_{l=1}^{\infty} h(k)h(l) \langle e(t-k)e(t-\tau-l) \rangle \\ &= \sigma \sum_{k=1}^{\infty} h(k)h(k-\tau) \end{aligned} \tag{3.44}$$

where $\langle \rangle$ denotes the expectation value.

We can define a noise transfer operator H in the same way that we define the system transfer operator G earlier:

$$H(q) = \sum_{k=1}^{\infty} h(k)q^{-k} \tag{3.45}$$

so the noise voltage signal becomes

$$v(t) = H(q)e(t) \tag{3.46}$$

and the system output $y(t)$ is

$$y(t) = G(q)u(t) + H(q)e(t) \tag{3.47}$$

If we now take our discrete time LTI model of Section 3.2 and add a noise signal

$$y(t) + a_1 y(t-1) + ... + a_n y(t-n) = b_0 u(t) + b_1 u(t-1) + ... + b_m u(t-m) + e(t) \tag{3.48}$$

the noise enters the difference equation directly as an error. We have the same parameter set θ as previously, but if we introduce the following operators:

$$\begin{aligned} A(q) &= 1 + a_1 q^{-1} + ... + a_n q^{-n} \\ B(q) &= b_0 + b_1 q^{-1} + ... + b_m q^{-m} \end{aligned} \tag{3.49}$$

then we can write the model of this system as

$$y(t) = G(q, \theta)u(t) + H(q, \theta)e(t) \tag{3.50}$$

where

$$G(q, \theta) = \frac{B(q)}{A(q)}; \quad H(q, \theta) = \frac{1}{A(q)} \tag{3.51}$$

The model schematic is shown in Figure 3.5. This is known as an *autoregressive model with exogeneous inputs*, ARX; the noise signal is the exogeneous input.

3.3.2 General Model Structure

This approach to modeling of the LTI system can be extended to a general structure, from which a family of models can be derived (see Ljung). The model structure is

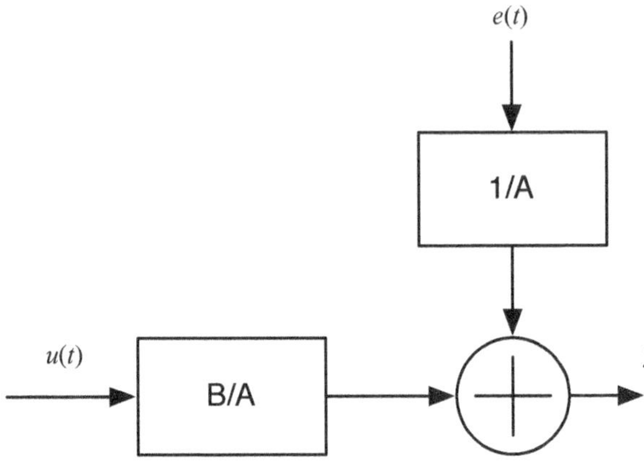

Figure 3.5 Schematic of the ARX model structure.

shown in Figure 3.6, and the model equation is

$$A(q)y(t) = \frac{B(q)}{F(q)}u(t) + \frac{C(q)}{D(q)}e(t) \tag{3.52}$$

Some common models that can be described by the general model equation are summarized in Table 3.1.

Table 3.1

Some Common LTI Models that Can Be Described by the General Model Structure.

Polynomials Used	Name of Model
B	FIR (Finite Impulse Response)
AB	ARX
AC	ARMA (Auto-Regressive Moving-Average)
ABC	ARMAX
BF	OE (Output Error)
BFCD	Box-Jenkins

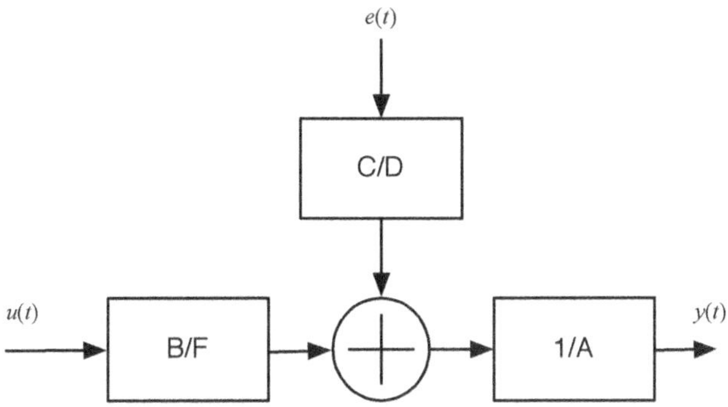

Figure 3.6 Schematic of the general model structure.

The FIR model is particularly interesting, as only the input signal is used to determine the output; such models are going to be more stable than recursive models, but will probably require more terms (greater memory) to achieve comparable accuracy.

3.4 MODEL BUILDING

3.4.1 Basis Functions

The solution for $y(t)$ may also be written in the form of a sum of basis functions $f_n(x)$, where these basis functions are fundamental solutions to the system:

$$y(t) = \sum_{n=0}^{N} a_n f_n\left(u(t)\right) \tag{3.53}$$

Common basis functions for linear systems are the trigonometric functions $sin(\omega t)$, $cos(\omega t)$, or the exponential form, which describe steady state signals as harmonic series: the Fourier series.

This formalism can be readily extended to nonlinear systems by using nonlinear basis functions. In this way, the nonlinearity is confined to the functions $f(x)$, and the solution is *linear in the parameters* a_n, which means that a wide array of linear solution techniques such as least squares can be used to extract the model parameters, even for the nonlinear model. We shall make much use of this approach in later chapters.

3.4.2 Least Squares Algorithm

Suppose that we have chosen a model structure, but we do not know the parameters; if we collect measured data over a large number of time samples, N, then we can choose the parameters θ that give us the closest approximation of the calculated value of the output, $\hat{y}$ to the measured data y, using the measured regression vector data $\phi(t)$. The measure for closeness is the average of the squares of the distances between the calculated and measured points, the L_2 norm:

$$Error = \frac{1}{N}\sum_{t=1}^{N}(y(t) - \hat{y}(t|\theta))^2 = \frac{1}{N}\sum_{t=1}^{N}\left(y(t) - \phi^T(t)\theta\right)^2 \tag{3.54}$$

The error is minimized by taking the derivative of the expression and equating to zero:

$$\frac{2}{N}\sum_{t=1}^{N}\phi(t)\left(y(t) - \phi^T(t)\theta\right) = 0 \tag{3.55}$$

hence, the set of parameters to be determined is found from the measured data set and the regression vector, through

$$\hat{\theta} = \frac{\sum_{t=1}^{N}\phi(t)y(t)}{\sum_{t=1}^{N}\phi(t)\phi^T(t)} \tag{3.56}$$

where we have used $\hat{\theta}$ to indicate that this is the set of parameters calculated from the measured data. This solution method is the least squares technique. As already noted, we shall make much use of this basic technique for model parameter extraction in later chapters, outlining limitations and enhancements to this method to overcome practical measurement issues such as noise, and mathematical limitations such as ill-conditioning of the data set.

3.4.3 Error Measures

The least squares estimator is a simple technique for estimating the parameters of a model given a set of data, through minimizing the difference between the measured output data and calculated output values over the whole data set. One thing we would like to know about our estimation is how good it is: what is the error between our estimate, the model, and the measured data? The mean squared error (MSE) measure is a commonly used error measure, and comes directly from the least squares technique. The MSE is usually normalized to measured data to yield the normalized mean square error (NMSE):

$$MSE = \frac{1}{N}\sum_{t=1}^{N}(\hat{y}(t) - y(t))^2 \tag{3.57}$$

where $\hat{y}$ is the estimated output, and y is the measured output.

The normalized MSE is:

$$NMSE = \frac{\sum_{t=1}^{N}(\hat{y}(t) - y(t))^2}{\sum_{t=1}^{N} y(t)^2} \tag{3.58}$$

and is often expressed in decibels: $10\log(NMSE)$.

The error may be weighted or filtered to give more prominence to certain sets of data, perhaps where the data collection is more accurate or where the model is designed to work best, given by the shape of a weighting function β.

$$wMSE = \frac{1}{N}\sum_{t=1}^{N}\beta(N,t)\,(\hat{y}(t) - y(t))^2 \tag{3.59}$$

For the PA models that we shall want to build later, we shall be interested in how the models perform in predicting the PA behavior when excited by digitally-modulated signals. In this respect, we shall want to see how well the model predicts such measures as error vector magnitude (EVM), which describes distortion generated by the PA in the information channel, and adjacent channel leakage ratio (ACLR), which describes the distortion outside the information channel. These are measures of the nonlinear performance of the model; we shall describe these in more detail in later chapters.

3.5 SYSTEM IDENTIFICATION

System Identification is essentially a mathematical framework for determining the dynamical properties of a given system. There are well-established methodologies used in many branches of science and engineering, including control engineering, chemical engineering, aerospace and mechanical engineering. The basic procedure to construct a model from observed data requires:

- a dataset, often time-sampled data;
- a model structure, such as FIR, or ARX;
- a criterion for selecting the best model, such as the minimization of the least squares error.

The experiment to generate the data set should be chosen with a view to producing data with the maximum information about the system under observation. For example, a single-tone test over a limited frequency and power range may cover the specification limits of the PA, but will not exercise the PA's dynamical behavior in the same way that a two-tone test or using a modulated signal can.

The model structure is chosen with a number of objectives in mind. Not least, is that the model should be able to represent the data accurately. Also the model parameters should be straightforward to determine; generally, models with large numbers of parameters can be difficult to fit uniquely. We may also find that increasing the number of model parameters causes the model error to increase, as the model structure can become ill-conditioned. Occam's Razor is always a good principle to follow.

The least squares estimation has been presented as a straightforward technique, easy to code, that can be used for estimating the model parameters in a linear system. It can be coded directly, as presented, or more robust techniques for overdetermined systems can be employed, such as QR factorization. More sophisticated parameter estimation techniques exist, such as recursive least squares (RLS), that can converge more rapidly, but are more expensive to implement. Using linear-in-parameters models in the nonlinear regime allows us to use these straightforward estimation techniques for model identification.

3.5.1 Nonlinear System Identification

The expressions for linear dynamical systems in continuous time and discrete time format can be written in a straightforward way for nonlinear dynamical systems, by replacing the linear combination of time derivatives or delays by a nonlinear functional description. The continuous time case:

$$y(t) = \mathrm{F}\left(u(t), \dot{u}(t), \ddot{u}(t), ..., \dot{y}(t), \ddot{y}(t), ...\right) \tag{3.60}$$

yields a multivariate nonlinear differential equation that must be solved to determine the model; and the discrete time expression:

$$y(t) = \mathrm{F}\left(u(t), u(t-1), ..., u(t-m), ..., y(t-1), y(t-2), ..., y(t-n)\right) \tag{3.61}$$

Using similar approaches to the linear case, we can obtain nonlinear analogues of the ARX and FIR transfer function models, where the model kernels are nonlinear functions.

The nonlinear model is a mapping of the past input and output data to the new output $y(t)$, defined by the set of parameters θ

$$\hat{y}(t|\theta) = \mathrm{g}\left(\{u^N, y^N\}, \theta\right) \tag{3.62}$$

where $\{u^N, y^N\}$ represents the past N input and output data samples, and g is a nonlinear mapping. The nonlinear mapping problem then comprises:

- the choice of the regression vector, $\phi(t)$;
- the choice of the mapping $\mathrm{g}(\phi, \theta)$ from the regression space to the output.

Often, the choice will be to use a basis function approach for the nonlinear mapping. Suitably chosen, this nonlinear model will confine the nonlinearity to the basis functions, and be linear-in-parameters so that we can use linear identification methods to determine the model, as outlined in Section 3.4.1.

$$\begin{aligned} \mathrm{g}(\phi, \theta) &= \sum_{k=1}^{n} \alpha_k g_k(\phi) \\ \theta &= [\alpha_1 \ ... \ \alpha_n] \end{aligned} \tag{3.63}$$

where g_k are the basis functions. A typical choice for nonlinear PA modeling is a polynomial basis, leading to a Volterra series dynamical model (see Chapter 6). This is a linear-in-parameters model, so we can follow the basic procedures of system identification to determine the model.

BIBLIOGRAPHY

There are several excellent texts on Linear Systems Theory; I have only scratched the surface here. A couple of excellent general background texts are:

- B. P. Lathi, *Signals and Systems*, Prentice-Hall, 1987
- T. Kailath, *Linear Systems*, Prentice-Hall,1980

In the context of linear and nonlinear model development, I have found the following text very useful:

- L. Ljung, *System Identification: Theory for the User*, Prentice-Hall, 2nd ed., 1999

This text supports a MATLAB® Toolbox.

Chapter 4

Modeling PA Nonlinearities

John Wood and Gayle Collins

As we have just seen in the previous chapter, a linear circuit produces an output that is a replica of its input: it introduces no distortion. Distortion is the addition of extra frequency components to the signal as it passes through the circuit; it is inherently a nonlinear phenomenon.

Distortion in the circuit context is not simply a change in the shape of the time-domain waveform of the signal. For instance, a linear filter will change the amplitude and phase of a signal dependent on the signal frequency; a multi-tone signal such as music may well have a different appearance after passing through a filter, but there will not be any new tones in the signal. The filter can be described by a linear transfer function. On the other hand, a PA running in compression is very nonlinear: harmonic and intermodulation tones appear at the output, new tones that are not present in the input signal. This amplifier is producing distortion.

With this viewpoint, all distortion is nonlinear, as linear circuits are incapable of producing it. The term 'nonlinear distortion' is frequently found: this seems to be a redundant phrase. The term 'linear distortion' is often presented, and in this context it is meaningless. But in image processing, for example, linear distortion is a useful term, describing the stretching of an image in one direction.

In this chapter we shall look at ways of describing the distortion produced by an RF power amplifier. This will include AM-to-AM and AM-to-PM distortion behavior, and we shall introduce a description of memory effects in RF PAs.

4.1 GAIN COMPRESSION IN THE PA

Figure 4.1 shows a typical power drive-up characteristic and gain response for an RF power amplifier at the fundamental frequency of operation. At low powers, we have the so-called linear region of operation; while this is a small-signal regime, it is unlikely to be linear for any practical RF or microwave PA. This is because the physical mechanisms governing charge control in the FET channel or the base of a BJT are not themselves linear functions. Even at low power, the PA will generate some low level of distortion.

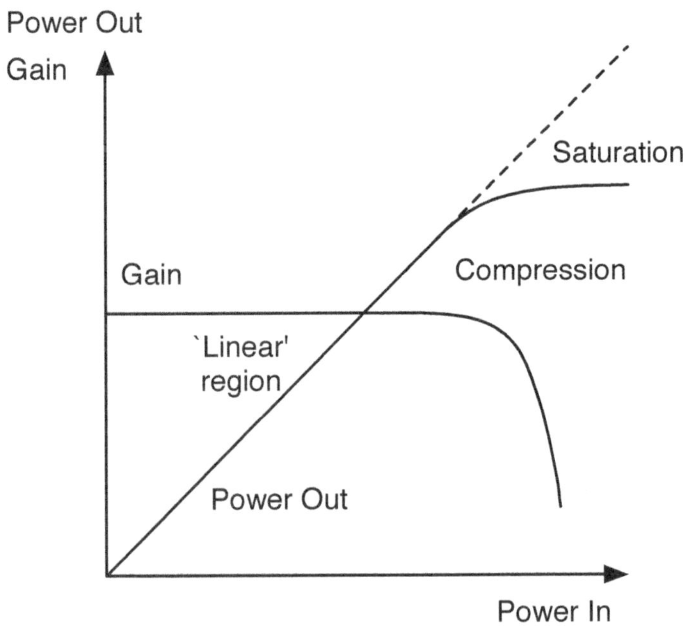

Figure 4.1 Power output and gain compression characteristics of a PA.

At higher powers, we can see that the output power and gain deviate significantly from the linear relationship at small signal. This is the compression region of operation, and at sufficiently high input drive, we will get no more power out of the PA: at this point, we are at the saturated power. In these regions of operation, the PA is very nonlinear. This compression behavior is also known as AM-to-AM conversion: by modulating (changing) the input signal amplitude, we affect or modulate the amplitude of the output signal in a nonlinear fashion.

The onset of this 'strong' nonlinearity is often claimed to be the 1-dB gain compression point: the (output) power at which the gain is one decibel below the small-signal value. The location of this point is somewhat subjective, since the small-signal gain is seldom ruler-flat, and, depending on the gate bias, may even exhibit some gain expansion. Even so, by the time the PA is at 1-dB compression, it will already be generating quite some distortion. The idealized PA device that we use to calculate the maximum linear power, efficiency, and so forth, allows a fundamental output voltage swing of twice the DC bias voltage, and a fundamental current amplitude of half the maximum current, I_{max} [1]. The real transistor has a knee region in its output $I-V$ characteristics, and as the RF voltage swings into this region, the current must follow this trajectory in the knee region, so the current (and the voltage) waveforms become nonsinusoidal: distortion is generated, in the form of harmonic products. This is illustrated schematically on the $I-V$ characteristics in Figure 4.2 and in the current waveform in Figure 4.3. The fundamental and harmonic loading will determine how these distortions are manifest in the desired signal.

4.1.1 Modeling the CW Input-Output Relationship

The output power is related to the input power by what at first sight appears to be a very simple function. Some elementary curve-fitting should allow us to model this relationship. Before we throw a math package at the problem, let's see what we are trying to do.

Our objective is to find a smooth function that approximates the measured input power–output power data. That is, a function that fits the data with the minimum of error. In fact, we want more than this: we don't want the model to fit well only at the data points themselves, but to fit the underlying function whose shape we don't know precisely, but one that we try and 'eyeball' as a smooth curve between the measured data points.

This function is our model, although we have yet to determine a class of functions from which to choose our model. The error measure allows us to choose from a candidate set of models the one that fits the data the best. We can define the error or cost function in a number of ways, though the most popular is a least-squares measure, outlined in Chapter 3.

We should expect the data to contain measurement noise, and without any further information, we assume this noise to be centered around the data and have

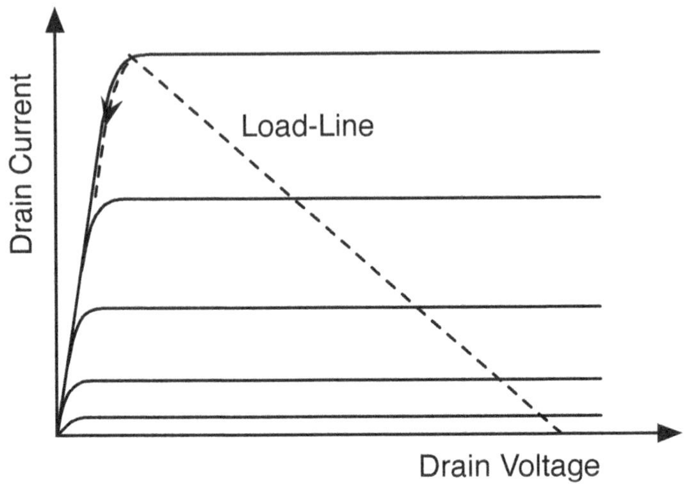

Figure 4.2 Output IV characteristics and load line indicating that at large signal amplitudes, the voltage enters the knee region, and the current must fall correspondingly.

a given variance; in other words, it is white noise. Because of this, we should not expect our model to go through all of the data points exactly. In that case we would be fitting the data plus noise, and not modeling the underlying input-output characteristic. This consideration eliminates simple spline functions from our candidate model set, as they are designed to connect the dots.

The input-output curve is clearly nonlinear, so we are free to choose any nonlinear function we like to fit this curve: polynomials, Bessel functions, Fourier series, logarithmic-exponential pairs, artificial neural networks, and so forth. All are worthy options, and all have been used in this context at some time or another. But let's focus on simplicity, with enough structure to allow us to explain the physical outcomes of this input-output relationship. We know that harmonic distortion is a result of this transfer characteristic, and we can obtain harmonic frequency components from polynomial products. All of which seems a long way round to choosing a polynomial as our representative example for modeling the input-output curve.

Nevertheless, this choice allows us to write the model as a series of nonlinear basis functions that are linear in the parameters of the model. Linear-in-parameters means that the nonlinearity is captured by the basis functions, and the model fit

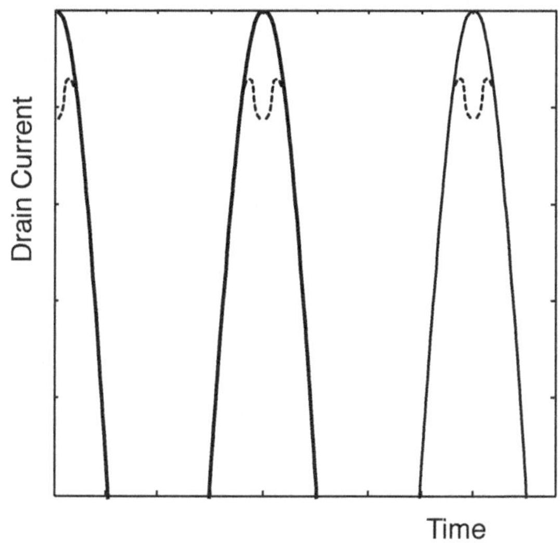

Figure 4.3 Sketch of the output current time-domain waveform, when the voltage amplitude is large enough to enter the knee region.

is determined by the coefficients, enabling linear mathematical techniques to be applied to solving this problem.

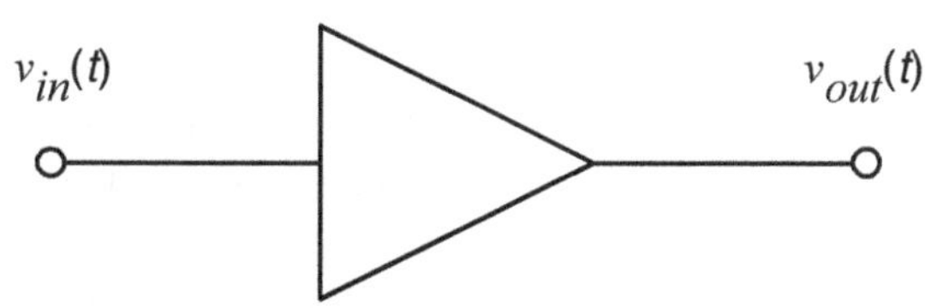

Figure 4.4 The schematic PA.

$$v_{out}(t) = a_1 v_{in}(t) + a_2 v_{in}^2(t) + ... + a_N v_{in}^N(t) = \sum_{n=1}^{N} a_n v_{in}^n(t) \tag{4.1}$$

Here we have written the polynomial in terms of the input and output voltage of the amplifier, as indicated in Figure 4.4. The values for the coefficients or parameters, a_n, and the degree of the polynomial, N, are the free parameters that define the model. The parameter a_1 can be identified as the linear gain. We can also write this expression in matrix form:

$$\mathbf{v}_{in}.\vec{A} = \vec{v}_{out} \tag{4.2}$$

$$\begin{pmatrix} v_{in}(t_1) & v_{in}^2(t_1) & \cdots & v_{in}^N(t_1) \\ v_{in}(t_2) & v_{in}^2(t_2) & \cdots & v_{in}^N(t_2) \\ v_{in}(t_3) & v_{in}^2(t_3) & \cdots & v_{in}^N(t_3) \\ \vdots & \vdots & \ddots & \vdots \\ v_{in}(t_T) & v_{in}^2(t_T) & \cdots & v_{in}^N(t_T) \end{pmatrix} \begin{pmatrix} a_1 \\ a_2 \\ \vdots \\ a_N \end{pmatrix} = \begin{pmatrix} v_{out}(t_1) \\ v_{out}(t_2) \\ v_{out}(t_3) \\ \vdots \\ v_{out}(t_N) \end{pmatrix} \tag{4.3}$$

If the number of data points, T, is the same as the polynomial degree, N, this means that the matrix $\mathbf{v_{in}}$ is square, and is known as a Vandermonde matrix. We can solve for the polynomial coefficients $\vec{A}$ by inverting $\mathbf{v_{in}}$ and multiplying by the output data vector $\vec{v}_{out}$. There are numerically efficient methods of solving the Vandermonde matrix [2].

The polynomial of degree N will then go through each data point. This is not a good idea, for a couple of reasons. First, we know the data is noisy, so the model cannot be correct; second, the resulting polynomial is unlikely to predict correctly the expected value of the output voltage at points between the data points: the interpolation of this polynomial is poor. Generally we try to use more data than fitting parameters to get a better fit for the function that underlies the data: an overdetermined problem. We can do this by collecting a lot of data, and by limiting the polynomial degree: a 5th- or 7th-degree is often sufficient for Class AB PAs, while 9th- or 11th-degree or maybe even higher can be required for Doherty and ET PAs.

The general case of an overdetermined system is the one we are interested in solving. For this general case, we can write $\mathbf{H}\vec{x} = \vec{y}$, where $\mathbf{H}$ is an $M \times N$ matrix, and the number of observations (rows) is much greater than the number of parameters or coefficients, $M > N$. There are potentially infinitely many solutions to this problem, and we find the best approximation to the vector $\vec{x}$ in the least squares sense. In many textbooks, you will find that the solution to this matrix equation can be written as:

$$\begin{aligned} \mathbf{H}\vec{x} &= \vec{y} \\ \mathbf{H}'\mathbf{H}\vec{x} &= \mathbf{H}'\vec{y} \\ \vec{x} &= (\mathbf{H}'\mathbf{H})^{-1}\mathbf{H}'\vec{y} \end{aligned} \tag{4.4}$$

These are known as the *Normal equations*, and the expression $(\mathbf{H}'\mathbf{H})^{-1}\mathbf{H}'$ is called the *pseudo-inverse*. This approach is often used, though strictly it is accurate only if the matrix $\mathbf{H}$ is of full rank: all the rows or columns are linearly independent of one another. In practice, with real data, we often find that $\mathbf{H}$ is rank-deficient; that is, the rank of $\mathbf{H}$, $R < N$, which results in the matrix inverse being ill-conditioned, and hence the solution may be inaccurate. In this case, we tend to favor more robust methods of solution of the matrix inversion.

One technique is *QR* factorization: the matrix $\mathbf{H}$ can be factorized into an orthonormal matrix $\mathbf{Q}$ and an upper triangular matrix $\mathbf{R}$, so that we can write

$$\begin{aligned} \mathbf{H}\vec{x} &= \vec{y} \\ \mathbf{QR}\vec{x} &= \vec{y} \\ (\mathbf{Q}'\mathbf{Q})\,\mathbf{R}\vec{x} &= \mathbf{Q}'\vec{y} \\ \mathbf{R}\vec{x} &= \mathbf{Q}'\vec{y} \end{aligned} \tag{4.5}$$

Because the matrix $\mathbf{Q}$ is orthonormal, $\mathbf{Q}'\mathbf{Q}$ is the identity matrix, and so the left hand side can be solved by back-substitution (Gaussian elimination) to find the unknown vector $\vec{x}$. In MATLAB, this can be achieved using the backslash '\' operator. In MATLAB, the pseudo-inverse function, *pinv*() can also be used for these overdetermined rank-deficient problems, as behind-the-scenes it performs the robust factorization.

$$\vec{x} = \mathbf{R}\backslash\,(\mathbf{Q}'\vec{y}) \tag{4.6}$$

The matrix $\mathbf{H}$ can also be written in the form [2]

$$\mathbf{H} = \mathbf{UWV}^T \tag{4.7}$$

where $\mathbf{U}$ and $\mathbf{V}$ are orthonormal matrices whose inverse is equal to their transpose, so

$$\mathbf{U}^T\mathbf{U} = \mathbf{VV}^T = \mathbf{I} \tag{4.8}$$

where $\mathbf{I}$ is the identity matrix, and all are $N \times N$ matrices. The matrix $\mathbf{W}$ is an $N \times N$ diagonal matrix whose elements, w_i, are known as the *singular values*. This is the singular value decomposition (SVD) of the matrix $\mathbf{H}$. In terms of the SVD, the solution to $\mathbf{H}\vec{x} = \vec{y}$ is

$$\vec{x} = \mathbf{VW}^{-1}\mathbf{U}^T\vec{y} \tag{4.9}$$

The inverse of **W** is also a diagonal matrix, found by replacing all of the elements w_i of **W** by their inverse. The SVD function is found in many commercial mathematical software tools. While these high-level functions are available in these commercial tools, making it easier for us to develop our model parameter extraction methods, when it comes to implementing them in a digital pre-distortion context, for example, in a FPGA, we will need to know much more about the underlying mathematical routines.

By the above means, we now have our polynomial model describing the input-output relationship of the PA, in terms of input and output voltage.

$$v_{out}(t) = \sum_{n=1}^{N} a_n v_{in}^n(t) \tag{4.10}$$

We can now apply a signal to this model and analyze the result. We use a cosinusoidal signal:

$$v_{in} = V_0 \cos(\omega_0 t + \varphi) \tag{4.11}$$

where V_0 is the magnitude of the input voltage, ω_0 is the angular frequency, and φ is an arbitrary phase offset of the signal. Either or both of V_0 and φ could be modulated qualities.

Putting (4.11) into (4.10), the output signal, $v_{out}(t)$, is:

$$\begin{aligned} v_{out}(t) &= a_1 V_0 \cos(\omega_0 t + \varphi) \\ &+ a_2 V_0^2 \cos^2(\omega_0 t + \varphi) \\ &+ a_3 V_0^3 \cos^3(\omega_0 t + \varphi) \\ &+ a_4 V_0^4 \cos^4(\omega_0 t + \varphi) \\ &+ a_5 V_0^5 \cos^5(\omega_0 t + \varphi) + ... \end{aligned} \tag{4.12}$$

The parameters $\{a_n\}$ in this model are real. Using the well-known trigonometrical identities, (4.12) becomes

$$\begin{aligned} v_{out}(t) &= \tfrac{1}{2}a_2V_0^2 + \tfrac{3}{8}a_4V_0^4 \\ &+ a_1V_0\cos(\omega_0 t+\varphi) + \left(\tfrac{3}{4}a_3V_0^3 + \tfrac{5}{8}a_5V_0^5\right)\cos(\omega_0 t+\varphi) \\ &+ \left(\tfrac{1}{2}a_2V_0^2 + \tfrac{1}{2}a_4V_0^4\right)\cos(2\omega_0 t+2\varphi) \\ &+ \left(\tfrac{1}{4}a_3V_0^3 + \tfrac{5}{16}a_5V_0^5\right)\cos(3\omega_0 t+3\varphi) \\ &+ \tfrac{1}{8}a_4V_0^4\cos(4\omega_0 t+4\varphi) + \ldots \\ &+ \tfrac{1}{16}a_5V_0^5\cos(5\omega_0 t+5\varphi) + \ldots \end{aligned} \tag{4.13}$$

The first line in this solution indicates a DC component of the output voltage, due to the even-order components of the transfer characteristic. This voltage will add to the applied bias voltage, causing a shift in the bias condition of the transistor. This is known as self-biasing. This changed bias condition will cause a change the nonlinear properties of the transistor. The second line contains the desired linear component of the output, $a_1V_0cos(\omega_0t+\varphi)$, plus some distortion contributions at the fundamental frequency, due to the odd-order components; this results in the compression or AM-to-AM effect. The remaining components of the output signal are harmonics due to the higher order components of the transfer curve.

We can hence identify three well-known forms of distortion from this simple model:

- self-biasing: a shift in the DC operating point, leading to changes in the detailed behavior of the transistor, such as changes in the gain;
- AM-to-AM compression: nonlinear gain;
- harmonic distortion: the generation of new frequency components at integer multiples of the input signal frequency.

If we now consider the possibility that either or both of V_0 and φ are modulated qualities, we can see that amplitude and phase distortion of the modulated signal will be present *in-band*, as a result of the $\frac{3}{4}a_3V_0^3 + \frac{5}{8}a_5V_0^5$ term at the fundamental. Amplitude and phase distortions of the modulated signal will also be present around the harmonics.

Further, there will also be modulation of the 'DC' terms, which are now not strictly DC but are amplitude-modulated signals at the baseband. This will cause modulation of the quiescent bias point of the transistor, essentially modulating the instantaneous gain of the PA about its quiescent value. For simplicity, if we assume a simple cosinusoidal modulation of the signal at frequency ω_m, then, from (4.13), the 'DC' term varies as (taking only the squared term for simplicity)

$$\delta V_{dc} \propto \cos^2(\omega_m t) \sim \cos(2\omega_m t) \tag{4.14}$$

This change in the output DC bias is reflected in a similar change in the drain (or collector) current, which is fed back to the input voltage of the transistor through the source (emitter) resistance. To first order, the transconductance of the device will be modulated with this signal. Thus, from (4.13), the linear gain term becomes

$$a_1(t) = a_1 + a_m \cos(2\omega_m t + \xi) \tag{4.15}$$

where a_m is the amplitude of the gain variation, and ξ accounts for any delay in the change in gain in the transistor as a result of the impressed baseband modulation. The linear component of the output signal then will contain an additional term due to this distortion term

$$v_{dist}(t)\big|_{\omega_0} = a_m \cos(2\omega_m t + \xi).V_0 \cos(w_m t) \tag{4.16}$$

where the ω_0 component is dropped for clarity. This product of cosines can be simplified to

$$v_{dist}(t)\big|_{\omega_0} = \tfrac{1}{2} a_m V_0 \left(\cos(3\omega_m t + \xi) + \cos(\omega_m t + \xi)\right) \tag{4.17}$$

This contains an up-mixed distortion component at three times the modulation rate, with some phase offset, as well as a delayed version of the original signal. This is manifestation of a *memory effect*, which we shall describe further in Chapter 5. This signal will appear in the adjacent channels to the desired modulation channel.

So even with this relatively simple model, a polynomial with real coefficients, we can observe and predict a number of physical, measurable phenomena. As we shall show in the sequel, we can use the polynomial model with complex coefficients to describe AM-to-PM effects, and to investigate phenomena in the envelope or modulation domain further.

But first, just a word or two of caution regarding the use of polynomial models. While they are relatively straightforward to identify from the measured data, care is needed in their implementation and use. We have already mentioned the desirability of an overdetermined data set from which to create the model; this should help to ensure that the model predicts a smooth curve between the data points. Despite this, the model may still exhibit oscillations in its response between the data points, if too few measurement points are used.

Another potential problem with polynomials is extrapolation: high-degree polynomials tend to do poorly at this, and predictions outside the range of the measured data can have large errors, with the polynomial shooting sharply away from the end of the measured data range. This will lead to errors, and probable non-convergence in simulation for inputs that are outside the data range. In fact, it is more problematical than this. During simulation, the simulator may make estimates of the solution on its path to the answer that are outside the measured data range; the model result for this input may have a Jacobian entry (derivative) whose value takes the simulation away from the true solution, leading to non-convergence, even though the correct result does lie within the measured data range.

4.2 PHASE CHARACTERISTIC: AM-TO-PM BEHAVIOR

The relative change in the transfer phase, $\angle s_{21}$, with power input is known as the amplitude modulation to phase modulation characteristic, AM-to-PM. The reference point for the phase is usually the small-signal value. The AM-to-PM characteristic for various amplifier types is illustrated in Figure 4.5. It is evident from this figure that the AM-to-PM is very dependent on the amplifier configuration, with the bias point playing a major role. Further, the phase change can occur at power levels that are well backed-off from the compression region. The mechanisms for AM-to-PM are the nonlinear input and output capacitances of the transistor, examples of which are shown in Figures 4.6 and 4.7.

In III-V MESFET-type devices, the input capacitance is a nonlinear function of the gate-source voltage, though it is quite smooth in the normal reverse-bias region of operation of the gate. For III-V HEMT devices, the depleted barrier layer acts as an insulating layer and the capacitance is nearly constant in the normal operating region of the transistor. When used in Class C, as the peaking amplifier in a Doherty, the transistor is switched from off to on during operation, and this introduces a nonlinear change in the input capacitance. The output capacitance of these FET devices is fairly constant with respect to the drain-source voltage, except at low voltages around the $I - V$ knee region, where the capacitance increases.

LDMOS FETs have a more pronounced change in the output capacitance at low values of drain voltage than the III-V FETs, resulting from the diffused n-drift layer. Being MOS-based, their input capacitance is relatively constant in normal operation, although in Class C operation, as the Doherty peaking device,

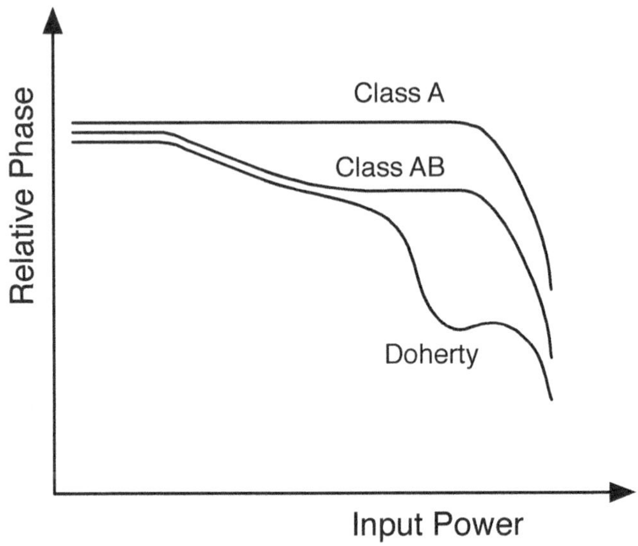

Figure 4.5 Stylized AM-to-PM curves for several amplifiers, covering bias types and architecture: Class-A, Class-AB, and (a rather poor) Doherty examples are shown.

the capacitance will sweep through the threshold region, again introducing a strong nonlinearity.

Heterojunction bipolar transistors (HBTs) are used as the PA device in many handsets, and high-voltage HBTs are becoming competitive for base-station applications. These devices have a nonlinear input capacitance, with a contribution from the diffusion charge in the base that governs the output current. The output capacitance is basically a reverse-biased p-n junction, which has a nonlinear, although smooth, voltage dependence.

Despite the physical origins and the detailed shape of the AM-to-PM characteristic, our task is to model its effect on the input-output characteristics of the PA. Again, we can use polynomial functions to do this. We have two choices in how to apply the model for the AM-to-PM:

1. Use a completely separate polynomial function for the AM-to-PM characteristic, and add the output signal vectorially to the output of the AM-to-AM function, to account for the absolute phase. The phase variation with the input signal

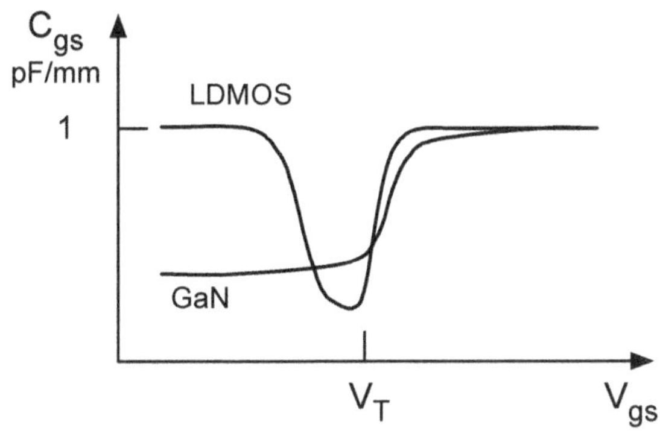

Figure 4.6 The relative variation of input capacitance with input voltage, for various power FET transistor types. A rule of thumb is about 1 pF/mm for many technologies.

can be written as

$$\varphi_{out}(t) = \sum_{n=1}^{N} q_n v_{in}^n(t) \tag{4.18}$$

The input signal, $v_{in}(t)$, is again given by equation (4.11); for convenience, the phase offset φ can be set to zero, because in the AM-to-PM characteristic we are looking for deviations from a reference phase at small signal. The output phase given by 4.18 can be added to the phase values $N\varphi$ in (4.13), to yield the complete form of the nonlinear output voltage $v_{out}(t)$. This approach allows considerable freedom in the choice of fitting function for the AM-to-PM characteristic, and we may choose to use an altogether different function from polynomials. The potential difficulty with this approach is the combining of the AM-to-AM and AM-to-PM effects at the output node, where care with causality is needed to describe the overall phase changes.

2. Use complex-valued coefficients in the same polynomial model that is used for the AM-to-AM model. Because we are defining the measures of distortion in polar form, AM-to-AM and AM-to-PM, we shall use the polar form for representing the complex coefficients:

$$a_i \rightarrow |a_i| \angle \psi_i = \text{Re}\left\{a_i e^{j\psi_i}\right\} \tag{4.19}$$

We take the real coefficients from our original polynomial and add a phase angle ψ to each argument. Using De Moivre's theorem, the squared term becomes 2ψ, the cubic term becomes 3ψ, and so on.

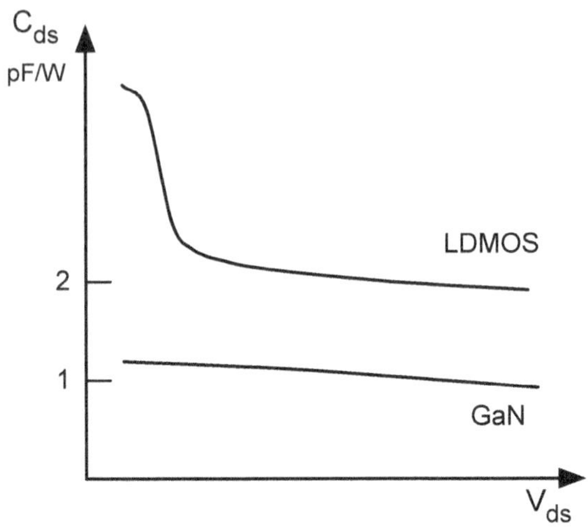

Figure 4.7 The relative variation of output capacitance with output voltage, for various power FET transistor types. The absolute values will depend on device size and power rating, but the relative capacitance per watt of output power is indicated. GaNFETs have a higher power density than LDMOS.

The polynomial relationship for $v_{out}(t)$ then becomes

$$\begin{aligned} v_{out}(t) &= \tfrac{1}{2}a_2V_0^2 + \tfrac{3}{8}a_4V_0^4 \\ &+ a_1V_0\cos(\omega_0 t + \varphi + \psi_1) \\ &+ \left(\tfrac{3}{4}a_3V_0^3 + \tfrac{5}{8}a_5V_0^5\right)\cos(\omega_0 t + \varphi + \psi_3 + \psi_5) \\ &+ \left(\tfrac{1}{2}a_2V_0^2 + \tfrac{1}{2}a_4V_0^4\right)\cos(2\omega_0 t + 2\varphi + 2\psi_2 + 2\psi_4) \\ &+ \left(\tfrac{1}{4}a_3V_0^3 + \tfrac{5}{16}a_5V_0^5\right)\cos(3\omega_0 t + 3\varphi + 3\psi_3 + 3\psi_5) \\ &+ \tfrac{1}{8}a_4V_0^4\cos(4\omega_0 t + 4\varphi + 4\psi_4) + \ldots \\ &+ \tfrac{1}{16}a_5V_0^5\cos(5\omega_0 t + 5\varphi + 5\psi_5) + \ldots \end{aligned} \quad (4.20)$$

The phase distortions ψ_n can be seen in the arguments of the cosine terms; the AM-to-PM effects are produced by the vector additions described by the magnitude and phase components.

So far we have been able to generate a relatively simple polynomial-based model from CW drive-up input and output data that describes AM-to-AM and AM-to-PM responses, and that can also be used to provide information on harmonic and DC bias responses in the system. The phase response outlined above is strictly a memory phenomenon, and can be used to describe the memory effects in the PA. We shall encounter this further in Chapter 5.

4.3 TWO-TONE ANALYSIS

The single-tone analysis of the previous sections has provided us with a polynomial model fitted to that data. This model predicts out-of-band responses at harmonics and DC. In this section we shall use this model to investigate the two-tone response, from which we will generate in-band distortion products, close to the input signals. In a linearization scheme, it is these in-band distortions that we need to remove.

Two-tone measurements are made to assess the in-band distortion performance of a transistor or PA: the intermodulation distortion products. These measurements are often carried out using a spectrum analyzer, which provides magnitude information only; relative phase information can be obtained using a vector network analyzer. It is possible to set up a nonlinear vector network analyzer (nVNA) to measure the absolute phase of the intermodulation products [3]; this measurement will be an excellent validation or data generator for this simple polynomial model.

If the two-tone input signal is two identical sinusoids, separated by a small frequency difference, we have:

$$v_{in}(t) = V\cos(\omega_1 t + \phi_1) + V\cos(\omega_2 t + \phi_2) \tag{4.21}$$

$$\omega_2 - \omega_1 \ll \omega_1, \omega_2 \tag{4.22}$$

where V is the magnitude of each signal, and we have allowed for arbitrary phase on each signal.

The two-tone expression can also be written in the following way, using a trigonometrical identity:

$$\begin{gathered} V\cos(\omega_1 t + \phi_1) + V\cos(\omega_2 t + \phi_2) = 2V\cos(\omega_m t + \phi_m)\cos(w_c t + \phi_c) \\ \omega_m = \tfrac{1}{2}(\omega_2 - \omega_1) \\ \omega_c = \tfrac{1}{2}(\omega_2 + \omega_1) \end{gathered} \tag{4.23}$$

This is a double-sideband, suppressed-carrier AM signal, with a carrier frequency of ω_c and a modulation frequency of ω_m. Note that the amplitude of the modulating signal is $2V$. In using our polynomial model of (4.10) with this two-tone signal, we must ensure that the total signal amplitude, $2V$, is no greater than the single-tone signal amplitude, V_0, for the model to remain within its range of validity.

To be more general, we can allow the amplitudes of the two tones to be different, V_1 and V_2:

$$v_{in}(t) = V_1 \cos(\omega_1 t + \phi_1) + V_2 \cos(\omega_2 t + \phi_2) \tag{4.24}$$

Substituting this value of $v_{in}(t)$ into the complex-valued polynomial (4.10) yields terms not only in the *degree* powers of the polynomial, but also mixing terms of various *orders*. If we write the polynomial coefficients in the following way:

$$a_i = |a_i|\, e^{j\psi_i} = |a_i|\, e^{j\omega\tau_i} \tag{4.25}$$

and take the real part of this analytic expression, this will permit the easy identification of the phase terms with the different orders.

After some pages of algebra using several of the trigonometrical identities, and collecting the terms of the different orders together [4], we get the following result for $v_{out}(t)$:

First-order

These are the purely linear terms in the output voltage of the amplifier, which are the ones we want for linear gain:

$$a_1 V_1 \cos(\omega_1 t + \phi_1 + \psi_{11}) \tag{4.26}$$

$$a_1 V_2 \cos(\omega_2 t + \phi_2 + \psi_{21}) \tag{4.27}$$

where $\psi_{ij} = \omega_i * \tau_j$, here and in the following.

Everything below here is distortion!

Second-order

$\frac{1}{2}a_2\left(V_1^2+V_2^2\right)$	DC bias shift
$a_2V_1V_2\cos\left((\omega_2-\omega_1)t+(\phi_2-\phi_1)+(\psi_{22}-\psi_{21})\right)$	IM2: baseband
$a_2V_1V_2\cos\left((\omega_2+\omega_1)t+(\phi_2+\phi_1)+(\psi_{22}+\psi_{21})\right)$	IM2: sum frequency
$\frac{1}{2}a_2V_1^2\cos\left(2\omega_1t+2\phi_1+2\psi_{12}\right)$	2nd harmonic
$\frac{1}{2}a_2V_2^2\cos\left(2\omega_2t+2\phi_2+2\psi_{22}\right)$	2nd harmonic

Third-order

$\frac{3}{4}a_3V_1^3\cos(\omega_1t+\phi_1+\psi_{13})$	AM-AM, AM-PM
$\frac{3}{4}a_3V_2^3\cos(\omega_2t+\phi_2+\psi_{23})$	AM-AM, AM-PM
$\frac{3}{2}a_3V_1V_2^2\cos(\omega_1t+\phi_1+\psi_{13})$	Cross-modulation
$\frac{3}{2}a_3V_1^2V_2\cos(\omega_2t+\phi_2+\psi_{23})$	Cross-modulation
$\frac{3}{4}a_3V_1^2V_2\cos((2\omega_1-\omega_2)t+(2\phi_1-\phi_2)+(2\psi_{13}-\psi_{23}))$	IM3 Lower
$\frac{3}{4}a_3V_1V_2^2\cos((2\omega_2-\omega_1)t+(2\phi_2-\phi_1)+(2\psi_{23}-\psi_{13}))$	IM3 Upper
$\frac{3}{4}a_3V_1^2V_2\cos((2\omega_1+\omega_2)t+(2\phi_1+\phi_2)+(2\psi_{13}+\psi_{23}))$	IM3 at H3
$\frac{3}{4}a_3V_1V_2^2\cos((2\omega_2+\omega_1)t+(2\phi_2+\phi_1)+(2\psi_{23}+\psi_{13}))$	IM3 at H3
$\frac{1}{4}a_3V_1^3\cos(3\omega_1t+3\phi_1+3\psi_{13})$	3rd Harmonic
$\frac{1}{4}a_3V_2^3\cos(3\omega_2t+3\phi_2+3\psi_{23})$	3rd Harmonic

Important fourth-order terms

$\frac{3}{8}a_4V_1^4 + \frac{3}{2}a_4V_1^2V_2^2 + \frac{3}{8}a_4V_2^4$	DC bias shift
$\frac{3}{4}a_4V_1^3V_2\cos\left((\omega_1-\omega_2)t+(\phi_1-\phi_2)+(\psi_{14}-\psi_{24})\right)$	IM2 : baseband
$\frac{3}{4}a_4V_1V_2^3\cos\left((\omega_2-\omega_1)t+(\phi_2-\phi_1)+(\psi_{24}-\psi_{14})\right)$	IM2 : baseband
$\frac{3}{4}a_4V_1^2V_2^2\cos\left((2\omega_2-2\omega_1)t+2(\phi_2-\phi_1)+2(\psi_{24}-\psi_{14})\right)$	IM4 : baseband

...plus terms at even-order harmonics and intermodulation products at higher frequencies.

...and important fifth-order terms

$\frac{5}{8}a_5V_1^5\cos(\omega_1t+\phi_1+\psi_{15})$	AM-AM, AM-PM
$\frac{5}{8}a_5V_2^5\cos(\omega_2t+\phi_2+\psi_{25})$	AM-AM, AM-PM
$\frac{15}{4}a_5V_1^3V_2^2\cos(\omega_1t+\phi_1+\psi_{15})$	Cross-modulation
$\frac{15}{4}a_5V_1^2V_2^3\cos(\omega_2t+\phi_2+\psi_{25})$	Cross-modulation
$\frac{15}{8}a_5V_1V_2^4\cos(\omega_1t+\phi_1+\psi_{15})$	Cross-modulation
$\frac{15}{8}a_5V_1^4V_2\cos(\omega_2t+\phi_2+\psi_{25})$	Cross-modulation
$\frac{15}{8}a_5V_1^2V_2^3\cos((2\omega_1-\omega_2)t+(2\phi_1-\phi_2)+(2\psi_{15}-\psi_{25}))$	IM3 Lower
$\frac{5}{4}a_5V_1^4V_2\cos((2\omega_1-\omega_2)t+(2\phi_1-\phi_2)+(2\psi_{15}-\psi_{25}))$	IM3 Lower
$\frac{15}{8}a_5V_1^3V_2^2\cos((2\omega_2-\omega_1)t+(2\phi_2-\phi_1)+(2\psi_{25}-\psi_{15}))$	IM3 Upper
$\frac{5}{4}a_5V_1V_2^4\cos((2\omega_2-\omega_1)t+(2\phi_2-\phi_1)+(2\psi_{15}-\psi_{15}))$	IM3 Upper
$\frac{5}{8}a_5V_1^3V_2^2\cos((3\omega_1-2\omega_2)t+(3\phi_1-2\phi_2)+(3\psi_{15}-2\psi_{25}))$	IM5 Lower
$\frac{5}{8}a_5V_1^2V_2^3\cos((3\omega_2-2\omega_1)t+(3\phi_2-2\phi_1)+(3\psi_{25}-2\psi_{15}))$	IM5 Upper

...plus terms at odd harmonics and higher intermodulation products.

4.3.1 Comments on the Two-Tone Analysis

We can see in the previous tables the richness of the distortion terms that can be identified, even for this relatively simple nonlinear model. This gives us some insight into the difficulties that await us in implementing a pre-distortion linearizer. The distortions that can be classified include:

- self-biasing: DC shift with contributions from even-order harmonic terms;
- AM-to-AM and AM-to-PM effects;
- harmonic distortion: generation of new frequency components at integer multiples of each of the two tones;
- intermodulation (IM) distortion: nonlinear mixing between the various fundamental and harmonic components of the signal, $n\omega_1$ and $m\omega_2$, leading to new frequency components in the spectrum;
- cross-modulation distortion: nonlinear mixing between the various fundamental and harmonic components of the signal, $n\omega_1$ and $m\omega_2$, leading to products at *existing* frequency components of the signal.

The first three items in the list are the same as we identified from the single-tone analysis, and the last two terms are specifically from the nonlinear mixing of the two tones and their harmonics. The values of n and m can be positive or negative integers, their sum giving the order of the distortion product. For example, the fourth-order terms $\{\omega_1, \omega_2, \omega_2, -\omega_2\}$ yield the second-order intermodulation product at $\omega_2 - \omega_1$. From this consideration, it can be seen in the tables that the intermodulation products contain components from higher-order mixing. Another example is the IM3 terms, which are produced by third-order and fifth-order mixing products.

The spectrum of products is shown in Figure 4.8, with the in-band distortion products highlighted in Figure 4.9.

4.3.2 Do We Need the Even-Degree Terms in the Polynomial?

Given that we are measuring the magnitude and phase of the output voltage of the amplifier as a function of the single-tone (or even two-tone) drive-up signal, then to get the best representation of the vector transfer response, it would seem clear that

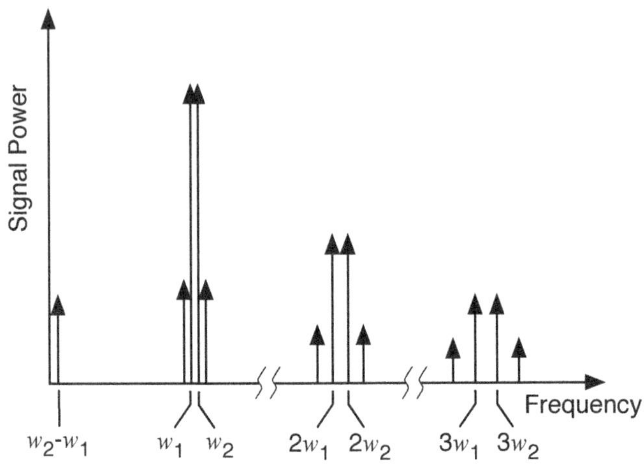

Figure 4.8 Output spectrum from PA model, showing harmonic and third-order intermodulation products.

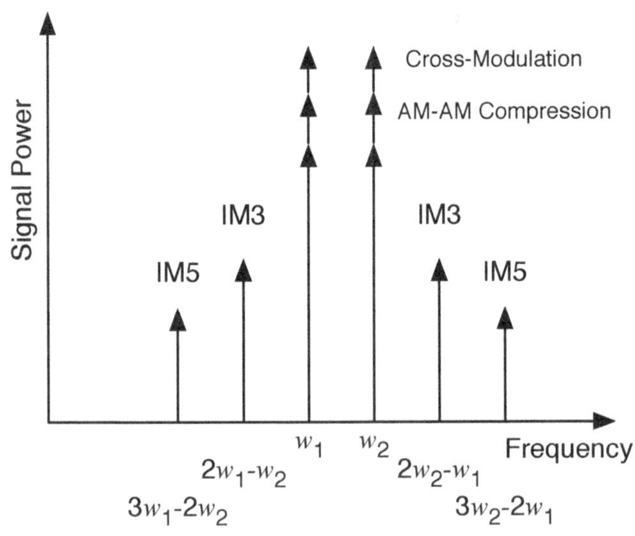

Figure 4.9 Two-tone output spectrum from PA model, showing the in-band intermodulation and cross-modulation products, and AM-to-AM compression effects.

we should use all the terms in the polynomial to hand. That will give us the best fit to the AM-to-AM and AM-to-PM curves that we measure.

On the other hand, when we measure a real signal out of the PA, this is likely to be a signal that is limited to the frequency band of interest: a band-limited signal around the carrier(s). In this case, we shall only detect the carrier and odd-order intermodulation signals. The limited pass band will not permit the *direct* detection of the even-order intermodulation products, the harmonic distortions, or the DC bias shift. All of these effects must either be inferred from the passband data, or neglected altogether. From this view, the even-degree terms would appear to be redundant. Indeed, there are many reports of successful pre-distortion models that are generated from the fitting odd-degree polynomial-based models to bandpass data. But these models also include terms to describe memory effects in the PA, to create the successful pre-distorter. We shall turn our attention to these memory effects in Chapter 5.

4.4 SUMMARY REMARKS

In this chapter, we have developed a simple nonlinear instantaneous model that uses nonlinear basis functions to describe the AM-to-AM and AM-to-PM characteristics of the PA. The model structure is linear-in-parameters, meaning that the nonlinearity is captured by the basis functions, and the model fit is determined by the coefficient, which can be found using well-established linear solution techniques based on least-squares principles.

In this example, the model is determined from the RF drive-up data: the gain compression and phase transfer characteristics. Using polynomials for the basis functions we are able to show that the model can predict the gain and phase behavior at RF, and also the distortion products at harmonic and intermodulation frequencies, including DC self-bias effects, and cross-modulation or in-band distortion.

While polynomials are straightforward to use, they have some drawbacks, particularly with interpolation and extrapolation of the function. Low-degree polynomials may not be able to capture the nonlinearity with sufficient fidelity, whereas high-degree polynomials may show oscillatory behavior between the measured data points. Orthogonal polynomial sets such as Hermite or Chebyshev polynomials offer better convergence and possibly fewer parameters. All polynomial models need care with extrapolation beyond the measured data set. More complicated functions can be used, within the linear-in-parameters framework; examples include combinations of *log/exp* functions, *multi-tanh* functions, and other trigonometric or hyperbolic series. These all possess their own features and adherents, although the polynomial approach remains popular.

The simple memoryless model described here is essentially a model of the complex gain function of the PA, dependent on the input power. This model can form the basis of a simple pre-distortion linearizer: by inverting this complex gain function we generate the inverse of the nonlinear function, and hence the cascade of the two effects is linear. The inverse function is often implemented as a look-up table of the complex gain, phase values, indexed by the input power.

References

[1] S. C. Cripps, *RF Power Amplifiers for Wireless Communications*, 2nd ed., Norwood, MA: Artech House, 2006.

[2] G. Golub and C. van Loan, *Matrix Computations*, 3rd ed., London, UK: Johns Hopkins University Press, 1996.

[3] L. Betts, D. T. Bespalko, and S. Boumaiza, "Application of Agilent's PNA-X nonlinear vector network analyzer and X-parameters in power amplifier design," White Paper, Agilent Technologies, Inc., Tech. Rep. 5990-7768EN, 2010.

[4] G. F. Collins and D. Runton, "Nonlinear design of power amplifiers," *Microwave Journal*, Vol. 50, No. 9, pp. 164–176, Sep. 2007.

Chapter 5

Memory Effects

5.1 INTRODUCTION

Memory is the ability to recall an event that happened some time ago. How does this apply to power amplifiers? *Memory effects* is the term used to describe the influence of the history of the signal on its present value. In other words, the values of the signal at past time instances contribute to the present value. This is illustrated in Figure 5.1. The input signals' value at times prior to the current instant t_0 contribute to the power amplifier output at t_0. The number of samples that contribute is known as the memory depth of the PA.

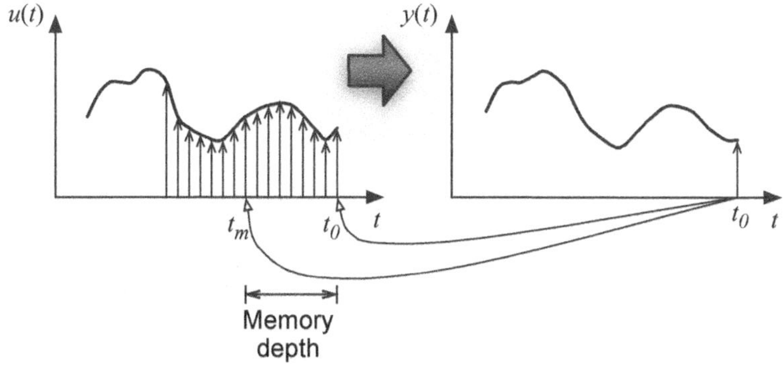

Figure 5.1 The action of past signals $u(t)$, up to some time t_m, on the present output of a power amplifier, $y(t)$.

This behavior should come as no surprise to an electrical engineer: a simple series R-C circuit exhibits memory effects, because the current flowing in the circuit depends on how much charge has already built up on the capacitor. Memory effects are a result of energy storage in the system, and are also a manifestation of the dynamical behavior of the system, and for linear systems can be described by the time derivative or time delay expressions, and system identification techniques outlined in Chapter 3. An example of how memory affects the output of a power amplifier is shown in Figure 5.2, where the rise and fall times of the amplifier are not the same, resulting in a different output voltage for the same input voltage, depending on whether we are on a rising or falling edge of the input signal. This phenomenon is also known as *hysteresis*.

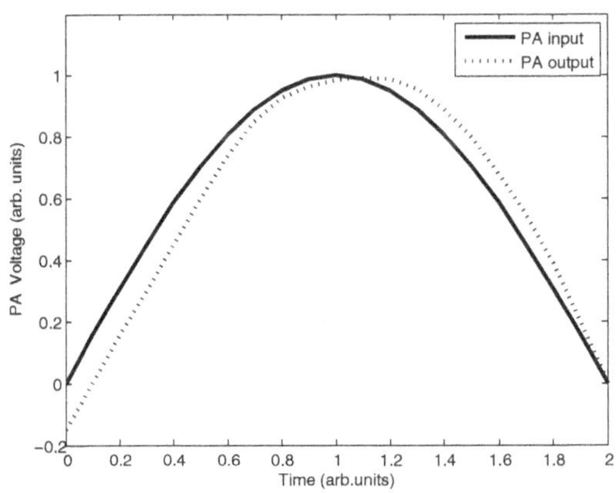

Figure 5.2 Memory effects action on the output signal of a power amplifier, manifest as a difference in the rising and falling slew rates of the PA.

For nonlinear systems such as a power amplifier, we see from Figures 5.1 and 5.2 that the history of the signal influences the output. The amount of the history that is needed is finite: after some time the influence of the past time sample becomes negligible. This is known as *fading memory*, and is a fundamental requirement for the realization of a nonlinear dynamical model [1]. The dynamic effects in a PA can occur over a wide range of timescales. Memory effects or dynamics that occur at about the same rate as the RF signal are known as short-term memory (STM) effects, and dynamical behavior that happens at a slower rate, close to the modulation rate, is called long-term memory (LTM) effects, though the classification is broad, and often the distinction is not clear, especially when simply "memory effects" are referred to. In this respect, memory effects has become something of a catch-all phrase for dynamical behavior in the power amplifier.

5.1.1 Short-Term Memory Effects

Short-term memory effects are characterized by time constants that roughly of the order of the period of the RF carrier frequency. The STM effects are essentially the RF frequency response of the amplifier, which may include bias- or signal-dependent effects as well as the passive behavior that we normally associate with the term frequency response. Components or properties that contribute to the STM, or frequency response, include:

- *Matching networks.* The input and output reactive matching networks contain passive components with dynamic properties, viz. capacitors, inductors, and transmission lines. These components may be part of the transistor product, that is, capacitors and bond-wire inductance built into the package, or on the printed-circuit board as external matching circuits. Multistage and integrated amplifiers may also contain interstage matching networks. The matching circuits' frequency response at the harmonics may also be of interest for linearization and efficiency enhancement.
- *Device capacitances.* Or, more correctly, the time delays associated with the build-up, transport, and decay of charge within the transistor. In the small-signal regime, these are represented by the capacitances and transit time model parameters. Under large-signal conditions, these memory effects are modeled as dQ/dt functions, and these can have significant dependence on the signal level, as evidenced by the variation of the transfer phase with power level in a PA: AM-to-PM effects. If we are identifying the STM from simulation rather than measurement of a real PA, then it is important that the transistor model is charge-conservative; otherwise, during circuit simulation, we will create an unphysical charge in the device [2], and the time behavior of this charge is indistinguishable from a true memory effect.

Generally, a good approximation to the small-signal STM can be obtained from measurement or simulation of the small-signal frequency response. This can be implemented in the PA model as a filter. The large-signal effects require more care (and properly-formulated models).

5.1.2 Long-Term Memory Effects

The long-term memory effects are characterized by time constants much longer than the period of the RF carrier frequency. This means that LTM includes those effects that are on or about the timescale of the envelope or modulation signal, and also those effects that take place on much longer timescales. Examples of LTM mechanisms include:

- *Thermal effects.* The flow and storage of thermal energy in semiconductors take place relatively slowly compared with the RF or envelope signal. Energy from a large signal that is dissipated as heat changes the local, temperature-dependent, physical parameters in the FET channel, such as the carrier mobility or velocity. This is then observed as a small change in gain that persists for a short time after the signal has passed, as the heat diffuses away: subsequent signals will experience the different gain for a short time, a memory effect of the earlier large signal.
- *Charge trapping.* Semiconductor materials and processing are not perfect, and the materials can contain defects that can trap the mobile charge carriers to release them later when the local potential conditions are different. Traps are present in the III-V semiconductors; hence, devices based on gallium arsenide (GaAs) or gallium nitride (GaN) will experience trapping effects. Generally, the trap densities are quite low, reflecting the high cleanliness and precision of the semiconductor device fabrication processes, but the effects can be noticeable in the distortion products.
- *Bias circuit effects.* The job of the bias line is to provide an unimpeded flow of current at DC, but prevent any RF signal from being impressed on the DC supply. A simple view is that the ideal bias line is a short circuit at DC, and an open circuit at RF. In practice, while it may be possible to achieve a very low DC resistance, this low impedance is maintained only over a limited frequency range, known as the video bandwidth (VBW). If this low impedance can be maintained over a frequency range that accommodates, say, the IM2, IM4 distortion products, then these distortions will be effectively shorted to ground and will have little impact. If the VBW is low, or the bias line impedance is high, then these distortion products will induce bias voltage variations at the modulation frequency. This 're-modulation' affects the bias-dependent properties of the power transistor, such as the gain, and thus constitutes a memory effect.

- *Control circuitry.* Many PA circuits include control features such as automatic bias and gain control. These control circuits generally have time constants associated with them that are long compared to the RF period, and so effectively introduce memory into the operation of the PA. While such control circuits improve the performance of the raw amplifier, they can introduce memory features that can be difficult to correct for.

The range of LTM can cover from tens or hundreds of milliseconds down to a few microseconds.

The distinction between short- and long-term memory effects is quite vague; some of the phenomena such as thermal effects have responses that can cover several orders of magnitude of time. The distinction can also be blurred when we consider the measurement of memory effects. Often, the PA characterization and modeling is done in the modulation domain: the measured signal is down-converted and demodulated into the I and Q data, for the model generation. Are memory effects on this timescale considered short term or long term? In practice, their distinction is not as important as their characterization and mitigation.

The elimination, or at least minimization, of memory effects in a PA requires that each of the causes is addressed specifically. We shall now describe the memory effect behaviors in more detail, and outline some techniques for mitigation of the effects in the power device and at the design stages of the PA. What is left, we will have to leave to the pre-distorter.

5.2 MATCHING NETWORKS

In the wireless infrastructure business, laterally-diffused MOS (LDMOS) high-power transistors account for about 95% of the market share (at the time of this writing), with other technologies, mainly III-V based, making up the difference. These other, newer, technologies may be anticipated to gain market share with time. A high-power LDMOS transistor, with the ceramic lid removed, is shown in Figure 5.3.

Apart from the transistor die, there are several other components present in the package: ceramic (low-temperature co-fired ceramic (LTCC)) and MOS capacitors, and arrays of bond wires that are designed to be the capacitance and inductance elements of pre-matching networks in the package. They are called pre-matching networks because their job is to raise the impedance presented at the package tab from a small value to something in the region of 2 to 5 ohms, to make the rest of the match to 50 ohms on the printed circuit board (PCB) easier to realize with practical surface mount components and microstrip lines. High-power RF transistors all have

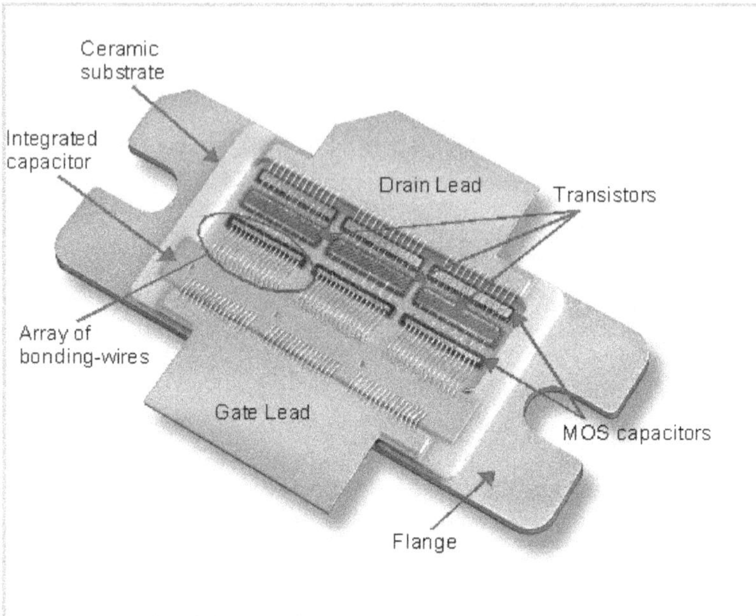

Figure 5.3 An example of a packaged high-power RF transistor, showing the in-package components. This is a 170-watt LDMOS transistor product designed to operate at 1.9 GHz. (Photograph courtesy of Freescale Semiconductor, Inc.)

this basic arrangement to a more or less sophisticated degree. Integrated RFIC or MMIC power amplifiers may have the inductor and capacitor components built onto the semiconductor die itself, and printed spiral inductors and metal-insulator-metal (MIM) capacitors; their roles are the same as those of the discrete components described above.

The memory effects due to these in-package networks arise from the phase response of these filter networks, and also RF feedback effects from the mutual inductance between the input and output bondwire arrays, and inductance from the source node of the FET through the semiconductor and package to the RF ground.

The matching networks, both in-package and on PCB, are generally designed for economy of size and part count. Single-tuned matching networks will give the best peak performance from the transistor, and broader bandwidth behavior can be obtained from double-tuned filter networks, at the cost of some gain. For low memory impact, the matching networks should be carefully designed to obtain the desired amplitude response for the PA, along with a phase response that is as close as possible to linear phase across the bandwidth to yield a constant group delay that can be accommodated fairly easily by the pre-distorter.

For a power amplifier using a digital pre-distortion system for linearization, the instantaneous bandwidth of the PA will need to be several times the nominal signal bandwidth, to accommodate the additional frequency components introduced by the pre-distorter (see Chapter 7). This wider DPD pass-band is often not as flat as the signal bandwidth, requiring equalization in the linearizer before the pre-distortion can be applied. As the signal bandwidth desired by the wireless service providers continues to increase, driven by the increased demand for data services, the PA design and the pre-distorter design become ever more challenging.

This is especially the case for Doherty PA design. In this PA architecture, impedance transformations are required for the load modulation of the main amplifier path. The impedance transformers at the output of the transistors are usually constructed from quarter-wavelength transmission lines, which are narrowband. The input power splitter is often realized using a 3-dB hybrid splitter component, or in microstrip using a Wilkinson splitter arrangement. Such approaches can be made wide to have a wide bandwidth, of up to an octave, although this is not currently a production Doherty PA design. Additionally, the path delay between the main and peaking amplifiers needs to be optimized for proper load modulation; this is generally achieved using a short piece of transmission line, which again is a frequency-dependent technique. A new design approach will be required for Doherty PAs if and when the total signal and pre-distortion bandwidth requirement exceeds about 20% of the carrier frequency.

Finally, a comment on matching network filter design. The classic microwave filter design approaches are reflective filters; that is, the out-of-band energy is reflected back to the signal source. If the signal source is a 50-ohm source, this presents no problem, but in our case the source is the power transistor whose output is usually far from 50 ohms, and comprises nonlinear output admittance as well. The transistor output can be subjected to reflected energies at harmonic impedances. The delay in the return of the reflected signal contributes a memory effect, and the nonlinear output admittance will generate further mixing products from the reflected and transistor output signals. These new frequency mixed components will contribute further to the distortion and to the bias-modulation effects described in Chapter 4, leading to additional nonlinear dynamical contributions to the overall signal output. This process illustrates one aspect of how changes in the passive networks of a PA can result in different nonlinear performance.

One remedy to this problem to use absorptive filters, wherein the out-of-band energy is absorbed rather than reflected. This approach has been used in multi-octave amplifiers, where the reflected harmonic energy degraded the output match of the PA [3], and using a low-pass filter that absorbed the higher-frequency components improved the performance of the amplifier. Such filters may be difficult to realize in the confines of the transistor package. An alternative approach may be to use the reflected energy constructively, by providing the appropriate load impedances at the harmonics to create a 'Class F/F^{-1}-*lite*' amplifier and scavenge an additional point or two of efficiency.

5.3 CHARGE STORAGE IN THE TRANSISTOR

The transistors themselves contribute to the memory effects through the charge storage and transport mechanisms in the semiconductor device. These behaviors are modeled through the reactive components in the small-signal models, illustrated in Figure 5.4 for the FET and bipolar transistor. The model components are derived from the two-port Y-parameters of the transistor, which in turn are transformed from high-frequency two-port S-parameter measurements. The two-port Y-parameters yield the input and output admittances and the forward and reverse transadmittances. The passive admittances are written in terms of real and imaginary components to give the resistive and capacitive elements of the model. The forward transfer controlled current source values are often written in terms of magnitude and phase, as the transconductance, g_m, and delay, τ, with the phase being given by $\tau * \omega$, where ω is the signal (angular) frequency. The gain could also be written in terms of real and imaginary components; this yields a transconductance, and a transcapacitance term, C_m, where $C_m = g_m/2\pi\tau$.

The reactive passive elements in the model describe the frequency response of the transistor, contributing to the short-term memory. The time-delay in the transadmittance describes the time lag between the application the gate voltage and the change in the drain current (in the FET). This can be thought of as the time taken for the charge to move along the channel of the device. This transit time delay is related to the transition frequency, f_T.

The III-V semiconductors have high electron mobility and velocity, and hence have short lag times and high f_Ts, compared to silicon-based FETs. The pseudomorphic high electron mobility transistor (pHEMT)[1], a FET whose conducting channel is built from a thin layer of very high mobility indium-gallium arsenide (InGaAs) semiconductor sandwiched between confining barrier layers of GaAs, has

1 A misnomer, it is really the saturated velocity of the electrons in the channel that determines the transition frequency.

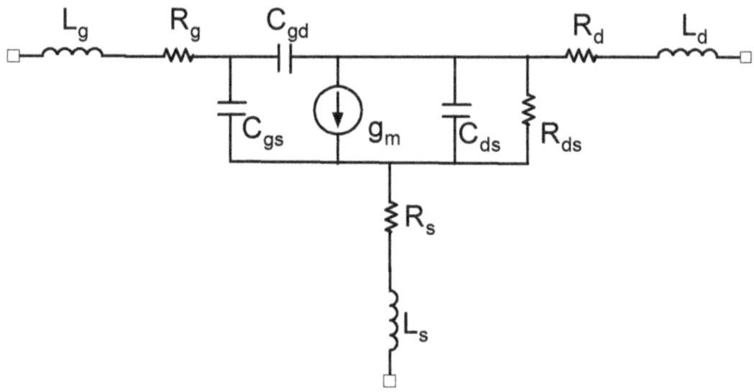

(a)

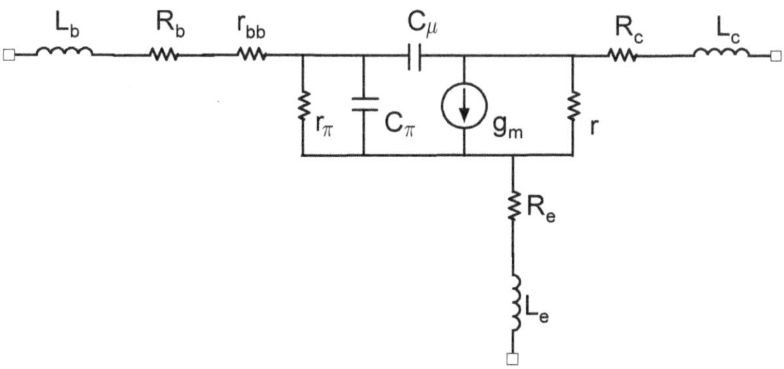

(b)

Figure 5.4 Small-signal equivalent-circuit models for (a) the field effect transistor, and (b) the bipolar junction transistor.

very high f_T. Gallium nitride (GaN) also has desirable high electron mobility and velocity for high-speed power transistors.

The elements of the small-signal models are voltage-dependent: this is *bias voltage*-dependent, when we are concerned with S-parameter measurements and other small-signal phenomena. The model elements may also exhibit some signal-level dependence: this is manifest as changes in the transfer phase of the PA, or AM-to-PM effects; much of the change in phase may be ascribed to changes in the in-phase behavior of the PA as it enters compression at high signal levels, causing changes in the local operating point of the device, and hence the local charge behavior. Nevertheless, the change in phase must be accommodated in the behavioral model and, eventually, in the pre-distorter.

The models shown in Figure 5.4 are intrinsic models. The complete transistor model also contains an extrinsic network that surrounds the intrinsic circuit, connecting it to the external world. The extrinsic network comprises resistors, capacitors, and inductors, that account for the series resistance of the semiconductor, contact resistances between the metallizations and the semiconductor, capacitances between the metallizations and the semiconductor, and inductances of metallizations, for example. Generally, we try to minimize the capacitances and resistances in the model and in the extrinsic network, to reduce the delays in the circuit, and hence improve the transit time and charging times of the capacitors. However, it is not possible to eliminate these effects.

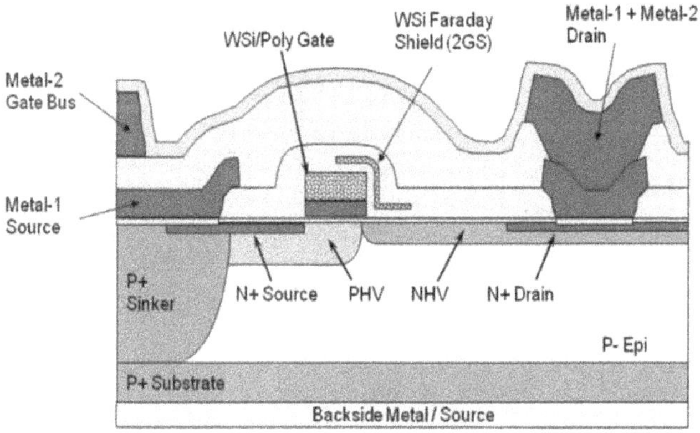

Figure 5.5 Schematic cross-section of an LDMOS RF power transistor, showing the channel under the gate, and the NHV drift region. (Figure courtesy of Freescale Semiconductor.)

An example of a cross-section of an LDMOS RF power transistor is shown in Figure 5.5. The output capacitance, C_{ds}, of this transistor is given largely by the reverse-biased $p - n$ junction formed between the laterally-diffused drift region and the epitaxial support layer. This is basically a feature of the transistor that enables the high output voltage swings necessary for a high-power transistor, and consequently the output capacitance for LDMOS transistors is comparatively high. This junction capacitance also exhibits significant drain voltage dependence. Several ingenious attempts have been made to reduce this capacitance, and its voltage dependence, without compromising the output power, gain, frequency response, or reliability of the device.

In the GaN power FET, the capacitances in the small-signal model are still present, and the drain-side drift region is also built into the device structure. But, the GaN FET is a heterojunction FET, and the conducting channel is electrostatically confined to a region close to the heterojunction, and so the confining $p-n$ structure of the LDMOS drift region is not required. This means that the junction capacitance of the LDMOS structure is absent, and the GaN FET output capacitance is defined mostly by the thickness of the barrier layer semiconductor above the heterojunction. This capacitance can be made smaller than the LDMOS output capacitance, and largely independent of the drain voltage. While this has other benefits for Doherty and envelope-tracking (ET)PA design, for example, it also means that there are fewer short-term memory effects in the device.

Similar limitations and compromises exist in other technologies, too. For example, the output capacitance of the heterojunction bipolar transistor (HBT), frequently used in handset PAs, can be reduced by reducing the collector region doping, at a cost of gain and frequency response. One of the features of the HBT device is that the emitter efficiency of the electron injection over the base-emitter junction is determined by the heterojunction band offset, allowing the base region to be heavily doped and therefore be of high conductivity. This means that the base resistance is low, leading to a reduced input time constant and hence lower memory than a conventional BJT would have.

While the small-signal models provide a convenient circuit-based description of the frequency response, it should be remembered that the transistors work by moving charge around, and so it is the 'dq/dt' effects that fundamentally describe these short-term memory effects. The inherent switching speed of the transistor, the inverse of f_T, depends on how quickly charge can be pushed through the channel of the FET or base of the HBT, and taken out again. In practice, FETs are limited by their switch-on time, and bipolar transistors are limited by their switch-off time: this switch-off delay is well-described in undergraduate texts [4].

5.4 THERMAL EFFECTS

Power transistors in PAs get very hot, as we all know. This is because the amplifier is not 100% efficient, and so some of the output energy is dissipated as heat. The heat energy flows through the semiconductor from the heat source, in the FET channel or HBT base, to the cold heat-sink usually found at the back of the die. The heat flow is governed by a diffusion equation, and can be described by a heat flow component, governed by the *thermal resistance* of the material and the temperature difference across this resistance, and a heat storage component dependent on the *thermal capacitance* of the material and the thermal energy. The thermal resistance and capacitance can be modeled as electrical analogs, and the concept of a thermal time constant as the product of this resistance and capacitance is very natural. Hence, it can be expected that the generated heat energy will take some time to flow away from the point of heat generation, and this is the source of thermal memory effects in the transistor.

The modulated signals that the RF power amplifier has to deal with may exhibit substantial changes in envelope amplitude over a short period of time. We may therefore expect to see some signal-dependent thermal memory effects in the transistor. For example, let's assume that the signal envelope has been above the average power level for some time; the energy dissipated by the PA in the transistor channel region (of an FET, or base in HBT) will cause a temperature rise in the channel. Consequently, the temperature-dependent charge transport properties such as electron mobility and velocity will be changed. The values of these parameters fall with increasing temperature around room temperatures and above, and so the gain of the transistor will fall as a result of the increased thermal energy dissipation. If the modulation envelope then drops to a low value, the gain remains at the lower value associated with the high dissipation condition until the channel temperature falls back towards the equilibrium value, because the time constant of the thermal energy dissipation is long compared with the modulation rate. We can see that the gain of the transistor is dependent upon the local history of the modulating signal envelope, and the thermal time constant of the semiconductor material used to make the power transistor.

While this behavior is no doubt present in all PAs under modulated signal drive to some extent, it is most easily observable with very time-dependent signals, especially those using time-division duplex (TDD) communications protocols, such as WiMAX or TD-LTE. In these cases, the up- and down-link operate at different times; the PA is being switched on and off at the frame rate or less of the signal, and therefore operates at alternately high power dissipation and no power dissipation, which causes the PA gain to change during the transmission period as the device heats up. Another signal that produces thermal memory effects is GSM: as shown in Figure 5.6, the power to the PA is ramped up and down rapidly at the beginning and end of each slot in the frame, again causing the PA gain to change during the data transmission time of the slot. This gain modulation effect can be significant in handset PAs.

These thermal effects are also exacerbated in multi-carrier signal environments, where individual carriers can be turned on and off to accommodate the signal traffic demands, resulting in rapid changes in the dissipated energy in the PA. As the digital modulation for wireless communications becomes more sophisticated, with

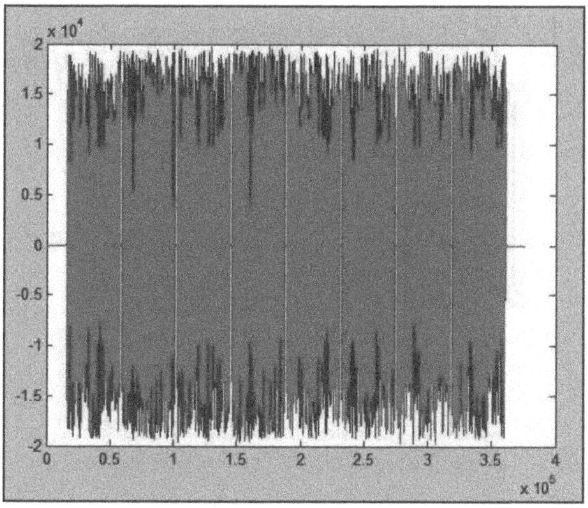

Figure 5.6 A single frame of a GSM signal with eight slots occupied, showing that the power output is dropped to zero between each slot. These lead to thermal memory effects in the PA.

time- and frequency-dependent signal changes during the frame or resource block in LTE, the thermal memory effects will have more of a contribution to the nonlinear behavior of the power transistor and amplifier, requiring more sophisticated pre-distortion algorithms and techniques to overcome the signal degradation.

The thermal memory effects can be modeled using the electrical R-C network analog of the thermal diffusion outlined earlier. This yields a simple one-pole equivalent circuit for predicting the temperature rise in a transistor, as shown in Figure 5.7. Here, the instantaneous power dissipation, resulting (mostly) from the product of the transistor output current and voltage, passes through a low-pass filter given by the thermal resistance and capacitance, and the voltage across the thermal resistance is the analog of the temperature rise in the device. The transistor output current can then be modified as a function of this temperature rise, based on the original work of Canfield *et al.* [5], to produce a self-consistent dynamic electrothermal model. Several commercially-available transistor models incorporate this sort of thermal model [2, 6, 7].

In a real transistor, the thermal time constant can vary over several decades, describing the various microscopic and macroscopic aspects of the thermal behavior. Such aspects include the rapid heating and cooling of the channel region in the FET, resulting from envelope-signal level variations, and the long time constant

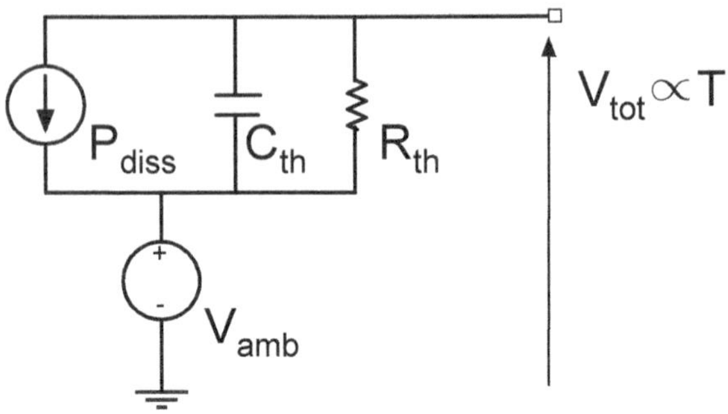

Figure 5.7 Simple thermal equivalent circuit model used in many dynamic electrothermal transistor models.

behavior due to the average power dissipation in the transistor. Thorough characterization and modeling of the thermal properties using pulsed measurement techniques have shown that the thermal behavior is best modeled with a thermal pole of fractional order [8]. A more practical implementation from a circuit model point of view is to use a multiple R-C ladder network to provide a suitable span of thermal time constants [2].

5.5 TRAPPING OF CHARGE

Traps in semiconductor materials are a result of imperfections in the crystal structure, causing local variations in the internal potential in the material. The imperfections arise mainly at interfaces between dissimilar semiconductors, where stresses and strains at the interface disrupt the perfect crystal structure, or at the semiconductor surface, where there are unsatisfied atomic bonds. In both cases, the local potential energy in the material is disrupted from its regular pattern. The disruptions in the local potential can be potential wells that can capture or trap a passing electron, thereby causing a change in the local current flow. The local potential well or trap can capture or release the electron depending on the values of impressed external electric fields or potentials, which may be DC or RF, the speed of change of such external fields, or on the material temperature, which again may have local variations in space and time.

The capture and release of the mobile charge takes place on timescales that are long, compared with the period of the RF signal, and can range from fractions of a second to hours. The trap behavior can be characterized by pulse measurements, used to provide an isothermal environment and eliminate thermal memory effects. Parker *et al.* [9] have observed interesting trap-mediated behavior of GaAs-based pHEMTs using pulsed IV and S-parameter measurements. For transistors biased at the Class A bias point, in the middle of the IV-plane, and the voltage pulses taken from this quiescent point, the measured output $I - V$ characteristics are classic textbook shape; but if the quiescent bias point is shifted to below the threshold voltage, a Class C bias, a significant droop is seen in the knee region of the $I - V$ characteristics, as shown in Figure 5.8. This droop is associated with traps that are empty under the Class C bias, and must be filled by electrons, reducing the conduction current.

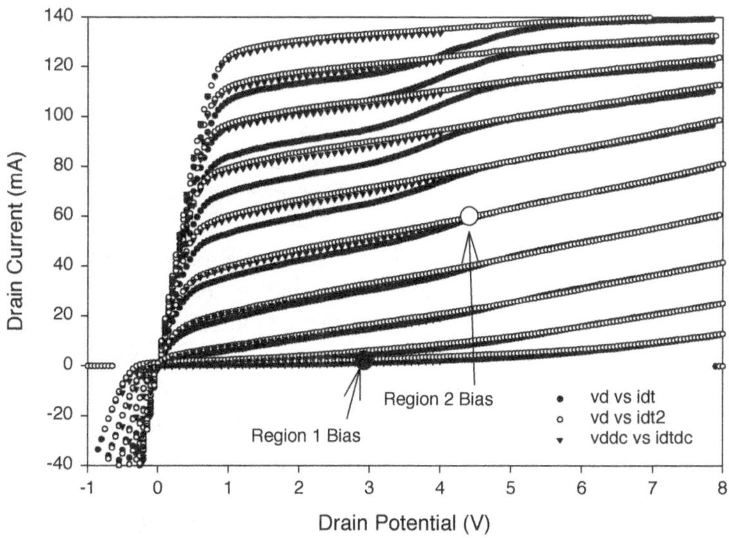

Figure 5.8 Output current-voltage characteristics of a GaAs pHEMT measured using pulsed IV techniques from different quiescent bias points [10].

Gallium Nitride FETs have also exhibited memory effects thought to be caused by trap-related phenomena, in particular, 'knee walk-out' where under RF drive and various quiescent bias conditions, the knee of the $I - V$ characteristics both drops in peak current and 'walks out' to higher drain voltages. This effect can take several seconds to appear, and is repeatable over many cycles. The result is a drop in output power as the load-line excursion is curtailed, and an increase in distortion as the voltage swing enters further into the knee region. Improvements in the quality of the semiconductor materials and interfaces in these devices has improved this behavior. Trap-related phenomena are time-dependent in this sense, too: as the semiconductor material and device processing technologies improve, resulting in purer materials and cleaner interfaces, the trap densities become smaller and more controllable, and the device performance becomes more repeatable and predictable.

Modeling of the trapping effects can be accomplished at the transistor-model level by including an energy storage network in the output network of the model: a simple approach by Brazil *et al.* models the traps using an L-R network, which is shown to be quite effective and easy to implement [11].

5.6 BIAS AND SUPPLY LINES

The term *bias* is used here for the voltage connections to the gate or base of the power transistor, and *supply* is used for the high-current connection at the output (drain or collector) terminal of the transistor.

The job of the bias and supply connections to the power transistor is to provide the appropriate DC operating conditions, and yet be invisible to the RF or modulating signal. For this purpose, we can turn to the classic quarter-wavelength DC supply connection that is illustrated in PA design textbooks, for its simplicity in doing this job. This high-impedance transmission line is connected to the transmission line at the transistor's output terminal (for example), which will be a relatively low impedance line of a few ohms characteristic impedance; the high impedance line does not load the transistor output significantly. The DC supply is attached at the other end of the high-impedance line, using traditional capacitive decoupling techniques to provide a low impedance at the RF and modulating signal frequencies; the quarter-wavelength line then transforms the low impedance to high impedance at the transistor output port, so the DC connection has minimal loading effect at RF.

The question now is how broadband is the low-impedance region provided by the capacitive decoupling of the DC supply beyond the quarter-wave transformer. Often, several capacitors of various values are placed in parallel to provide broadband decoupling, with the smallest value capacitor closest to the PA connection, and progressively larger capacitors, providing lower impedance at low frequencies, as we move away from the PA towards the DC supply. The capacitors that are used at RF are not an ideal capacitance: they possess parasitic resistive loss and inductance due to the size and shape of the component. The capacitance and inductance produce a self-resonant frequency, below which the component behaves as a capacitor of the appropriate value, but above this frequency the component looks inductive. Large-value capacitors have low self-resonant frequencies, hence the need to parallel these with lower values capacitors, with higher self-resonant frequencies, to produce a broadband low impedance path to ground for the RF signals. This results in a L-C ladder structure, a low-pass filter, for the decoupling network. The bandwidth of this filter is the video bandwidth (VBW).

The goal in the design of the decoupling network is to make the VBW wide enough so that the even-order intermodulation products generated by the nonlinear transistor, which appear at frequencies close to DC, see the low impedance path to ground and therefore cannot impress any significant distortion voltage or current onto the bias voltage or supply current. As mentioned earlier, the distortion on the bias and supply lines modulates the gain and other properties of the transistor, generating further distortion products in the output signal. This is a nonlinear memory effect. The wider the video bandwidth, the wider the modulation signal IM2 and higher-order products that can be eliminated. Generally, the VBW at the bias and supply sides of the transistor should be of similar bandwidth to be effective.

In the context of a typical high-power LDMOS PA, some comments on what we mean by 'low' and 'high' impedance are in order. The optimum output resistance for the power transistor may be around 2 to 5 Ω at the package terminal. The 'low' impedance that we need to create in the supply line over the video bandwidth frequencies should be an order of magnitude or more smaller than this value to be effective as a short circuit: impedances of 0.5 Ω or less are difficult to realize in practice, and so the suppression of these nonlinear memory effects may be limited. To help in presenting the low DC impedance at the drain contact of the transistor, special packages with extra pins for the DC connections have been developed [12], and techniques for including a small decoupling capacitor inside the package have been devised [13]. On the other hand, the 'high' impedance of the quarter-wave line only needs to be an order of magnitude or so greater than the output resistance, $20-100\ \Omega$ to be effective. In practice, this range of impedance may be realized with a smaller transformation than 90°, and shorter lines are often used, saving space and reducing loss.

Doherty PAs can introduce more nonlinear memory effects through the action of the peaking amplifier. This is commonly a regular power transistor operated in Class C mode, which is in itself a nonlinear model of operation. The peaking amplifier switches on only during the peak excursions of the signal, and it is switched on hard in order to supply the necessary power to perform the load modulation of the main amplifier and drive the load. During these switching pulses, the current demand is high, and any losses in the supply line will generate a distortion voltage: a memory effect.

The bias and supply circuits are where the PA designer has the opportunity to influence the impact of memory effects on the performance of the PA, and hence influence how easy or difficult the PA is to linearize with a given pre-distortion algorithm. Enhancing the video bandwidth both in terms of its frequency extension and by minimizing the supply line impedance will reap big rewards.

5.7 INCLUDING MEMORY EFFECTS IN THE NONLINEAR MODEL

A simple but intuitive model that includes the memory effects described in the chapter is shown in Figure 5.9. Here, the nonlinearity of the transistor or PA is assumed to be memoryless or static, dependent only on the instantaneous value of the input voltage. The short-term memory effects due to the input and output matching networks, and the input and output reactive elements of the transistor, are represented by passive filter blocks. The long-term memory effects are modeled by another filter network that is a feedback network from the output of the nonlinearity to the input: the LTM affected the value of the input signal by a delayed output signal. It is important to note that the output signal goes around this feedback loop only once: there is only one delayed instance of the signal. This LTM model is quite flexible, as we can add parallel paths to represent specific long-term memory behavior independently, such as thermal and trapping effects.

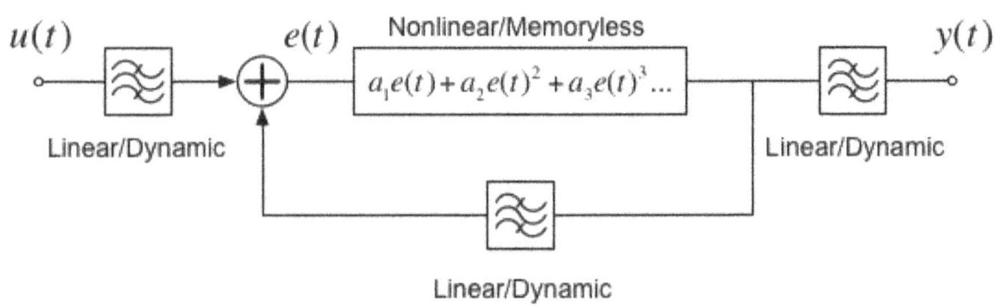

Figure 5.9 A simple block diagram model of a nonlinear PA indicating how the short- and long-term memory effects can be included.

The memory effects are a manifestation of time-dependent behavior, and can be described by time-dependent differential equations or difference equations. The differential equations describe the rate of movement of charge, through reactive elements and trapping effects, or the movement of thermal energy through the semiconductor: these are the dq/dt and dQ/dt effects in the device. We can use the time-dependent differential equations describing these phenomena directly in our model, or, as shown in Chapter 3, use the time-sampled equivalents describing the nonlinear dynamical behavior

$$y[t] = f_{NL}\left(u[t], u[t-1], u[t-2], ..., u[t-m]\right) \tag{5.1}$$

where f_{NL} is the nonlinear function, and m the memory depth. This is essentially the System Identification approach to behavioral model development that we met in Chapter 3, so our nonlinear modeling approach is already memory-capable. The development of nonlinear dynamical models is described in the next chapter, where the focus is primarily on using Volterra series for the model. The model structure in Figure 5.9 can be recast in the form of a Volterra series, with a little bit of work.

References

[1] S. Boyd and L. O. Chua, "Fading memory and the problem of approximating nonlinear operators with Volterra series," *IEEE Trans. Circuits Syst.*, Vol. 32, No. 11, pp. 1150–1161, Nov. 1985.

[2] P. Aaen, J. Plà, and J. Wood, *Modeling and Characterization of RF and Microwave Power FETs*. Cambridge, UK: Cambridge University Press, 2007.

[3] J. Breitbarth and D. Schmelzer, "Absorptive near-Gaussian low pass filter design with applications in the time and frequency domain," *IEEE MTT-S Int. Microwave Symp. Dig.*, vol. 3, Fort Worth, TX, June 2004, pp. 1303–1306.

[4] B. G. Streetman, *Solid State Electronic Devices*. Englewood Cliffs: Prentice-Hall International, Inc, 1990.

[5] P. C. Canfield, S. C. F. Lam, and D. J. Allstot, "Modeling of frequency and temperature effects in GaAs MESFETs," *IEEE J. Solid State Circuits*, Vol. 25, No. 1, pp. 299–306, Feb. 1990.

[6] A. E. Parker and D. J. Skellern, "A realistic large-signal MESFET model for SPICE," *IEEE Trans. Microw. Theory Tech.*, Vol. 45, No. 9, pp. 1563–1571, Sep. 1997.

[7] I. Angelov, H. Zirath, and N. Rorsman, "A new empirical nonlinear model for HEMT and MESFET devices," *IEEE Trans. Microw. Theory Tech.*, Vol. 40, No. 12, pp. 2258–2266, Dec. 1992.

[8] A. E. Parker and J. G. Rathmell, "Broadband characterization of FET self-heating," *IEEE Trans. Microw. Theory Tech.*, Vol. 53, No. 7, pp. 2424–2429, Jul. 2005.

[9] A. E. Parker and J. G. Rathmell, "Bias and frequency dependence of FET characteristics," *IEEE Trans. Microw. Theory Tech.*, Vol. 51, No. 2, pp. 588–592, Feb. 2003.

[10] A. E. Parker and D. E. Root, "Pulse measurements quantify dispersion in PHEMTs," *Proc. 1998 URSI Int. Symp. On Signals, Systems, and Electronics*, Pisa, Italy, Sep. 1998, pp. 444–449.

[11] I. Kallfass, H. Schumacher, and T. J. Brazil, "Multiple time constant modeling of dispersion dynamics in hetero field-effect transistors," *IEEE Trans. Microw. Theory Tech.*, Vol. 54, No. 6, pp. 2312–2320, June 2006.

[12] C. Blair and D. Fowlkes, "Method of manufacturing a power transistor module and package with integrated bus bar," U.S. Patent 8 258 014 B2, 2011.

[13] H. Ladhani, J. Jones, and G. Bouisse, "Improvements in the instantaneous bandwidth capability of RF power transistors using in-package high-k capacitors," *IEEE MTT-S Int. Microwave Symp. Dig.*, Baltimore, MD, June 2011, pp. 1–4.

Chapter 6

Nonlinear Dynamical Models

In the previous two chapters we have reviewed the two fundamental cornerstones of nonlinear dynamical modeling: nonlinear function fitting techniques and the system dynamics or memory effects. Here we shall combine these principles and descriptions to create a framework for building nonlinear dynamical models. We will build our models in the time domain, the natural home for nonlinear dynamical phenomena. The framework that we shall use is the Volterra series. This is a well-established mathematical description for nonlinear dynamical systems, with a considerable amount of supporting literature and methodologies (see the bibliography at the end of this chapter) and, as we shall see, it is fairly straightforward to implement as a modeling technique. A more detailed description of the Volterra series formalism is presented next.

We shall also focus on sampled data models. As presented in Chapter 3, much of the measured and simulated data that we shall be using to build our models is discrete-time sampled data; we shall apply the System Identification techniques outlined also in Chapter 3 as the basis of our model-building methodology. The Volterra series can be written as a *linear-in-parameters* model, enabling straightforward linear identification techniques to be used.

While the main focus of this chapter is on nonlinear behavioral modeling using polynomial or Volterra series and simplifications thereof, taking advantage of the linear-in-parameters formulation for ease of solution, we have also included a section on non-Volterra methods. These methods include the use of artificial neural networks, a general nonlinear modeling approach that uses nonlinear optimization methods to arrive at a solution.

6.1 INTRODUCTION TO VOLTERRA SERIES

Vito Volterra (see Figure 6.1) was an Italian mathematician whose pioneering work on integral and differential equations led to the development of the branch of mathematics known as functional analysis. Volterra studied under Enrico Betti at the University of Pisa, graduating with a Doctor of Physics degree in 1882, and in his thesis on hydrodynamics he developed independently some of the fluid mechanical equations that had been presented earlier by Stokes. He was appointed Professor of Rational Mechanics at Pisa in 1883, and immediately began working his theory of *functionals*: functions that depend on a continuous set of values of another function. The theory of functionals led to the development of new fields of analysis, including the solution of differential and integral equations. In 1884, Volterra himself began to work on the solution of integral equations, which has become his lasting contribution: "the integral equation of the Volterra type." His first paper on the theory of functionals was published in 1887, and his collected works were published in the volume *Theory of Functionals and of Integro-Differential Equations* in 1936, a few years before his death.

The current usage of the Volterra theory of functionals is due to Norbert Wiener, a professor at MIT, who became interested after contact with Volterra's student, Paul Levy. Wiener applied his knowledge of Brownian motion to the integration of the Volterra functionals, identifying the kernels of the functional using white noise. Wiener used Volterra series for a system analysis of the effect of radar noise on a receiver circuit. Wiener also hypothesized that a time-invariant nonlinear dynamical system could be modeled by a series of Volterra functionals, provided that the nonlinearities in the system were not too 'strong.' In the context of PA modeling, a strong nonlinearity would be a discontinuity in the response of the input signal, caused by clipping of the waveform, for instance. It is often stated that the (PA) system must be *weakly nonlinear* for it to be amenable to Volterra analysis: what this means is that the system response must be continuous, and therefore can be represented by a finite series of contributing terms. Sometimes, weakly nonlinear is interpreted as a *small* number of terms, no more than cubic(!), but this is unnecessarily artificial and constrictive.

The development of Wiener's methods are well presented by Schetzen [1], who provides a thorough introduction to the mathematical underpinnings and use of Volterra series. Wiener's work on the application of Volterra theory to nonlinear circuits and communications systems was picked up and applied by researchers at the Rome Air Development Center in the 1960s [2], and sparked further interest leading to some classic publications on the application of Volterra methods to the analysis of nonlinear systems [3–5].

Figure 6.1 Vito Volterra (b.1860, d. 1940).

The Volterra techniques have recently been "rediscovered," as distortions in communications circuits have again become a significant design concern. High-power, high-frequency devices are not inherently very linear, and the complexity and level of integration of modern wireless communications systems means that the distortion has to be addressed at the design stage of the circuits and systems [6–8]. The memory effects in the power amplifiers in these systems are a big problem, and the Volterra series techniques have been widely applied to the behavioral modeling of the RF power amplifiers, and to their linearization using pre-distortion techniques [9–11].

In the following sections, we show how a Volterra series expansion for a nonlinear dynamical system can be developed in a number of ways. These developments are more or less heuristic, but are intended to show that the Volterra series model is a fairly natural development from linear systems, from nonlinear instantaneous model descriptions, and from a simple prototype nonlinear dynamical system model that we could extend to describe a more complicated system such as an RF power amplifier. In this way, we shall see how the Volterra series is a straightforward, though complicated, expression for describing a general (weakly) nonlinear dynamical system, and we can then proceed to some practical implementations and some techniques of identifying the components of the Volterra series model.

6.1.1 Volterra Series as a Development of a Taylor Series

One way of building up a Volterra series expression is as a development of a nonlinear instantaneous model that can be described by a Taylor series expansion. We take as our prototype the polynomial model from Chapter 4 to describe a nonlinear instantaneous system:

$$y(t) = a_0 + \sum_{n=1}^{N} a_n u^n(t) \tag{6.1}$$

The polynomial coefficients a_n are found from the Taylor series of the input-output relationship, expanded around some operating point u_0

$$\begin{aligned} y\left(u(t)\right) &= y(u)|_{u=u_0} + \frac{1}{2!}\frac{dy}{du}\bigg|_{u=u_0}(u-u_0) + \frac{1}{3!}\frac{d^2y}{du^2}\bigg|_{u=u_0}(u-u_0)^2 + ... \\ &= a_0 + a_1 u + a_2 u^2 + ... \end{aligned} \tag{6.2}$$

Now, instead of the instantaneous relationship, let's consider the output $y(t)$ as a function of $u(t)$ and the values of u at some previous times, thus describing the memory effects:

$$y(t) = f\left(u, u_1, u_2, , u_n\right) \tag{6.3}$$

where

$$u = u(t), u_1 = u(t-\tau_1), u_2 = u(t-\tau_2), ... \tag{6.4}$$

The Taylor series expansion for this is

$$\begin{aligned} y\left(u(t)\right) &= y(u)|_{u=u_0} + \frac{1}{2!}\frac{dy}{du}\bigg|_{u=u_0}(u-u_0) + \frac{1}{2!}\frac{dy}{du_1}\bigg|_{u=u_0}(u_1-u_0) + ... \\ &+ \frac{1}{3!}\frac{d^2y}{du^2}\bigg|_{u=u_0}(u-u_0)^2 + \frac{1}{3!}\frac{d^2y}{du_1^2}\bigg|_{u=u_0}(u_1-u_0)^2 \\ &+ \frac{1}{3!}\frac{d^2y}{du du_1}\bigg|_{u=u_0}(u-u_0)(u_1-u_0) + ... \end{aligned} \tag{6.5}$$

which is a *multinomial* series. The memory terms in $(u_1 - u_0)$ and the so-called Volterra *cross-terms* in $(u-u_0)(u_1-u_0)$ can be readily observed in this expression.

The Volterra series can be thought of as a Taylor series with memory. The nonlinearities described by such a series must satisfy some smoothness criterion for the series to be convergent. This is another way of saying that the series will approximate the true value of the function to within some specified tolerance, that is, the truncation error becomes smaller than this tolerance value.

6.1.2 Volterra Series as a Development of a Linear System

In Chapter 3, we reviewed some of the properties of linear systems. In particular, the output from a linear dynamical system can be described by the convolution of system transfer response function, $h(t)$, with the input signal, $u(t)$, and is given by

$$y(t) = \int_0^t h(t-\tau)u(\tau)d\tau \tag{6.6}$$

If we again take as our prototype nonlinear dynamical model the polynomial model from Chapter 4, we note that the nonlinearity is described by the higher-order contributions in u^2, u^3, and so on. Adapting this approach to the convolution integral (6.6), we can write these higher-order terms as

$$y(t) = \int_{-\infty}^{+\infty}\int_{-\infty}^{+\infty} h_2(\tau_1, \tau_2)u(t-\tau_1)u(t-\tau_2)d\tau_1 d\tau_2 \tag{6.7}$$

for the second-order term, and

$$y(t) = \int_{-\infty}^{+\infty}\int_{-\infty}^{+\infty} \cdots \int_{-\infty}^{+\infty} h_n(\tau_1, \tau_2, ..., \tau_n)u(t-\tau_1)u(t-\tau_2)...u(t-\tau_n)d\tau_1 d\tau_2 ... d\tau_n \tag{6.8}$$

for the general n^{th}-order term. The integral limits have been extended to $[-\infty, +\infty]$ for generality in these expressions, although we would expect the usual time-invariance and causality criteria to apply. The complete model is the sum of all of these expressions, from the linear convolution term to the nth-order expression, yielding a polynomial model including the dynamics.

The Volterra *kernels*, $h_n(\tau_1, \tau_2, ..., \tau_n)$, can be thought of as the coefficients of each component of this nonlinear dynamical model expression, analogous to the coefficients of the instantaneous polynomial.

6.1.3 Volterra Series from the Cascade of Linear and Nonlinear Components

This development of a Volterra series expression is taken from a presentation by Steve Maas [12, 13], and shows how a *nonlinear transfer function* can be inferred from the cascade of a linear component, such as a filter, and a nonlinear memoryless component, such as the polynomial model from Chapter 4, as shown in Figure 6.2.

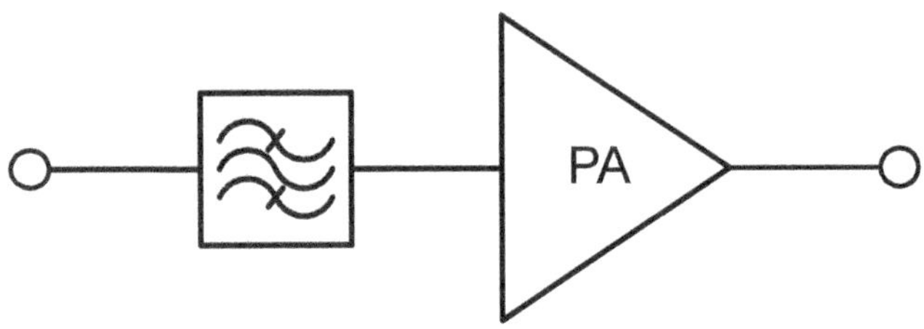

Figure 6.2 Cascade of a linear filter and an instantaneous nonlinearity, described by a polynomial series, to create a Volterra series description of this system.

The input-output relationship of the nonlinear component in Figure 6.2 is given by the polynomial

$$y(t) = f\left(v(t)\right) = a_0 + \sum_{n=1}^{N} a_n v^n(t) \tag{6.9}$$

and the filter block can be described by its frequency-domain transfer function, $H(\omega)$, where

$$V(\omega) = H(\omega)U(\omega) \tag{6.10}$$

The input signal to this cascade network, $u(t)$, is written in terms of the signal frequency components using the Fourier transform:

$$u(t) = \sum_{k=-\infty}^{k=+\infty} U[k]e^{j\omega_k t} = \sum_{\substack{k=-K \\ k\neq 0}}^{k=+K} U_k e^{j\omega_k t} \tag{6.11}$$

Here we limit the summations by considering the input signal to be band-limited, and removing the $k = 0$ term ensures there is no DC component in the output signal, which can then be written as

$$v(t) = \sum_{\substack{k=-K \\ k \neq 0}}^{k=+K} H(\omega_k) U_k e^{j\omega_k t} \tag{6.12}$$

and substituting for $v(t)$ in (6.9) gives:

$$y(t) = \sum_{n=1}^{N} a_n \left[\sum_{\substack{k=-K \\ k \neq 0}}^{k=+K} H(\omega_k) U_k e^{j\omega_k t} \right]^n \tag{6.13}$$

which can be expanded as:

$$y(t) = \sum_{n=1}^{N} a_n \sum_{\substack{k_1=-K \\ k_1 \neq 0}}^{k_1=+K} \dots \sum_{\substack{k_n=-K \\ k_n \neq 0}}^{k_n=+K} H(\omega_{k_1}) \dots H(\omega_{k_n}) U_{k_1} \dots U_{k_n} e^{j(\omega_{k_1} + \dots + \omega_{k_n})t} \tag{6.14}$$

We can generalize this expression by writing the product of the linear transfer functions as a *nonlinear transfer function* for the whole system:

$$H_n\left(\omega_{k_1}, \dots, \omega_{k_n}\right) = H(\omega_{k_1}) \dots H(\omega_{k_n}) \tag{6.15}$$

By generalizing this product term in this way, we can account for a number of phenomena that were not present in the simple block diagram, but are commonly found in nonlinear systems, such as feedback and memory effects.

This frequency-domain expression, (6.14), can be converted back into the time domain using the Fourier transform to obtain

$$y(t) = \sum_{n=1}^{N} a_n \int_{-\infty}^{+\infty} \dots \int_{-\infty}^{+\infty} h_n(\tau_1, \dots, \tau_n) u(t - \tau_1) \dots u(t - \tau_n) d\tau_1 \dots d\tau_n \tag{6.16}$$

which, with the inclusion here of the polynomial coefficients a_n, is the same expression for the Volterra series as was developed in Section 6.1.2, from a heuristic expansion of the linear convolution integral.

6.1.4 The Higher-Order Kernels

We have already seen, in Chapter 3 and the foregoing, the idea of a linear system being described by the one-dimensional convolution integral, (6.6). If we let $u(t)$

become a delta function, $\delta(t)$, then this equation yields the impulse response of the linear system, $h_1(\tau)$, given by the first-order Volterra kernel. And from here, we can obtain the frequency response of the system by the Fourier transform.

The second-order Volterra kernel, $h_2(\tau_1, \tau_2)$, characterizes the second-order response of the system through (6.7), and the obvious question is: "What is the physical significance of the second-order Volterra kernel?" Letting $u(t)$ become the impulse function again, we can see from (6.7) that the impulse response of the second-order Volterra integral is

$$\begin{aligned} y(t) &= \int_{-\infty}^{+\infty}\int_{-\infty}^{+\infty} h_2(\tau_1, \tau_2)\delta(t-\tau_1)\delta(t-\tau_2)d\tau_1 d\tau_2 \\ &= h_2(t,t) \end{aligned} \tag{6.17}$$

In fact, this does not characterize the second-order system completely, but only along the line $\tau_1 = \tau_2$. To get a full picture of the second-order kernel, we need to drive the system in a different way, to extract the two-dimensional impulse response. We use two different input signals, and the input signal is now written as

$$u(t) = u_1(t) + u_2(t) \tag{6.18}$$

and the response $y(t)$ is given by (6.7), as before

$$\begin{aligned} y(t) &= \int_{-\infty}^{+\infty}\int_{-\infty}^{+\infty} h_2(\tau_1, \tau_2)u(t-\tau_1)u(t-\tau_2)d\tau_1 d\tau_2 \\ &= \mathbf{H}_2[u(t)] \end{aligned} \tag{6.19}$$

where $\mathbf{H}_2[u(t)]$ is the second-order Volterra operator. Now

$$\begin{aligned} y(t) &= \mathbf{H}_2[u(t)] = \mathbf{H}_2[u_1(t) + u_2(t)] \\ &= \mathbf{H}_2\{u_1(t), u_1(t)\} + 2\mathbf{H}_2\{u_1(t), u_2(t)\} + \mathbf{H}_2\{u_2(t), u_2(t)\} \\ &= \mathbf{H}_2[u_1(t)] + 2\mathbf{H}_2\{u_1(t), u_2(t)\} + \mathbf{H}_2[u_2(t)] \end{aligned} \tag{6.20}$$

The $\mathbf{H}_2[u_n(t)]$ terms represent the second-order response to two copies of the same input signal delayed with respect to each other, as expressed by (6.7), and the term

$\mathbf{H}_2\{\,.\,\}$is a bilinear Volterra operator, with the following functional representation

$$\mathbf{H}_2\{u_1(t), u_2(t)\} = \int_{-\infty}^{+\infty}\int_{-\infty}^{+\infty} h_2(\tau_1, \tau_2)u_1(t-\tau_1)u_2(t-\tau_2)d\tau_1 d\tau_2 \qquad (6.21)$$

The bilinear operator is linear in u_1 when u_2 is held constant, and *vice versa*. Rearranging the expression for the bilinear operator in terms of the second-order Volterra operator

$$2\mathbf{H}_2\{u_1(t), u_2(t)\} = \mathbf{H}_2[u_1(t) + u_2(t)] - \mathbf{H}_2[u_1(t)] - \mathbf{H}_2[u_2(t)] \qquad (6.22)$$

is the basis of the interpretation of $h_2(\tau_1, \tau_2)$ as a two-dimensional impulse response. By choosing the values of τ_1 and τ_2 to cover the time axes for $t > 0$ (assuming causality) and calculating or measuring $\mathbf{H}_2[\ ]$ for the input signals u_1 and u_2, which may be copies of the same signal applied at times τ_1, τ_2, we can generate the surface that represents the two-dimensional impulse response of the second-order system. This is illustrated in Figure 6.3.

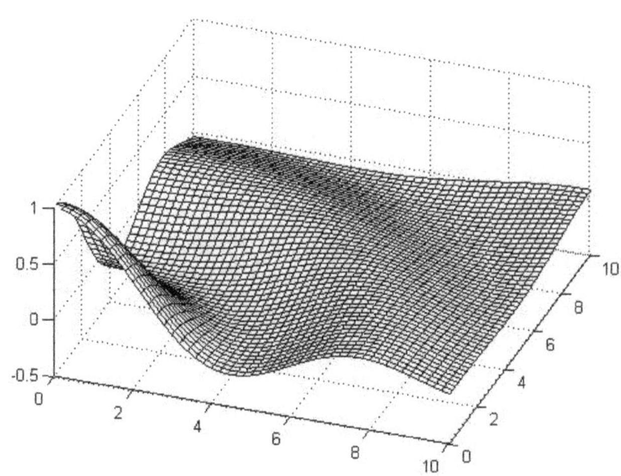

Figure 6.3 Schematic representation of a two-dimensional impulse response for a hypothetical second-order Volterra system.

This impulse response can be transformed into the frequency domain, using a two-dimensional Fourier transform, to give a frequency response of the second-order system, which is the output of the system in response to inputs at two different frequencies:

$$Y(\omega) = H(\omega_1, \omega_2)U(\omega_1)U(\omega_2) \tag{6.23}$$

where $H(\omega_1, \omega_2)$ is the second-order Volterra transfer function, and is a nonlinear transfer function.

This can be interpreted fairly straightforwardly. If a linear system is excited by two signals at different frequencies, the system output will contain only those two frequencies. If a second-order system is excited by signals at two frequencies, the system output will include signals at those two frequencies and, in addition, it will contain components at the sum and difference frequencies $\omega_1 + \omega_2$ and $\omega_1 - \omega_2$, arising from the second-order nonlinearity.

The higher-order kernels can be interpreted in a similar fashion, and are described in detail by Schetzen [1], but they are much more difficult to draw.

Third-order and higher odd-order distortions are of particular interest for RF PAs, as they produce intermodulation products within the passband of the system, adding in-band distortion products. We shall discuss these odd-order distortion products in more detail in the context of the narrowband approximation to the Volterra series, in which the model can be determined from the analytic signal representation of the modulation envelope signal. The narrowband approximation will be outlined in more detail in Section 6.4, "Narrowband Approximation."

6.1.5 Discrete-Time Expressions for Volterra Series

As noted earlier, the data that we shall use for the model generation, whether obtained from measurement or from simulation of the PA, will be discrete time-sampled data. We cast the continuous-time, fading-memory Volterra series expression in a discrete, time-sampled form:

$$y(t) = f_{poly}\left(u(t), u(t-1), ...u(t-M)\right) \tag{6.24}$$

where f_{poly} represents the Volterra multinomial function, t is the sampling instant, and M is the memory depth of the model, measured as a multiple of the sampling time steps, and known as "delays" or "taps." This expression can be expanded in the orders of the multinomial series:

$$\begin{aligned} y(t) &= \sum_{m_1=0}^{M} h_1(m_1)u(t-m_1) \\ &+ \sum_{m_1=0}^{M}\sum_{m_2=0}^{M} h_2(m_1,m_2)u(t-m_1)u(t-m_2) \\ &+ \sum_{m_1=0}^{M} \dots \sum_{m_n=0}^{M} h_n(m_1,\dots,m_n)u(t-m_1)\dots u(t-m_n) \\ &= \sum_{n=1}^{N}\sum_{m_1=0}^{M} \dots \sum_{m_n=0}^{M} h_n(m_1,\dots,m_n)\prod_{j=1}^{n} u(t-m_j) \end{aligned} \tag{6.25}$$

where N is the polynomial degree and M is the memory depth. When m_i is zero, this expression yields the instantaneous or memoryless response. As is usual for classic Volterra series, the memory depth is the same for each order of the polynomial.

6.1.6 Some Remarks on Volterra Series Modeling of Power Amplifiers

We have seen that the Volterra series approach has been used as a general representation of a nonlinear dynamical system [1–5]. In general, the Volterra Series description used in these analyses is applied to time-invariant and causal nonlinear systems; this is not a limitation from the point of view of modeling practical nonlinear systems such as power amplifiers. Nevertheless, these works do have the following limitations, identified by Boyd and Chua [14]: the input signals are only defined as non-zero over a finite time interval, and the Volterra series approximation only holds and is valid over this compact subset of the input space, on this time interval. Power amplifiers tend to be used in a much more arbitrary manner, and so these limitations could be a problem. Boyd and Chua [14] demonstrated that the Volterra series approximation could be shown to be valid for input signals defined over infinite time, producing a general model that was also valid for all times, provided that the approximation or model has *fading memory*. This means that the memory effects of the system to be approximated, and hence the model, have a finite memory time: the memory effects fade to zero after some time. This is a very practical condition, and clearly power amplifiers fall into this class of systems, so the Volterra series approach is very appropriate for modeling of power amplifiers, as demonstrated by several notable works [8–11, 15, 16].

The Volterra series representation of nonlinear dynamical system such as a power amplifier is still just a model, and therefore suffers from all of the limitations and compromises that we need to make in constructing a model. These trade-offs include accuracy and speed, complexity and ease of identification, and so forth.

As we have seen, the Volterra series is a Taylor series with memory. The accuracy of a Taylor series model is best close to the point around which the approximation is being made, u_0, and degrades as we get further away. In generating the Volterra model of a PA, we would normally center the model at or close to the average signal power, and use a modulated signal to generate the data for the model construction. This signal will typically exercise the PA into compression, and, being a digitally-modulated signal, will also present low signal powers to the PA. We use the average power point as the center of approximation as the PA is then operating in the typical use conditions, so the thermal environment and associated memory effects will be modeled properly. For communications signals with high peak-to-average power ratio, typical of modern wireless signals, the compression region may be several dB away from the average power point of the PA, and we need to ensure the model will be accurate at this distance from the approximation point.

The accuracy of the Volterra series can be improved by increasing the number of terms in the polynomial series. In other words, the model is approximating the actual data to a smaller tolerance. Whereas increasing the number of terms in a Taylor series is a straightforward exercise, with a Volterra series the cross-terms and memory terms as illustrated in (6.5) cause the number of terms in the series to increase dramatically as the polynomial degree and memory depth are increased. This is shown diagrammatically in Figure 6.4.

This is one reason why Volterra Series modeling has seen only limited application in the past. A cubic model of modest memory was considered manageable, and 5th-degree barely tractable. This is perhaps also why Volterra series approximations have been considered to be applicable only to weakly-nonlinear systems, of low polynomial order, as mentioned earlier. As a further attempt to reduce the number of terms when modeling PAs, the even-order terms are often ignored. The usual reason given for this is that the even-order mixing components are found at the even harmonics and at the difference frequencies at baseband, close to DC, and therefore fall outside the frequency band of interest. One problem with this approach is that these baseband components generate modulation on the PA bias and supply lines, creating memory effects as described in Chapter 5. It would seem prudent to include these terms in our model.

With the computational power now available in modern laptop computers, such limitations on the Volterra series polynomial degree are artificial. Volterra

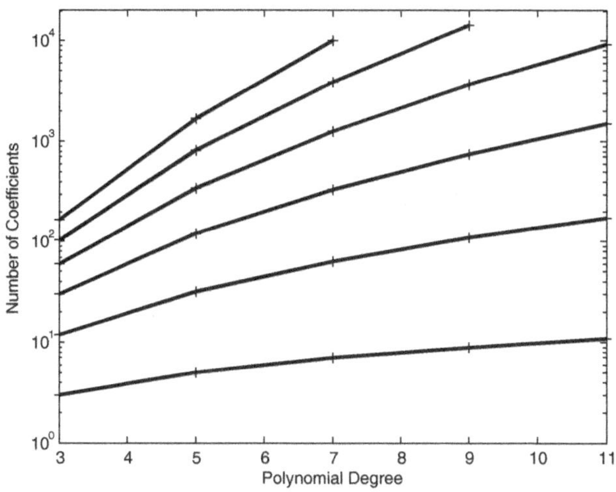

Figure 6.4 The number of coefficients in a Volterra Series increases dramatically with increasing polynomial degree and memory depth.

series models for power amplifiers can be created in a few seconds, having nonlinearity up to 15th-degree, including even-order terms, and several memory delays. To do this, we employ some sophisticated 'pruning' techniques to limit the number of coefficients in the polynomial to a manageable level, around 100 being typical, without any significant loss of model fidelity. We shall describe some of these pruning techniques in more detail in Section 6.3.

Model Parameter Identification

A corollary of having a large number of model coefficients or parameters is that we need to make a large number of measurements to be able to identify these coefficients accurately. Schetzen [1] states that a pth-order Volterra system needs 2^p independent measurements to identify the pth-order Volterra kernel, although in practice some redundancy means that around $2^p - p$ are needed. Once the pth-order kernel has been identified, then the $(p-1)^{th}$-order kernel can be found. The pth-order identification is transparent to the $(p-1)^{th}$-order signals; this enables the unique identification of each kernel order of the system. The complete identification of all the parameters in this system will require of the order of

$$\sum_{p=1}^{P} 2^p - p$$

measurements. This can be costly if each measurement is set up and recorded individually. This is another reason why Volterra series have not been popular in the past.

The work of Wiener [17] showed that the kernels could be identified using white Gaussian noise. We can make use of this result in the modeling of PAs for wireless communications. The spreading code of the WCDMA modulation makes the signal appear noise-like: this improves the interference properties of the signal, but we can use this approximation to band-limited white noise for model parameter identification. We also take the approach that we shall solve an overdetermined system of equations to find the model parameters. The statistical distribution of the noise-like WCDMA signal allows us to capture a large number data samples, much greater than the number of coefficients in the Volterra series model of the PA, and use least-squares techniques to find these coefficient values and identify the model. We touched upon this in Chapter 4, in the context of finding the coefficients of *linear-in-parameters* nonlinear models, and we shall discuss this approach in more detail in Section 6.5, where we present some practical solution techniques for Volterra-based models.

6.2 SOME EXAMPLES OF VOLTERRA MODELS FOR POWER AMPLIFIERS

To overcome the problems with the complexity of the Volterra series, both in terms of its implementation and in the identification of the kernels' coefficients, several simplified structures and models have been proposed over the years. Here we shall present a brief overview of these historical approaches, some of which still see practical application today.

6.2.1 A Nonlinear Feedback Model

A nonlinear dynamical model with feedback to represent a power amplifier was presented by Pedro and coworkers in 2003 [18], and is shown in Figure 6.5. In this model structure, we can identify the cascade linear filters with the input and output matching networks and capacitances of the power transistor, and the feedback network represents long-term memory effects such as thermal and bias contributions, which affect future inputs to the transistor. The transistor itself is modeled as a memoryless nonlinearity, arising from the nonlinear transfer function of the device transconductance function. While this model has a reassuring basis

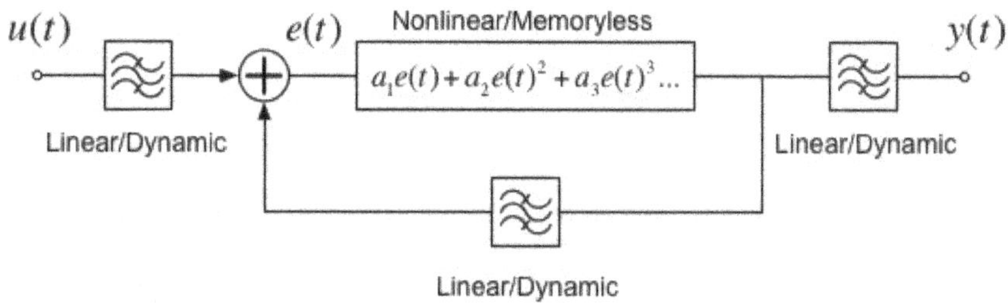

Figure 6.5 A nonlinear feedback model of a power amplifier, cascading linear dynamic system and a static band-pass nonlinearity, and including a feedback path for long-term memory effects.

in the physical structure of the power amplifier, it can be shown to map onto the Volterra series description of (6.25) [10], although its recursive (feedback) construction makes it difficult to extract the model parameters. This problem was overcome using a low-pass approach, but maintaining the original topology of the model [19].

6.2.2 Wiener and Hammerstein Models

Following from our introduction to Volterra series using a cascade of a linear filter and a static PA nonlinearity, here we present a generalized form using a pair of dual models. The Wiener model is the cascade of a linear dynamical system, such as a filter, with a static nonlinearity, such as an instantaneous polynomial. The Hammerstein model places these components in the reverse order. The models are shown in Figures 6.6 and 6.7.

Again, following the development of the model input-output relationship described in Section 6.1.2, we can derive expressions for the output signal from these two models, using a discrete-time input signal. For the Wiener model, the output signal is

$$y(t) = \sum_{n=0}^{N} a_n \left[\sum_{m=1}^{M} h(\tau_m) u(t - \tau_m) \right]^n \tag{6.26}$$

Here, the filter coefficients or kernels are integrated into the power series. These coefficients are therefore no longer linear in parameters, and the model is

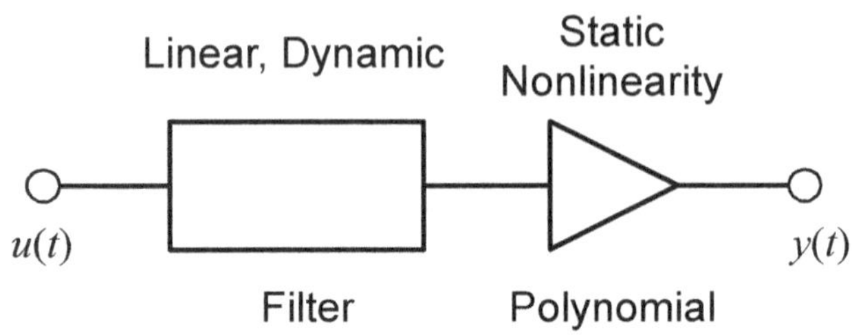

Figure 6.6 Wiener model structure: cascade of a linear dynamic system and static nonlinearity.

more difficult to identify: fully nonlinear optimization techniques are used. This leads to an additional complexity in the model generation. The relationship to a full Volterra series model can be summarized as follows:

$$h_n(\tau_1, ..., \tau_n) = a_n h(\tau_1)...h(\tau_n) \tag{6.27}$$

The nth-order Volterra kernel is given by the product of the filter coefficients over the range of samples or frequencies of interest.

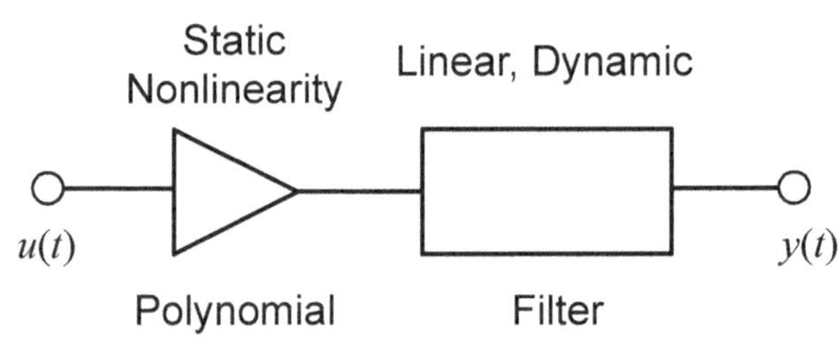

Figure 6.7 Hammerstein Model structure: cascade of a static nonlinearity with a linear system.

The Hammerstein Model, on the other hand, has a much simpler expression for the output signal:

$$y(t) = \sum_{m=1}^{M} h(\tau_m) \left[\sum_{n=0}^{N} a_n u^n(t - \tau_m) \right] \tag{6.28}$$

where the filter coefficients can now be easily identified, and are often included with the polynomial coefficients, and the whole model can be seen to be linear in the parameters $h(\tau_m)a_n$. This makes the identification of the Hammerstein model quite straightforward. The relationship between the Hammerstein kernels and the Volterra kernels is:

$$\begin{aligned} h_n(\tau_1, ..., \tau_n) &= a_n h(\tau) \quad \text{for } \tau_1 = ... = \tau_\text{n} \\ &= 0 \qquad\quad\ \ \text{otherwise} \end{aligned} \tag{6.29}$$

Mathematically, this structure is similar to the 'Memory Polynomial' model, outlined below.

The Wiener-Hammerstein model combines the two model structures, placing a linear filter on each side of the static nonlinearity. This closely mimics the structure of a power amplifier, where the two filters can be associated with the input and output matching networks, and the amplifier is represented by the instantaneous nonlinearity, perhaps in the form of static AM-to-AM and AM-to-PM curves. While this model can be overlaid on the physical structure of the PA, the problems of coefficient or model parameter extraction arising from the Wiener model are still present, and the overall model does not account for any of the long-term memory effects associated with thermal behavior of the PA or the bias circuits.

6.2.3 Memory Polynomial Model

The memory polynomial model is an elegant and simple structure for incorporating memory effects into the static nonlinear polynomial model. The model consists of a delay line and polynomial function, as illustrated in Figure 6.8. The input signal passes through the delay line, and at each tap, the signal is input to the polynomial function. As in the classic Volterra series, in the memory polynomial structure all the polynomials have the same degree. The total number of taps in the delay line defines the memory depth of the model. The output is taken as the sum of the outputs from the polynomial functions, and is given by

$$y(t) = \sum_{n=1}^{N} \sum_{m=0}^{M} a_{nm} u(t - \tau_m) |u(t - \tau_m)|^{n-1} \tag{6.30}$$

The memory polynomial is often described as being a Volterra series model that includes only the "diagonal" terms. In other words, the second-degree terms include only $u^2(t), u^2(t-1), \ldots, u^2(t-n)$, and the third-degree terms include only $u^3(t), u^3(t-1), \ldots, u^3(t-n)$, and there are no cross-terms such as $u(t) * u(t-1) *$

$u(t-2)$; each polynomial is a function of one delayed signal only. The mathematical structure of the memory polynomial model can be shown to be the same as the Hammerstein model.

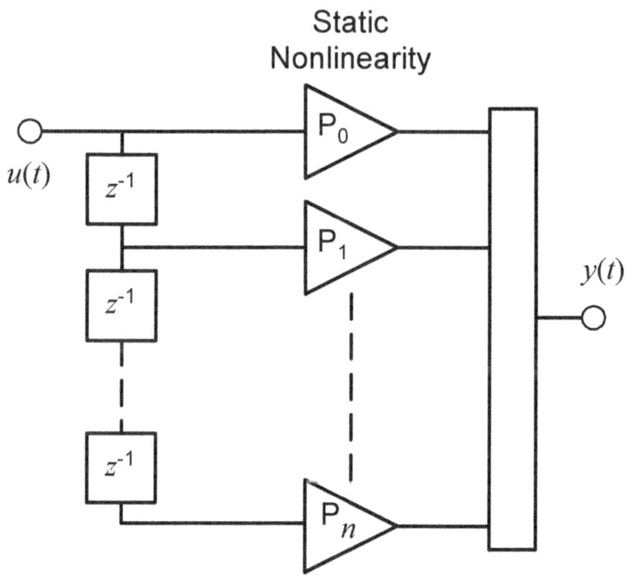

Figure 6.8 Memory Polynomial model: the polynomials are each fed by delayed versions of the input signal.

More sophisticated memory polynomial models have been proposed and demonstrated [20], in which the memory depth decreases as the polynomial order increases. The assumption is that the higher-order terms do not contribute significantly to the memory effects. That is, terms such as, say, $u^{11}(t-9)$ have coefficients that are very tiny, and so do not contribute much to the overall accuracy of the model. In fact, because these terms are so small, they can be difficult to determine accurately in the numerical precision of the calculation, and can therefore introduce noise and even numerical instability into the model solution. By using this 'fading memory polynomial' approach, we are forcing these coefficients to be zero, and hence removing the need to identify and calculate these components of the model. This approach reduces the number of parameters in the model, without affecting the accuracy of the model predictions, and having the potential advantage of greater numerical stability.

A *Generalized* Memory Polynomial structure was introduced by Morgan [21] in the context of a digital pre-distorter. The generalized structure allows for an arbitrary number of additional delays to be included in each of the regular memory polynomial expressions, as follows:

$$\begin{aligned} y(t) &= \sum_{n=1}^{N}\sum_{m=0}^{M} a_{nm}u(t-\tau_m)|u(t-\tau_m)|^{n-1} \\ &+ \sum_{n=1}^{N}\sum_{m=0}^{M}\sum_{k=1}^{K} b_{nmk}u(t-\tau_m)|u(t-\tau_m-\tau_k)|^{n-1} \\ &+ \sum_{n=1}^{N}\sum_{m=0}^{M}\sum_{k=1}^{K} c_{nmk}u(t-\tau_m)|u(t-\tau_m+\tau_k)|^{n-1} \end{aligned} \qquad (6.31)$$

This is effectively adding some upper and lower diagonals to the simple memory polynomial main diagonal terms, and including a limited number of cross term delays into the expressions. Note that Morgan introduces a change of summation index, to be "over the diagonal," which I have not included explicitly in the above representation.

6.2.4 Parallel Cascade

The Volterra series expressions of (6.16) for continuous time or (6.25) for discrete time can be mapped onto a bank of nonlinear filters, with each filter representing one order of the Volterra series, as shown in Figure 6.9. This structure is known generically as a parallel cascade and has been used in, for example, biological applications [22], as well as for RF power amplifier modeling [23]. The Volterra filters can be implemented directly as the convolution algorithm, or more practically by using band-pass filters for each of the harmonics, with a static nonlinearity, as shown in Figure 6.10. This example is using a Wiener-Hammerstein model of reach order of the Volterra series, though simpler filters could be used, such as Hammerstein, which are easier to identify. Korenberg [22] developed a formal technique for the identification of the parallel cascade elements and estimation of the Volterra kernels, for nonlinear models of biological systems. In these systems the memory depth can be very large. He showed that a discrete-time nonlinear system with finite memory (which could be a power amplifier) that can be described by a Volterra series of finite order can be represented by a finite number of parallel paths of cascaded linear and nonlinear elements. In his example, he used a dynamic linear system, which could be a filter, followed by a static nonlinear system such as a

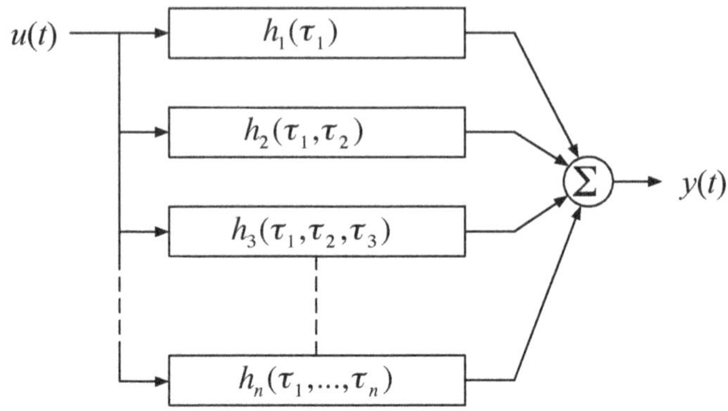

Figure 6.9 Parallel Cascade of Volterra filters of each order.

polynomial, for each cascade path. This is then a parallel cascade of Wiener models. Korenberg's analysis provided an upper limit for the number of cascade paths to represent a Volterra series of a given order, and also a procedure for estimating the Volterra kernels. Evaluating the parallel cascade can be a much quicker method of estimating the Volterra kernels than traditional functional analysis, especially as the polynomial order and memory depth increase. The key is that the nonlinear functions in the cascade paths are static polynomials, which are rapid to evaluate, and increasing the model order simply means adding more parallel cascade paths. This method can also be readily extended to model multiple-input, multiple-output (MIMO) systems, which are becoming popular for improving transmitter system diversity.

6.2.5 Orthogonal Basis Functions

In the modeling approaches described so far, we have dealt only with simple polynomial basis functions. Our model approximation can be written in the following form, for an instantaneous response

$$y(t) = \sum_{n=0}^{N} a_n \varphi_n(t) \tag{6.32}$$

where the basis function $\varphi_n(t)$ is a polynomial

$$\varphi_n(t) = u^n(t) \tag{6.33}$$

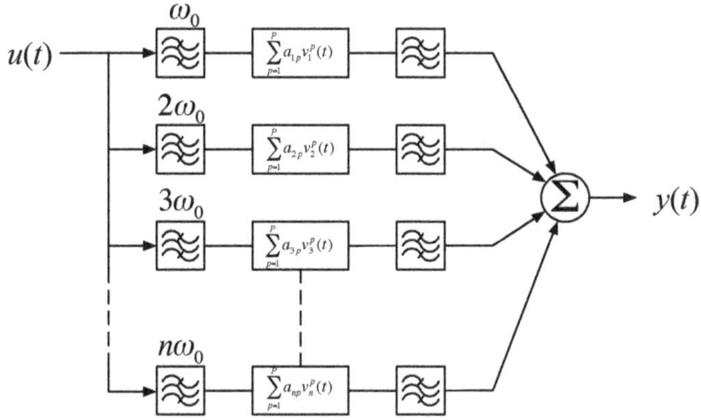

Figure 6.10 Using a cascade of filters and static nonlinearities to implement the cascade of Volterra filters. This example is a Weiner-Hammerstein model.

We can think of each polynomial order, n, as one dimension of the model, with the total number of dimensions being equal to the polynomial degree, N. A multi-dimensional system is best described using coordinates that are orthogonal to each other; this gives the most compact description. In the same way, the most compact model description is when the basis functions are orthogonal.

Orthogonal bases are already familiar concepts to electronic engineers, with Fourier series being perhaps the most well known. A Fourier series is an expansion of a periodic function in terms of (an infinite sum of) sines and cosines, making use of the orthogonality property of the sines and cosines, which can be expressed as:

$$\begin{aligned} &\int_{-\pi}^{\pi} \sin(mx)\sin(nx)dx = \pi \quad \text{if } m = n, \ 0 \text{ otherwise} \\ &\int_{-\pi}^{\pi} \cos(mx)\cos(nx)dx = \pi \quad \text{if } m = n, \ 0 \text{ otherwise} \\ &\int_{-\pi}^{\pi} \sin(mx)\cos(nx)dx = 0 \end{aligned} \tag{6.34}$$

Thus, the periodic function or signal $f(x)$ can be represented as a sum of sinusoidal and cosinusoidal harmonic components:

$$f(x) = \tfrac{1}{2}a_0 + \sum_{n=1}^{\infty} a_n \cos(nx) + \sum_{n=1}^{\infty} b_n \sin(nx) \tag{6.35}$$

with the Fourier coefficients a_n, b_n being found in the usual way, exploiting the orthogonal property.

The orthogonal set of basis functions form a compact representation of the function or signal $f(x)$. We can use an orthogonal basis of polynomials in the same way, thereby minimizing the number of basis functions and hence coefficients that must be found to describe the model.

The orthogonal basis is determined by the probability distribution function of the signal that is used for identification of the model. In principle, any signal can be used, although finding a closed form expression for generating the polynomial series for an arbitrary distribution may be impossible. We stick instead with some more familiar signals.

If the input signal $u(t)$ used for the model identification is sinusoidal, in other words, a delta-function probability density function *pdf* in the frequency domain, the orthogonal polynomial basis is the set of Chebyshev polynomials:

$$y(t) = \sum_{k=1}^{K} a_k T_k\left(u(t)\right) \tag{6.36}$$

where K is the model order, and $T_k(x)$ is the kth Chebyshev polynomial, with

$$\begin{aligned} T_0(x) &= 1 \\ T_1(x) &= x \\ T_2(x) &= 2x^2 - 1 \\ T_3(x) &= 4x^3 - 3x \quad \text{etc.} \end{aligned} \tag{6.37}$$

If the input signal is a zero-mean, real-valued Gaussian random variable with unit variance, in other words, white noise, the orthogonal polynomial basis that is derived is the Hermite polynomial set:

$$\begin{aligned} H_0(x) &= 1 \\ H_1(x) &= 2x \\ H_2(x) &= 4x^2 - 2 \\ H_3(x) &= 8x^3 - 12x \quad \text{etc.} \end{aligned} \tag{6.38}$$

If the signal $u(t)$ has a uniform *pdf* over a given interval, such as, for example, a band-limited communications signal such as WCDMA or LTE, the resulting orthogonal basis is the set of Shifted Legendre polynomials, defined over that interval, which is usually normalized to [0,1]. They have the following orthogonality relationship:

$$\int_0^1 \bar{P}_m(x)\bar{P}_n(x) = \frac{1}{2n+1}\delta_{mn} \tag{6.39}$$

and

$$\begin{aligned} \bar{P}_0(x) &= 1 \\ \bar{P}_1(x) &= 2x - 1 \\ \bar{P}_2(x) &= 6x^2 - 6x + 1 \\ \bar{P}_3(x) &= 20x^3 - 30x^2 + 12x - 1 \quad \text{etc.} \end{aligned} \tag{6.40}$$

These Legendre polynomials are useful in that their values are bounded, as shown in Figure 6.11 for the first five Shifted Legendre polynomials, and therefore the resulting model identification does not suffer from problems of ill-conditioning. We shall comment further on this topic in Section 6.5.

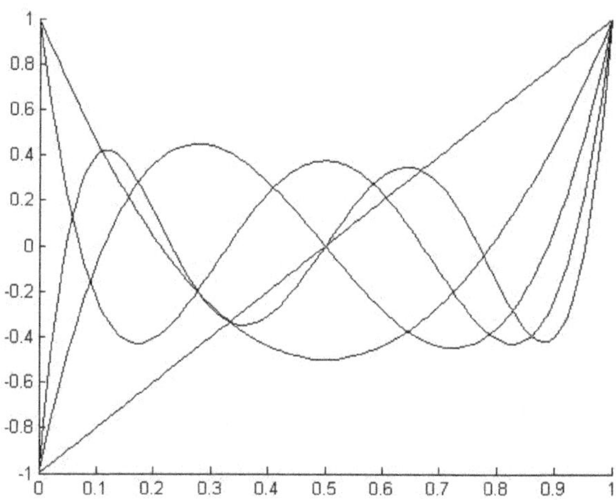

Figure 6.11 The first five Shifted Legendre polynomials calculated on the interval x:[0, 1].

The above expressions for Chebyshev, Hermite, and Shifted Legendre polynomials can all be adapted for dynamical models by including the delay and cross terms in the Volterra series. Recall that Norbert Wiener used a noise signal to identify a Volterra series model for a nonlinear dynamical system, which he described in terms of an orthogonal basis, called the Wiener G-functionals. The signal that he used was a real-valued Gaussian noise input obtained from Brownian motion. The

G-functionals were determined iteratively using the Gram-Schmidt procedure to produce the orthonormal set. The Hermite polynomials above are the instantaneous form of the Wiener G-functionals.

When we consider the dynamical component of the model, expressed in delayed values of the input signal, $u(t-1), u(t-2)$ and so forth, it is evident that we have another degree of orthogonality that we can use to minimize the model dimension: the values of the delay itself. We can determine a characteristic delay time between the input and a delayed version of itself as the interval that minimizes the autocorrelation of the signals,

$$R = \int_{-\infty}^{+\infty} u(t)u(t-\tau)dt \tag{6.41}$$

and $R \to 0$ as $\tau \to$ characteristic delay time. When the autocorrelation is zero, the two versions of the signal are orthogonal.

An alternative technique for determining the characteristic delay time to give orthogonal signals that is applicable to digitally modulated signals is a statistical method known as mutual information. This method uses the statistical probability density of the data at sample times in the data set to determine whether the signal and its delayed version are orthogonal:

$$I_M(\tau) \propto \sum_{y(t)} \sum_{y(t-\tau)} p_2\left(y(t), y(t-\tau)\right) - \sum_{y(t)} p_1\left(y(t)\right) - \sum_{y(t-\tau)} p_1\left(y(t-\tau)\right) \tag{6.42}$$

where I_M is the mutual information for a delay of τ, p_1 is the probability density of the data sample y at times t and $t-\tau$, and p_2 is the joint probability of the original and delayed samples. The mutual information exhibits a minimum value at the optimal delay time.

6.3 ADVANCED VOLTERRA SERIES MODELS

The Volterra series description of a nonlinear dynamical system can rapidly expand to several hundred coefficients as the order of the nonlinearity and memory depth increase beyond five or so, as indicated in Figure 6.4. There are several difficulties with such a complex model. Finding the values of all of these coefficients is difficult to do accurately. Many of the coefficients will have only a small influence on the model output, and so large variations in the values of such coefficients cannot be detected accurately. This is an ill-conditioned problem that is difficult to solve accurately; we shall investigate some practical techniques for overcoming ill-conditioned model descriptions in Section 6.5, Techniques for Building Volterra Models of Power Amplifiers. Further, such models, even if we can extract the coefficient values from the data, may be poor at predicting the system behavior. The poorly-identified coefficients just add statistical noise to the model output, which can mask some of the behaviors of interest.

A simpler model may capture the underlying system behavior better, and hence have a better predictive performance. Hence the search for ways of simplifying the Volterra series, to create a model that still captures the nonlinear and dynamic effects that the Volterra series describes so well, but retains only the most significant terms. We met some examples of simplified Volterra models in the previous section. Here, we describe a formal mathematical technique for pruning the Volterra series, to retain only the level of nonlinearity and dynamics necessary for accurate modeling.

The technique is known as a "dynamic kernels" method, where the model is partitioned into a static or memoryless nonlinearity, and a dynamic nonlinear component. This model uses a change of variable to a "dynamic deviation" from the static condition to describe the excitation. This approach was first described by Filicori *et al.* and applied to the modeling of field effect transistors (FETs) [24]. This model was called the nonlinear integral model (NIM). It was later adopted by Ngoya *et al.*, who recast the model formalism to a complex baseband description, calling it a "Modified Volterra Series," and applied this technique to the modeling of RF power amplifiers [25]. Filicori and Ngoya both found that reasonable accuracy and good convergence of the dynamic kernels models could be obtained by truncating the dynamic part of the model to only first order. Ngoya and coworkers later extended the general approach to describe long-term memory effects, in their "Nonlinear Impulse Response Model" [26]. A review of the complex envelope modeling by Ngoya and co-workers can be found in Chapter 3 of [15].

The drawback with this dynamic kernels description is that the model is no longer a linear-in-parameters structure, with some of the coefficients being included inside the nonlinear dynamical terms of the model. This makes it much more difficult to identify and extract these coefficients. This problem was solved by Zhu in 2006 [27], by recasting the dynamic kernels description and recovering the linear-in-parameters structure. The resulting model is now simple to extract using the familiar linear systems methods outlined earlier, and the levels of nonlinearity, memory depth, and dynamical order are under the explicit control of the user. Zhu called his modeling approach "Dynamic Deviation Reduction" (DDR).

6.3.1 Dynamic Deviation Model

The basic assumption is that the system can be described by a static nonlinearity and a dynamic nonlinear contribution. The model is derived by introducing a *dynamic deviation* between the input signal $u(t)$ at the current instant and the input signals over the memory depth M of the system:

$$e(t, m_j) = u(t - m_j) - u(t) \tag{6.43}$$

where m_j is the jth memory delay, $j = 1, 2, M$.

If we take the regular discrete-time expression for the Volterra series, (6.25),

$$y(t) = \sum_{n=1}^{N} \sum_{m_1=0}^{M} \dots \sum_{m_n=0}^{M} h_n(m_1, ..., m_n) \prod_{j=1}^{n} u(t - m_j) \tag{6.44}$$

and substitute for $u(t - m_j)$ using the dynamic deviation, we get

$$y(t) = \sum_{n=1}^{N} \sum_{m_1=0}^{M} \dots \sum_{m_n=0}^{M} h_n(m_1, ..., m_n) \prod_{j=1}^{n} (u(t) + e(t, m_j)) \tag{6.45}$$

The product of the sum of the instantaneous signal and the dynamic deviation yields the simplification. If we take, for example, the cubic case $n = 3$, this product becomes

$$(u + e)^3 = u^3 + 3u^2e + 3ue^2 + e^3 \tag{6.46}$$

and we can see that we have generated the static term $u^3(t)$, and some mixed product terms. We can pull the static term, which we shall call $y_s(t)$, out of the full Volterra expression

$$y_s(t) = \sum_{n=1}^{N} u_n(t) h_n(0, 0, ..., 0) \tag{6.47}$$

This is equivalent to Filicori's F_{DC} ([24], equation(3)), and Ngoya's y_{dc} ([25], equation (4)). This static term is not necessarily the DC condition, although Filicori chose that for the NIM model of the FET; Ngoya chose the large-signal sinusoidal drive tone as the static condition for his power amplifier model. The static term reduces to an instantaneous polynomial

$$y_s(t) = \sum_{n=1}^{N} a_n u^n(t) \tag{6.48}$$

with a_n as the coefficients of this polynomial.

The dynamic part of the expression can also be recast, noting that from (6.46) as the powers of the $u(t)$ decrease, the powers of the dynamic deviation increase, and we can write

$$y_d(t) = \sum_{n=1}^{N}\sum_{r=1}^{n} u^{n-r}(t)\begin{pmatrix} n \\ r \end{pmatrix}\sum_{m_1=0}^{M}\cdots\sum_{m_n=0}^{M} h_n(m_1,...,m_n)\prod_{j=1}^{r} e(t,m_j) \tag{6.49}$$

where we have introduced an extra variable, r, which we use to decrease the powers of $u(t)$, and provide the correct products of $e(t, m_j)$. The value of r also sets the number or order of dynamics present in expression. The term

$$\begin{pmatrix} n \\ r \end{pmatrix} = \frac{n!}{r!(n-r)!} \tag{6.50}$$

generates the coefficients of the various product terms, as in (6.46), for example.

This is still not too convenient for parameter extraction, and some rearrangement is necessary. Zhu does this rearrangement by example, showing how first- and second-order dynamical expressions can be recast to create linear-in-parameters expressions [27]. The first-order dynamical term can be obtained by setting $r = 1$ in (6.49)

$$y_{d=1}(t) = \sum_{n=1}^{N} u^{n-1}(t)\sum_{m_1=0}^{M}\hat{h}_n(m_1)e(t,m_1) \tag{6.51}$$

where $\hat{h}_n$ is the nth-order kernel that contains one delay term, and the coefficient has been dropped for simplicity. Because the rest of the terms of $\hat{h}_n$ are the instantaneous or undelayed terms, we can write

$$\hat{h}_n(m_1) = h_n(0,...0,m_1) \tag{6.52}$$

where we have $n-1$ zero-delay terms.

Substituting for $e(t, m_1)$ from (6.43), and rearranging the $u^n(t)$, we get

$$\begin{aligned} y_{d=1}(t) &= \sum_{n=1}^{N} u^{n-1}(t) \sum_{m_1=0}^{M} \hat{h}_n(m_1) \left[u(t-m_1) - u(t)\right] \\ &= \sum_{n=1}^{N} \sum_{m_1=0}^{M} \hat{h}_n(m_1) u^{n-1}(t) \left[u(t-m_1) - u(t)\right] \\ &= \sum_{n=1}^{N} \sum_{m_1=0}^{M} \hat{h}_n(m_1) \left[u^{n-1}(t)u(t-m_1) - u^n(t)\right] \\ &= \sum_{n=1}^{N} \sum_{m_1=0}^{M} \hat{h}_n(m_1) u^{n-1}(t) u(t-m_1) - \sum_{n=1}^{N} \hat{h}_n(0) u^n(t) \end{aligned} \tag{6.53}$$

where the last term is an instantaneous quantity and can be rolled into the static term, $y_s(t)$, leaving

$$y_{d,r=1}(t) = \sum_{n=1}^{N} u^{n-1}(t) \sum_{m_1=0}^{M} \hat{h}_n(m_1) u(t-m_1) \tag{6.54}$$

This is the dynamic deviation reduction expression for the dynamic term with the dynamic order, r, equal to 1.

Generalizing this expression to arbitrary dynamical order, the dynamic deviation reduction (DDR) description of the Volterra series is

$$\begin{aligned} y(t) &= y_s(t) + y_d(t, r) \\ &= \sum_{n=1}^{N} h_{n,0}(0, ..., 0) u^n(t) \\ &+ \sum_{n=1}^{N} \sum_{r=1}^{n} u^{n-r}(t) \sum_{m_1=1}^{M} \cdots \sum_{m_r=m_{r-1}}^{M} h_{n,r}(0, ..., 0, m_1, ..., m_r) \\ &\cdot \prod_{j=1}^{r} u(t-m_j) \end{aligned} \tag{6.55}$$

In this expression, $h_{n,r}(0, ..., 0, m_1, ..., m_r)$ is the nth-order Volterra kernel, with the first $n - r$ terms equal to zero, and only r non-zero entries, corresponding to the input signal product with r delay terms

$$u^{n-r}(t) u(t-m_1) ... u(t-m_r) \tag{6.56}$$

We can see that the structure of this re-arranged Volterra series expression retains the same form as (6.44), and so is linear-in-parameters. The coefficients of this Volterra series can found using the same linear methods as we have outlined previously. But the number of coefficients to be found is now much reduced, and is controlled by our choice of the dynamic order, r. This yields a more tractable and stable model, without significant loss of accuracy.

6.4 NARROWBAND APPROXIMATION: BASEBAND VOLTERRA MODELS

In the foregoing we have described fairly traditional approaches to the modeling of nonlinear dynamical systems using Volterra series. The objective has been to produce generalized models that are accurate over a range of frequencies and timescales. If we are using signals whose bandwidth is small compared to the carrier frequency, as is the case with modern wireless communications signals such as WCDMA, or LTE, then we can make a narrowband approximation. It is the modulated envelope that carries the information in the signal; we assume that useful information can be recovered from only the *first zone* envelope, around the fundamental carrier frequency. Information around the harmonics is ignored. This is effectively placing a band pass filter at the output of the PA. This is illustrated schematically in Figure 6.12.

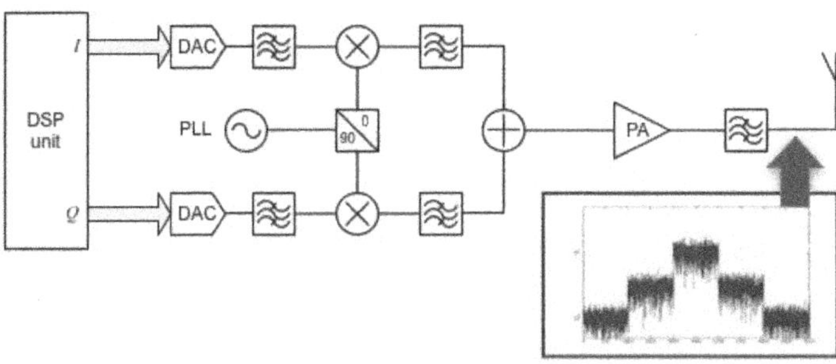

Figure 6.12 A schematic RF transmitter indicating the first zone envelope signal, after the band-pass filter at the PA output.

The amplitude- and phase-modulated RF signal at the fundamental is given by

$$y(t,\tau) = \hat{y}(\tau)\cos(\omega_0 t + \varphi(\tau)) = \operatorname{Re}\left\{\hat{y}(\tau)e^{j\omega_0 t}e^{j\varphi(\tau)}\right\} \tag{6.57}$$

We can simplify even further, by realizing that the carrier signal frequency is redundant from the modeling perspective: we are interested primarily in the nonlinear effects of the PA on the information or envelope signal:

$$\tilde{y}(\tau) = \hat{y}(\tau)e^{j\varphi(\tau)} \tag{6.58}$$

which we express in terms of its complex baseband representation,

$$\tilde{y}(\tau) = \mathrm{Re}\left\{\hat{y}(\tau)e^{j\varphi(\tau)}\right\} = \frac{\hat{y}(\tau)e^{j\varphi(\tau)} + \hat{y}^*(\tau)e^{-j\varphi(\tau)}}{2} \tag{6.59}$$

When we substitute this expression into the product term of the Volterra series, (6.25),

$$y(t) \quad = \quad \sum_{n=1}^{N} \sum_{m_1=0}^{M} \ldots \sum_{m_n=0}^{M} h_n(m_1, \ldots, m_n) \prod_{j=1}^{n} u(t - m_j) \tag{6.60}$$

we get products of the form

$$\prod_{j=1}^{n} \left(u(t - m_j) + u^*(t - m_j)\right) \tag{6.61}$$

ignoring the factor 2, which is rolled into the kernels or coefficients.

The value of representing the model in this way is that we can then build the Volterra series model of the nonlinear PA directly from the measured or simulated I & Q input and output data that describe the modulation signal. This approach will be outlined in Section 6.5. The narrowband approximation and complex baseband representation can also be applied to the dynamic deviation models outlined in Section 6.3.

Let's illustrate this narrowband, complex envelope approach using the third-order nonlinear products, as we know that 3rd-order (and higher odd-order) systems produce intermodulation products within the passband of the system, adding in-band distortion products. This can be shown by using a 3rd-order kernel, but by exciting with only two signals, at ω_1 and ω_2, the kernel then being written as $H(\omega_1, \omega_1, \omega_2)$, for example. Table 6.1 illustrates how the frequency components arise.

The in-band terms of interest are the distortion terms at the two input frequencies ω_1 and ω_2, the cross-modulation terms, and the intermodulation terms that generate new frequency components. All these terms arise when one of the three signal frequency components is negative: the conjugate signal. For simplicity, we have considered the cases when the third component in $H(\omega, \omega, \omega)$ is negative; all other cases are simply duplicates of this.

Table 6.1
The Frequency Components that Arise from the Nonlinear 3rd-Order Volterra Transfer Function.

Transfer Function	Output Frequency	Comment
$H(\omega_1, \omega_1, \omega_1)$	$3\omega_1$	3rd harmonic
$H(\omega_1, \omega_1, -\omega_1)$	ω_1	distortion at ω_1
$H(\omega_1, \omega_1, \omega_2)$	$2\omega_1 + \omega_2$	Intermodulation near 3rd harmonic
$H(\omega_1, \omega_1, -\omega_2)$	$2\omega_1 - \omega_2$	In-band intermodulation
$H(\omega_1, \omega_2, -\omega_1)$	ω_2	Cross-modulation at ω_2
$H(\omega_1, \omega_2, \omega_2)$	$2\omega_2 + \omega_1$	Intermodulation near 3rd harmonic
$H(\omega_1, \omega_2, -\omega_2)$	ω_1	Cross-modulation at ω_1
$H(\omega_2, \omega_2, -\omega_1)$	$2\omega_2 - \omega_1$	In-band intermodulation
$H(\omega_2, \omega_2, \omega_2)$	$3\omega_2$	3rd harmonic
$H(\omega_2, \omega_2, -\omega_2)$	ω_2	distortion at ω_2

We can write similar expressions in the time domain. All third-order baseband products can be expressed as one of the following products

$$\begin{gathered} u(t-m_1)u(t-m_2)u^*(t-m_3) \\ u(t-m_1)u^*(t-m_2)u(t-m_3) \\ u^*(t-m_1)u(t-m_2)u(t-m_3) \end{gathered} \tag{6.62}$$

where $m_1, m_2, m_3 = 0 : M$ represent delay terms, including zero delay, and M is the memory depth. We see that to obtain the third-order baseband products, one of the three terms is the conjugate. Similarly, for the fifth-order baseband products, two out of the five terms would be conjugates, leading to ten terms in all for the combinations:

$$\begin{gathered} u(t-m_1)u(t-m_2)u(t-m_3)u^*(t-m_4)u^*(t-m_5) \\ u(t-m_1)u(t-m_2)u^*(t-m_3)u(t-m_4)u^*(t-m_5) \\ u(t-m_1)u^*(t-m_2)u(t-m_3)u(t-m_4)u^*(t-m_5) \\ u^*(t-m_1)u(t-m_2)u(t-m_3)u(t-m_4)u^*(t-m_5) \\ u(t-m_1)u(t-m_2)u^*(t-m_3)u^*(t-m_4)u(t-m_5) \\ u(t-m_1)u^*(t-m_2)u(t-m_3)u^*(t-m_4)u(t-m_5) \\ u^*(t-m_1)u(t-m_2)u(t-m_3)u^*(t-m_4)u(t-m_5) \\ u(t-m_1)u^*(t-m_2)u^*(t-m_3)u(t-m_4)u(t-m_5) \\ u^*(t-m_1)u(t-m_2)u^*(t-m_3)u(t-m_4)u(t-m_5) \\ u^*(t-m_1)u^*(t-m_2)u(t-m_3)u(t-m_4)u(t-m_5) \end{gathered} \tag{6.63}$$

We shall illustrate this model development using a third-order nonlinearity with a memory depth M of 2 delays, shown in Table 6.2

Table 6.2
The Time Delay Components that Arise from a 3rd-Order Nonlinearity and Memory Depth of 2.

$u(t-m_1)$	$u(t-m_2)$	$u^*(t-m_3)$
$u(t)$	$u(t)$	$u^*(t)$
$u(t)$	$u(t)$	$u^*(t-1)$
$u(t)$	$u(t)$	$u^*(t-2)$
$u(t)$	$u(t-1)$	$u^*(t)$
$u(t)$	$u(t-1)$	$u^*(t-1)$
$u(t)$	$u(t-1)$	$u^*(t-2)$
$u(t)$	$u(t-2)$	$u^*(t)$
$u(t)$	$u(t-2)$	$u^*(t-1)$
$u(t)$	$u(t-2)$	$u^*(t-2)$
$u(t-1)$	$u(t-1)$	$u^*(t)$
$u(t-1)$	$u(t-1)$	$u^*(t-1)$
$u(t-1)$	$u(t-1)$	$u^*(t-2)$
$u(t-1)$	$u(t-2)$	$u^*(t)$
$u(t-1)$	$u(t-2)$	$u^*(t-1)$
$u(t-1)$	$u(t-2)$	$u^*(t-2)$
$u(t-2)$	$u(t-2)$	$u^*(t)$
$u(t-2)$	$u(t-2)$	$u^*(t-1)$
$u(t-2)$	$u(t-2)$	$u^*(t-2)$

We have used only unique combinations of m_1, m_2, m_3, eliminating redundancies. For example, the terms

$$\begin{array}{ccc} u(t-1) & u(t-2) & u^*(t) \\ u(t-1) & u(t-2) & u^*(t-1) \\ u(t-1) & u(t-2) & u^*(t-2) \end{array}$$

and

$$\begin{array}{ccc} u(t-2) & u(t-1) & u^*(t) \\ u(t-2) & u(t-1) & u^*(t-1) \\ u(t-2) & u(t-1) & u^*(t-2) \end{array}$$

are identical, and so this latter set are omitted from the table. We often write the table in terms of the delay values only, as a shorthand, which looks much cleaner and clearer (Table 6.3).

Table 6.3

Table 6.2 Written in Terms of the Values of the Delays.

m_1	m_2	m_3
0	0	0
0	0	1
0	0	2
0	1	0
0	1	1
0	1	2
0	2	0
0	2	1
0	2	2
1	1	0
1	1	1
1	1	2
1	2	0
1	2	1
1	2	2
2	2	0
2	2	1
2	2	2

This gives us a total of 18 coefficients to find, just for the cubic component of the model. The use of the pruning techniques as outlined in Section 6.3, "Advanced Volterra Series Models," will still be needed to produce a model with a manageable number of coefficients to find.

6.5 TECHNIQUES FOR BUILDING VOLTERRA MODELS OF POWER AMPLIFIERS

We shall build our models from data collected in the modulation domain, using I and Q (IQ) input and output modulation data from the PA. We shall use the narrowband approximation and complex envelope representation of the IQ data as described in Section 6.4. The gain and phase compression of the amplifier, its AM-to-AM and AM-to-PM responses of the modulation signal, completely describe the PA's nonlinear dynamical behavior. Also, using the IQ data is representative of how the PA models are identified in DPD applications.

Measured data can be obtained using a vector signal generator (VSG) to create the modulated RF signal, and a vector signal analyzer (VSA) to capture the output RF signal from the PA and demodulate to the IQ domain. A schematic of a typical measurement system is shown in Figure 6.13.

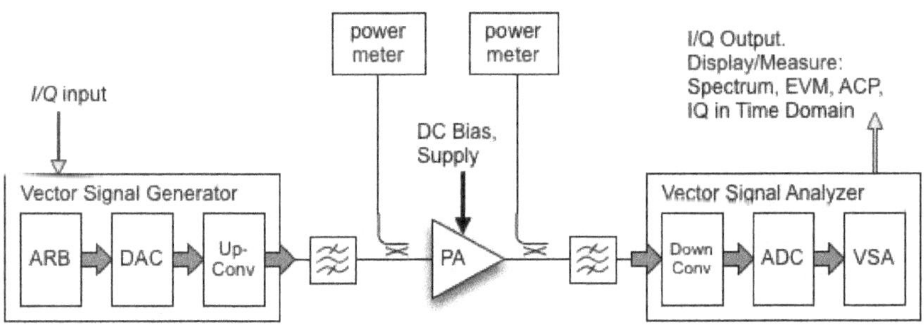

Figure 6.13 System for measuring the input and output IQ signals of a power amplifier. The input and output powers are also measured.

The VSG will have an arbitrary waveform generator (ARB) that is used to generate the modulation signal. The IQ data are often read in from a text file that has been created using Mathworks MATLAB or an RF system simulator. The IQ data is generally over-sampled compared to the native chip rate of the modulation, at a rate that is several times the signal bandwidth. For example, a four-carrier WCDMA signal may be sampled at 61.44 MSps, which is 16 times the chip rate of WCDMA, 3.84 MSps. These, and higher, data rates are typical of modern DPD systems, and the PA model that we build must operate at the same sample rates if it is to be useful in the design of a DPD system.

Typically the VSA will have only a single input port. This allows us to measure the PA output signal directly, and we then have to measure the input to the PA by removing the amplifier and connecting the VSA directly at that point.

Having a dual-input VSA would save the disconnection and reconnection, and would probably also enable us to estimate the delay between the captured signals more easily. The VSA should have as wide a demodulation bandwidth as possible. This is to enable the capture of the intermodulation products and adjacent channel data that arise from the nonlinearity of the PA, and will help up build an accurate model.

The demodulation bandwidth determines the sampling rate of the captured data. We may need to resample this raw data to match up to the input signal sampling rate. This can be done using, for instance, the MATLAB *resample* command; remember to use a wide enough window for the resampling to be accurate. This will depend on the (over-) sample rate of the original data.

The power meters collect the input and output average power information. It is important to use a power sensor that is capable of accepting modulated signals and calculating the average power. Not all power sensors are built alike. Many modern power sensors and software will also be able to calculate and display the CCDF and hence an estimate of the peak-to-average power ratio of the signal. We can use the peak power or PAPR information to estimate the compression level of the PA, to determine how hard it is being driven into the nonlinear regime of operation.

Other instrumentation can be added to this basic test bench, such as volt- and ammeters on the PA supplies, to enable the calculation of the energy efficiency of the PA. We may add a fast digital sampling oscilloscope to view the RF and modulation signals, although the resolution of these instruments is currently not high enough to enable capture of the digital data for modeling; this situation could change very quickly, though.

We can also obtain the PA input and output IQ data from simulation: this will be a Circuit Envelope or Envelope Transient simulation, where the PA is driven with an RF signal at the carrier frequency and analyzed using a harmonic balance nonlinear simulator, and the modulated signals are calculated using a transient simulation at the modulation sampling rate. The two simulators are coupled to ensure consistency of the solution.

The models for the PA that we are going to build will be based on Volterra series solutions as described in the preceding sections of this chapter. The overall flow for producing a Volterra-based model of the PA from IQ data is shown in Figure 6.14. We shall outline the activities in the major blocks in this diagram in the following sub-sections, but first, we shall give an overview of the procedure to indicate how these various activities are inter-related and dependent upon one another.

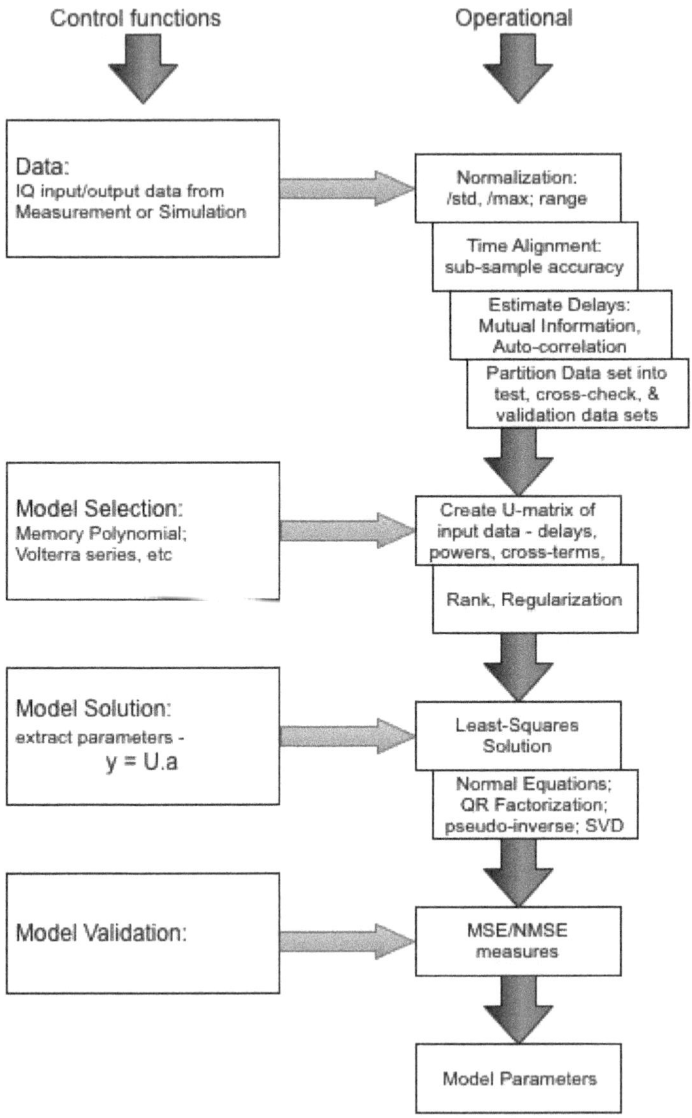

Figure 6.14 Flow of activity for creating a behavioral model of a power amplifier.

The Volterra models, as we have seen, are essentially basis function descriptions that are linear in the parameters or coefficients, a_n, with the dynamical nonlinearity described by the polynomial functions $\varphi(t - m_j)$, which are functions of the PA input IQ signal, $u(t)$:

$$y(t) = \sum_{n=0}^{N} a_n \varphi_n(t - m_j) \tag{6.64}$$

where $y(t)$ is the measured or simulated PA output IQ data at time t.

This linear-in-parameters feature allows us to use least-squares techniques of the general form

$$y = \mathbf{U}a$$

to determine the solution for a. Least squares methods generally use an inverse or *pseudo-inverse* to find x. Such methods rely on the matrix $\mathbf{U}$ being reasonably well-conditioned so that the error in the solution

$$\mathbf{U}^{-1}y - \hat{a}$$

is small for the estimate $\hat{a}$. If we simply use the raw measured voltage data for the input and output signals, then, given the PA will have gain, and that the Volterra polynomials will produce very large numbers for the high-order terms, the matrix $\mathbf{U}$ will be ill-conditioned: the values of the entries in the matrix will span many orders of magnitude. A more mathematical term for this is that its eigenvalues will span a range of several orders of magnitude, or that the condition number of the matrix is large. This means that even if the inverse of the matrix can be calculated, small changes or inaccuracies in the matrix entries, due to noise in the measurements or model errors in the simulation, can result in large errors in the estimate of the solution $\hat{a}$. With a large condition number, it may not even be possible to calculate the matrix inverse within the mathematical accuracy of the computer.

We can overcome these problems by applying a number of techniques. These include: scaling or normalization of the data, to avoid the generation of large numbers at high polynomial orders; regularization or rank-reduction of the matrix to reduce the condition number; using more robust mathematical techniques for finding the matrix inverse or pseudo-inverse; or combinations of these methods. We will describe these approaches in more detail next.

6.5.1 Data Pre-conditioning

We collect time records of the input IQ and corresponding output IQ signals, $\mathbf{u}$ and $\mathbf{y}$

$$\begin{aligned} \mathbf{y} &= \begin{bmatrix} y(1) & y(2) & \cdots & y(K) \end{bmatrix}^T \\ \mathbf{u} &= \begin{bmatrix} u(1) & u(2) & \cdots & u(K) \end{bmatrix}^T \end{aligned} \tag{6.65}$$

These raw IQ data from simulation or measurement will usually be in the form of a text file time record of the complex IQ values, in volts. These instantaneous values may be a few volts or smaller at the input to the PA or transmitter chain, to several tens of volts at the output of the PA. Different PAs will have different gains, and so for a robust modeling process we should normalize these signals to produce a standard range of values, typically between 0 and 1.

Several possibilities are open to us: dividing each value by the maximum value in the record, dividing by the standard deviation, using the norm, or more sophisticated scaling and compression algorithms. Typically the input and output IQ data will be obtained in the digital domain, from an analog-to-digital convertor (ADC). The ADC has a fixed range, and we should avoid going over-range and saturating the ADC. This suggests that a simple normalization by dividing by the maximum value in the signal record may well be good enough. The drawback with this approach is that the time record should contain enough data samples for us to be confident of including a large value, but it is statistically unlikely that we shall include the peak of the signal. Short time records of IQ data for cellular wireless signals may well have different maximum values, potentially leading to different models for the same PA. One way around this is to scale to the maximum value in the record, plus some guard value of, say, 1 bit of resolution in the ADC, and use this as the scaling for all other signals.

Part of the normalization process should also include calculation of the average input and output powers of the signals, and hence the power gain of the PA. These values can also be found from the power meter data. Knowing the average power is necessary for using the model in a digital pre-distortion system, where the input signal level to the PA will need to be adjusted to accommodate the linearization of the system. The PA model will need to be accurate over a range of input powers.

The input and output data records need to be aligned in time, so that a given $y(t)$ value corresponds with the $u(t)$ signal that caused it. Time misalignment arises from practical concerns such as cable lengths, and so forth, and also because of the group delay of the amplifier chain. Because the time delay is unlikely to be an integer multiple of the data sampling time, subsample alignment of the input and output data records is necessary. To accomplish this, the alignment is usually done in two steps. First, a coarse alignment is performed using cross-correlation; this will align the two signals to the nearest sample. This can be done easily and quickly in MATLAB, for example. The second step is often done in the frequency domain. The coarsely-aligned time-domain IQ data are transformed into the frequency domain using the Discrete Fourier Transform (DFT). In the frequency domain, the angle between the two signals corresponds to the delay. For each input-output pair of samples, the angle will be different; we find the average angle ϕ_{diff} using least-squares technique over the whole data set, to give the required rotation of the output signal to align the two signals:

$$\text{rot} = e^{j2\pi\phi_{diff}} \tag{6.66}$$

The time-aligned output signal in the discrete-time domain is then obtained using the inverse DFT, producing the required subsample alignment.

Once the input and output data have been aligned, we usually split the data into two, or even three sets, depending on the model identification and validation method that we use. One set of input/output data is the training data: this is used to identify the model parameters. If we are using a model identification technique that converges to a solution, such as a recursive least-squares calculation, we may choose to use a second set of data to cross-check that the model parameters will give an accurate solution. This will allow us to tell whether the model we obtain is general, in other words, that it is accurate for unseen data, or whether the model is simply mimicking the training data. In the latter case, we will see that the error measure increases, and the model is known as overtrained. This process of checking the model validity during the identification or convergence procedure is called *cross-validation*. It is especially useful if we are using fully nonlinear models such as artificial neural networks (ANNs), described in more detail in Section 6.6.

The final set of data is used to validate the generality of the model, once the parameters have been identified and the parameter set is converged. This is a set of data that has not been used in the model identification, and so if the model that we have built is sufficiently general, the error measure should be similar to the one obtained during the identification or training process. If the error measure is higher, then the model structure itself may not be good enough, or we have overtrained the model.

6.5.2 Model Selection

The next step in our system identification procedure for obtaining the nonlinear model is the selection of the model structure itself. In this example, this will be any one of the Volterra-based models we have described in the foregoing, such as a memory polynomial model, modified Volterra series, or DDR-Volterra series. Once we have chosen the model structure, we need to create the matrix of multinomial terms of the input signal that correspond to the model parameters a_n at each time step, such as

$$y(t) = \sum_{n=1}^{N} a_n \sum_{m_1=0}^{M} \dots \sum_{m_n=0}^{M} h_n(m_1, \dots, m_n) \prod_{j=1}^{n} u(t - m_j) \tag{6.67}$$

We take the time-aligned and normalized time records of the input and corresponding output signals, $\tilde{\mathbf{u}}$, $\tilde{\mathbf{y}}$

$$\begin{aligned} \tilde{\mathbf{y}} &= \begin{bmatrix} \tilde{y}(1) & \tilde{y}(2) & \cdots & \tilde{y}(K) \end{bmatrix}^T \\ \tilde{\mathbf{u}} &= \begin{bmatrix} \tilde{u}(1) & \tilde{u}(2) & \cdots & \tilde{u}(K) \end{bmatrix}^T \end{aligned} \tag{6.68}$$

and from $\tilde{\mathbf{u}}$ we build a matrix of the multinomial terms for the Volterra Series signals. We call this matrix $\mathbf{U}$. By way of illustration of the construction of the $\mathbf{U}$-matrix, we shall take a simple example Volterra Series model structure: a cubic nonlinearity with two memory terms, neglecting even-order terms. Even so, we shall not write out the full matrix longhand, as even this modest model is quite large, as indicated in Table 6.4.

On each row of the $\mathbf{U}$-matrix, corresponding to the kth-time sample, we will have: the instantaneous linear and polynomial terms (corresponding to the first row of Table 6.4); the delayed linear terms, $u(t-1)$ and $u(t-2)$; and the cubic delay terms given by each row of Table 6.4:

$$u(k) \quad u^2(k)u^*(k) \quad u(k-1) \quad u(k-2) \quad u(k)^2 u^*(k-1) \quad u(k)^2 u^*(k-2) \quad \dots \tag{6.69}$$

where the u^* terms are the conjugates of the IQ terms, and we have dropped the tilde overmark for clarity.

The $\mathbf{U}$-matrix is large and non-square: the number of rows is equal to the number of time samples, and the number columns is equal to the number of coefficients whose values we need to find. It is likely to be rank deficient, and some regularization is often necessary. This is often done during the parameter estimation or model solution step.

6.5.3 Model Solution: Parameter Estimation

Since the Volterra-based model that we are using is linear-in-parameters, of the form

$$\mathbf{y} = \mathbf{U}\mathbf{a} \tag{6.70}$$

we can use least-squares techniques to determine the parameters a. Because we have an overdetermined system, the $\mathbf{U}$-matrix is non-square and does not have a simple inverse, so we cannot use the following approach directly to estimate the coefficients $\hat{a}$

$$\hat{a} = \mathbf{U}^{-1} y \tag{6.71}$$

A straightforward solution technique that is often seen uses the *Normal equations* to estimate the coefficients $\hat{a}$

$$\left(\mathbf{U}^H \mathbf{U}\right)^{-1} \mathbf{U}^H y = \hat{a} \tag{6.72}$$

where $\mathbf{U}^H$ represents the Hermitian (complex) transpose of $\mathbf{U}$. The matrix:

$$\left(\mathbf{U}^H \mathbf{U}\right)^{-1} \mathbf{U}^H$$

is called the *pseudo-inverse*, often denoted by $\mathbf{U}^+$.

To obtain the pseudo-inverse, we have to invert a (number of coefficients) by (number of coefficients) matrix. This is a mathematically intensive and costly exercise, if we can do it at all. This matrix is often ill-conditioned, even with scaling and normalization of the input data, as the range of values of the polynomial orders can span several orders of magnitude. The resulting estimate for the parameters $\hat{a}$ can often be quite poor.

We can turn to regularization techniques to help obtain a solution to these ill-conditioned problems. A good and very accessible review of such techniques and their implementation is presented by Neumaier [28]. The condition (number) of the matrix $\mathbf{U}^H\mathbf{U}$ can be improved by modifying its values in a controlled manner. A simple way of doing this is by adding some value to the diagonal entries

$$\mathbf{U}^H \mathbf{U} \to \mathbf{U}^H \mathbf{U} + h^2 I$$

where I is the identity matrix. This new matrix has a condition number

$$\left(h^2 + \|\mathbf{U}\|^2\right) / h^2 \tag{6.73}$$

which gets smaller as h^2 increases. The estimate for the parameters $\hat{a}$ can now be written as

$$\hat{a} = \left(\mathbf{U}^H \mathbf{U} + h^2 I\right)^{-1} \mathbf{U}^H y \tag{6.74}$$

This technique is known as Tikhonov regularization, or ridge regression. The trick is in finding an appropriate value for h^2. Neumaier [28] provides a recipe that is based on a smoothing matrix derived from the original $\mathbf{U}$ matrix.

The pseudo inverse for a singular matrix is found using the singular value decomposition (SVD) technique. Briefly, the SVD of an $m \times n$ matrix $\mathbf{U}$ with rank r can be written as

$$\mathbf{U} = \mathbf{A}\Sigma\mathbf{B}^T \tag{6.75}$$

where $\mathbf{A}$ and $\mathbf{B}$ are orthogonal, and Σ is an $m \times n$ diagonal matrix with r non-zero entries given by the positive square roots of the eigenvalues of $\mathbf{U}^H\mathbf{U}$. The diagonal entries of Σ, called the singular values of $\mathbf{U}$, are written in descending order.

The pseudo-inverse of $\mathbf{U}$, the *Moore-Penrose pseudo-inverse*, is then given by

$$\mathbf{U}^+ = \mathbf{B}\Sigma^+\mathbf{A}^T \tag{6.76}$$

where Σ^+ is obtained by transposing Σ and inverting all of the non-zero diagonal entries; any diagonal values in Σ that are zero remain zero in Σ^+.

A solution of $\hat{a}$ found from the SVD is also a solution of our original set of ill-conditioned equations. The SVD technique is a very robust method of solving these ill-conditioned problems. The technique is also useful in that examination of the matrix Σ will tell us which of the parameters of the model have the most significant contributions, and which ones can effectively be ignored. This gives us another handle on reducing the number of parameters in the model, and essentially pruning the Volterra series approximation.

Another robust method for solving these ill-conditioned equations is orthog onal matrix triangularization, known as QR factorization or QR decomposition. For the case of an overdetermined system, then the $m \times n$ $\mathbf{U}$ matrix, whose columns are linearly independent, as we would expect since they correspond to the model parameters, can be factored into

$$\mathbf{U} = \mathbf{QR} \tag{6.77}$$

where the matrix $\mathbf{Q}$ is an $m \times n$ matrix with orthonormal columns, and $\mathbf{R}$ is an $n \times n$ upper triangular matrix that is invertible. The orthonormal matrix $\mathbf{Q}$ preserves the norm or distance in any transformation, so the QR factorization can be thought of as rotating (or reflecting) the $\mathbf{U}$ matrix until we arrive at the point where the set of linear equations can be solved by back-substitution in the triangular matrix $\mathbf{R}$. As $\mathbf{Q}$ is orthonormal,

$$\mathbf{Q}^H\mathbf{Q} = I \tag{6.78}$$

and we can write

$$y = \mathbf{U}a = \mathbf{QR}a \tag{6.79}$$

or

$$\mathbf{Q}^H y = \mathbf{Q}^H\mathbf{QR}a = \mathbf{R}a \tag{6.80}$$

and since $\mathbf{R}$ is invertible, our estimate for the parameters is

$$\hat{a} = \mathbf{R}^{-1}\mathbf{Q}^{H}y \tag{6.81}$$

which is relatively straightforward to solve: since $\mathbf{R}$ is triangular, the parameters can be found from back-substitution in the vector $\mathbf{Q}^{H}y$.

This procedure avoids the inversion of the $\mathbf{U}^{H}\mathbf{U}$ in the Normal equations, and the associated numerical instabilities. The orthonormal matrix $\mathbf{Q}$ can be found using a stable form of the Gram-Schmidt procedure [29] from the $\mathbf{U}$ matrix, or Householder transformations with pivoting, which will work even when $\mathbf{U}$ is rank-deficient; the MATLAB function *qr* for QR decomposition uses this latter method, is very stable and accurate, and less expensive than the SVD in terms of the number of calculations required.

The methods outlined above for solving these least squares problems can be found as built-in functions in mathematical software such as MATLAB. These functions are useful for model prototyping in MATLAB, and are generally quite robust in terms of dealing with ill-conditioned and rank-deficient matrices, for example. Moler provides a good description of these MATLAB implementations in Chapter 5 of [30]. A potential drawback with using these built-in functions is that because they are so robust, it can be easy to miss potential problems with the measured data that result in ill-conditioned or rank-deficient matrices. It is certainly worthwhile understanding how these methods are implemented, so that the most appropriate technique for the given data set can be used, and that you know what you are doing! A good reference for the implementation of these matrix methods is Golub and Van Loan [29].

6.5.4 Model Validation

Now that we have our model parameters, how can we tell how good our model is? We compare the model predictions with the data; this is where we use the set of validation data that was not used in the model identification. The comparison can be done in the time domain or frequency domain. Plotting both the data and model predictions will give us a visual sense of whether our model is a good fit for the data, but generally we prefer to use some objective measure.

A simple error measure such as the absolute value of the difference between the data and model signals is generally a poor indicator, as this tends to amplify zero-crossing errors. In other words, signals that are very similar but appear to be slightly misaligned in time will generate large sample-by-sample errors. Perhaps the time alignment is not quite accurate.

A typical measure of the goodness of fit is the mean squared error (MSE), which is averaged over the data set of N points in the time domain:

$$\mathrm{MSE} = \frac{1}{N}\sum_{1}^{N}(data - model)^2 \tag{6.82}$$

or the normalized version of this. The result is usually expressed in logarithmic form, as $10\log(\mathrm{NME})$ dB.

In the frequency domain, we can compare the model and data spectra. An interesting metric was proposed by Wisell *et al.* [31]. As the model is designed to predict the dynamic nonlinearities of the power amplifier, a metric based on these nonlinear components was suggested. The nonlinear behavior of the PA driven by the modulated signal is manifest in the spectral leakage into the adjacent communication channels: the intermodulation products. The adjacent-channel power ratio (ACPR) is a measure of the PA nonlinearity, and Wisell proposed that the error in the prediction of the ACPR should be a good metric for the model; this is called the adjacent channel error power ratio (ACEPR). The power in the spectrum is integrated over the adjacent channel for the data and model prediction, and the difference between them is again usually expressed in logarithmic form (dB).

We can also quantify the model accuracy in the code or modulation domain, by calculating the error vector magnitude (EVM) for the data and model, and again comparing the two values. In this case the difference is usually expressed in terms of percent.

The above-mentioned measures are generally the most often-quoted metrics for model accuracy. There are other metrics that we can use in the estimation of the goodness of fit of the model.

6.6 NON-VOLTERRA MODELS

Although we have focused on polynomial or Volterra series models, there are several other model types that could be used for nonlinear dynamical modeling of power amplifiers. These include piecewise polynomial or spline functions, rational polynomial functions, radial basis functions (RBFs), artificial neural networks (ANNs), and so forth. By modeling, we mean using the above functions to approximate the underlying but unknown function that describes the measured data: function approximation. A good general description of the use of these functions to approximate measured data, applied to the case of modeling power transistors, can be found in Chapter 7 of [32].

Spline functions provide a smooth and continuous curve through the data, though usually the degree of the polynomial used in the spline is limited, cubic being typical. This limits the order of nonlinearity that can be accommodated. Memory effects can be included, though not usually with the built-in functions, and the addition of memory effects makes the matching of the spline functions across the boundaries quite difficult.

Rational functions are an alternative to the simpler polynomial functions and offer the potential advantage that they can be constructed to have better, more controllable extrapolation properties. This is a possible advantage for application in modeling of pre-distortion linearizers, which have gain expansion. Memory can be incorporated quite straightforwardly into the rational function formulation, for example in the form of a memory polynomial function [33]:

$$y(t) = \frac{\sum_{n=1}^{N_N} \sum_{m=0}^{M_N} a_{nm} u(t-\tau_m)|u(t-\tau_m)|^{n-1}}{1 + \sum_{n=1}^{N_D} \sum_{m=0}^{M_D} b_{nm} |u(t-\tau_m)|^{n}} \tag{6.83}$$

After some rearrangement of this expression, multiplying through by the denominator polynomial and transferring the product terms to the right-hand side, we obtain

$$y(t) = \sum_{n=1}^{N_N} \sum_{m=0}^{M_N} a_{nm} u(t-\tau_m)|u(t-\tau_m)|^{n-1} - y(t) \sum_{n=1}^{N_D} \sum_{m=0}^{M_D} b_{nm} |u(t-\tau_m)|^{n} \tag{6.84}$$

which is an auto-regressive model that can be solved using the usual least-squares techniques to find the parameters a_{nm}, b_{nm}.

Radial basis functions and artificial neural networks, in contrast to the models described so far, are fully nonlinear models in which the model parameters are included in the nonlinear basis functions. Therefore, we cannot use linear least-squares techniques to identify those model parameters, and we have to use nonlinear optimization techniques to extract them.

The radial basis function including memory terms is given by

$$y(t) = \sum_{i=1}^{N} \sum_{m=0}^{M} a_{i,m} \exp\left[-\left(\frac{\|u(t-\tau_m) - c_i\|}{r_i} \right)^2 \right] \tag{6.85}$$

Radial basis functions have been used for power amplifier modeling with some success [34], in terms of achieving small errors, NMSE, and ACEPR.

6.6.1 Artificial Neural Networks

Artificial neural networks have been used as general-purpose nonlinear function approximators or modeling tools in a wide range of applications, including microwave transistor modeling [35], RF behavioral modeling [36, 37], and RF power amplifier and transmitter modeling [38, 39]. The ANN approximates the discrete measured data with a smooth function that is infinitely differentiable, enabling the accurate modeling of distortion products.

An artificial neural network is a regular structure comprising: an input layer, which defines the inputs to the ANN from the outside world; an output layer, which provides the output signals, and one or more so-called hidden layers, which contain the nonlinear processing elements, or neurons. The hidden layers, by convention, are not connected directly to the outside world. The structure of an artificial neuron is shown in Figure 6.15. The neuron takes several input signals and adds them together; this weighted sum of inputs is then passed to a nonlinear transfer function or activation function, which generates the neuron output signal. This nonlinear transfer function is often a sigmoid function, which is a nonlinear curve that eventually reaches a saturated value. It is often represented by a hyperbolic tangent $tanh(x)$ function.

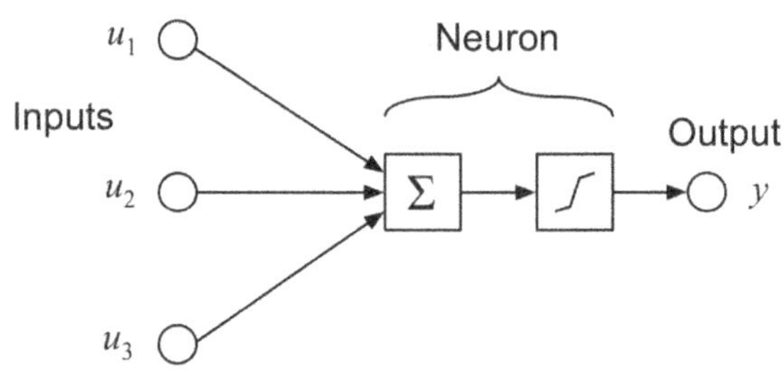

Figure 6.15 A simple artificial neuron structure.

The structure of a simple ANN with M inputs and one output is shown in Figure 6.16. Each input u_m is connected to every neuron through a weight, w_{mk}; at each neuron, the weighted inputs are summed together, and, together with an offset or bias value b_k, passed to the nonlinear transfer function. The output of each neuron, x_k, is thus a nonlinear function of the weighted sum of the input signals:

$$x_k = \tanh\left(\sum_{m=1}^{M} w_{mk}u_m + b_k\right) \tag{6.86}$$

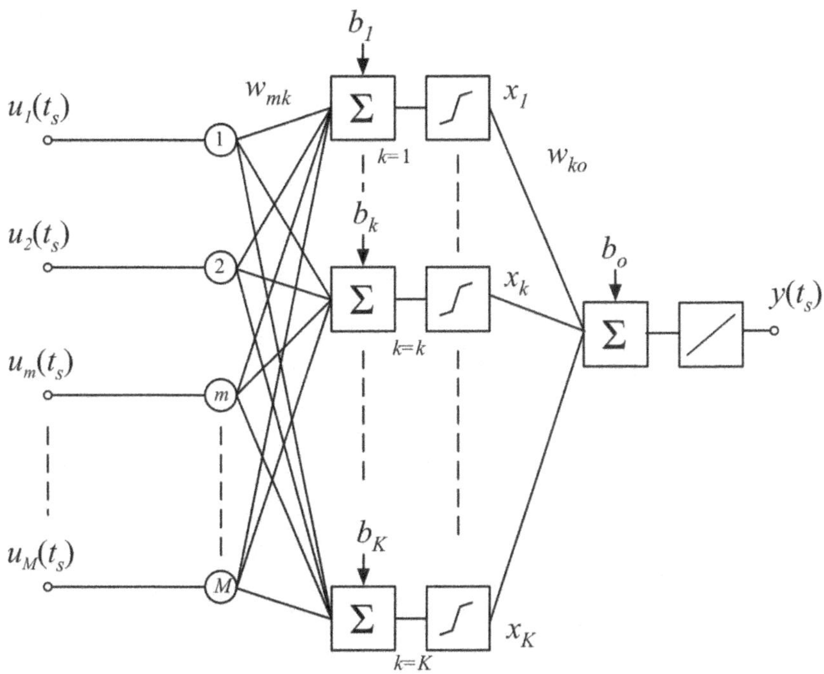

Figure 6.16 A single hidden-layer artificial neural network. This network has M inputs, K hidden-layer neurons, and one output node.

The weighted outputs from each neuron are then summed at the output node. The transfer function of the output is usually linear. The overall ANN function is then given by:

$$y(t) = \sum_{k=1}^{K} \left\{ w_{ko} \tanh \left(\sum_{m=1}^{M} w_{mk} u_m + b_k \right) + b_o \right\} \tag{6.87}$$

The hidden layer neurons provide the nonlinearity, and the weighting factors and biases at each neuron determine the shape or nature of the overall transfer function of the ANN. The ability of an ANN to approximate a given function depends on the values of the weights and biases. These values are determined in a nonlinear optimization process, commonly using a back-propagation method to adjust the weights through their input sensitivities, determined from the previous optimization cycle [40]. One of the main objectives of the optimization is to obtain a set of weights and biases that enable the ANN to predict the correct outputs for inputs that it has not seen before: *generalization*. These inputs can be other parts of the data set used for cross-validation, or signals of a similar class to the modeled data.

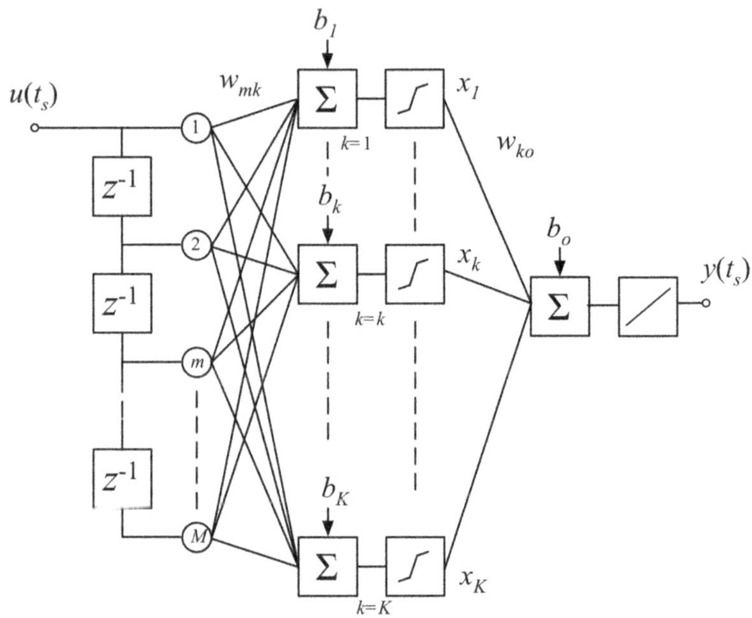

Figure 6.17 The time-delay artificial neural network, shown here using a single hidden layer ANN. This network has M inputs, corresponding to the M delays, or a memory depth, M.

Artificial neural networks have the potential of achieving high accuracy in function approximation of measured data. The *universal approximation theorem* [41] states that a single layer ANN can approximate any nonlinear function to an arbitrary degree of accuracy, provided that there are enough neurons. Unfortunately, this does not place a limit on the number of neurons that we will need for a given problem, and so a certain amount of trial and error remains in the function-fitting process, although the addition and removal of neurons can be made systematic. One problem with single-layer ANNs is that the neurons can interact with one another. In other words, adjusting the weights to improve the approximation for one region of the data can make it worse in another region (the *waterbed effect*). This problem can be overcome by using two hidden layers: the first layer operates only on localized regions of the data, and the second layer stitches these approximations together.

The artificial neural networks described so far are general nonlinear function approximators. For power amplifier modeling, we need to include memory effects. This is achieved by considering the nonlinear FIR structure for the PA model that we have focused on:

$$y(t) = \mathrm{F}\left(u(t), u(t-\tau_1), u(t-\tau_2), ..., u(t-\tau_m)\right) \tag{6.88}$$

By passing the input signal $u(t)$ through a delay line, we can generate the set of delayed inputs in $u(t)$; these delays become the inputs to a time-delay ANN (TD-ANN), as illustrated in Figure 6.17.

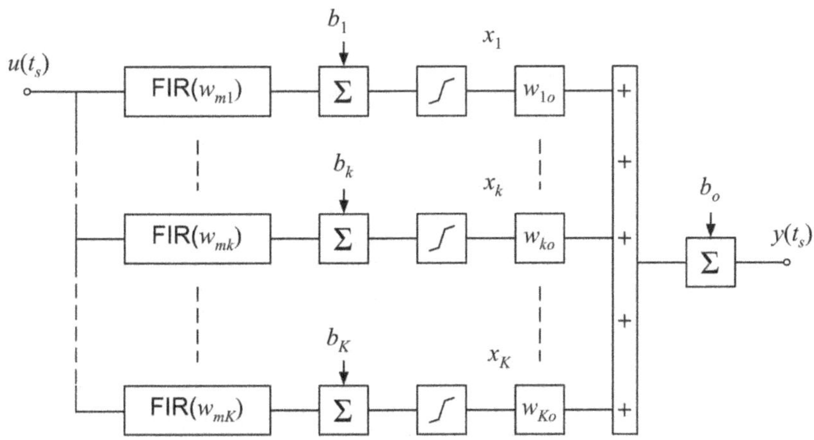

Figure 6.18 The time-delay ANN, reconfigured to resemble a Volterra series parallel cascade of filters.

This time-delay ANN can provide a similar level of model accuracy to Volterra series models, and has the potential advantage of not being limited to the case of weak nonlinearity, allowing the possibility of including heavily saturated or switching PA effects into the model. The relationship of the TD-ANN to Volterra Series has been investigated by Pedro & Mass [10], who concluded that the TD-ANN can be recast in the form of a Volterra Series parallel cascade approximation, as shown in Figure 6.18. Thus, the foundation and fundamentals of Volterra series approximations are seen to support this form of nonlinear dynamical modeling using ANNs as well.

BIBLIOGRAPHY

The classic texts on Volterra Series that I turn to for reference are:

- Martin Schetzen, *The Volterra and Wiener Theories of Nonlinear systems*, Malabar, FL: Krieger (recently reprinted in 2005), and
- Wilson Rugh, *Nonlinear System Theory - the Volterra/Wiener Approach*, Baltimore, MD: The Johns Hopkins University Press (1981)

which is now available online. Rugh's text is particularly accessible, as promised in his preface: "As to style I eschew the trappings of rigor and adopt a more mellifluous tone."

Other scientific disciplines have applied the Volterra approach well before the recent interest in using this technique for power amplifier modeling. In particular, the work of Michael Korenberg in the biological field is worthy of attention. In the context of behavioral modeling of PAs using Volterra-based approaches, books and papers by Professors Pedro and Carvalho and co-workers from the University of Aveiro, and by Professor Boumaiza from the University of Waterloo provide an excellent basis for study. Indeed, the review of microwave and RF power amplifier behavioral modeling approaches by Pedro and Maas is a classic reference on this topic [10]. Also, close reading of the work of Professor Filicori and co-workers, Professor Ngoya and Dr. Soury and their associates, and Professor Zhu, on the development of the dynamic deviation models will reward the attentive reader. As noted earlier, an accessible tutorial review of the mathematical techniques for solving the ill-conditioned equations that we encounter in solving the Volterra series models has been presented by Neumaier [28].

References to specific citations in this chapter on these topics are found below.

References

[1] M. Schetzen, *The Volterra and Wiener Theories of Nonlinear Systems*. Malabar, FL: Krieger, 2006.

[2] J. W. Graham and L. Ehrman, "Nonlinear system modeling and analysis with applications to communications receivers," Tech. Report, Rome Air Development Center, Tech. Rep. RADC-TR-73-178, 1973.

[3] E. Bedrosian and S. O. Rice, "The output properties of Volterra systems (nonlinear systems with memory) driven by harmonic and Gaussian inputs," *Proc. IEEE*, Vol. 59, pp. 1688–1707, 1971.

[4] W. J. Rugh, *Nonlinear Systems Theory: The Volterra/Wiener Approach*. Baltimore, MD: Johns Hopkins University Press, 1981.

[5] S. Boyd, L. O. Chua, and C. A. Desoer, "Analytical foundations of Volterra series," *IMA Journal of Mathematical Control and Information*, Vol. 1, No. 3, pp. 243–284, 1984.

[6] S. Maas, *Nonlinear Microwave Circuits 2nd ed.* Norwood, MA: Artech House, 2003.

[7] P. Wambacq and W. Sansen, *Distortion Analysis of Analog Integrated Circuits.* Boston, MA: Kluwer Academic, 1998.

[8] J. Vuolevi and T. Rahkonen, *Distortion in RF Power Amplifiers.* Norwood, MA: Artech House, 2003.

[9] J. Wood and D. E. Root, (eds.) *Fundamentals of Nonlinear Behavioral Modeling for RF and Microwave Design.* Norwood, MA: Artech House, 2005.

[10] J. C. Pedro and S. A. Maas, "Comparative overview of microwave and wireless power-amplifier behavioral modeling approaches," *IEEE Trans. Microw. Theory Tech.*, Vol. 53, No. 4, pp. 1150–1163, Apr. 2005.

[11] A. Zhu and T. J. Brazil, "Optimal digital Volterra pre distorter for broadband RF power amplifier linearization," 31st *European Microwave Conf.*, London, England, Sept. 2001, pp. 1–4.

[12] S. Maas, "Volterra methods for behavioral modeling," *IEEE MTT-S Workshop on Fundamentals of Nonlinear Behavioral Modeling*, Philadelphia, PA, June 2003.

[13] J. Wood, "Volterra methods for behavioral modeling," in *Fundamentals of Nonlinear Behavioral Modeling for RF and Microwave Design*, J. Wood and D. E. Root, (eds.) Norwood, MA: Artech House, 2005, Ch. 2.

[14] S. Boyd and L. O. Chua, "Fading memory and the problem of approximating nonlinear operators with Volterra series," *IEEE Trans. Circuits Syst.*, Vol. 32, No. 11, pp. 1150–1161, Nov. 1985.

[15] E. Ngoya and A. Soury, "Envelope domain methods for behavioral modeling," in *Fundamentals of Nonlinear Behavioral Modeling for RF and Microwave Design*, J. Wood and D. E. Root, (eds.) Norwood, MA: Artech House, 2005, Ch. 3.

[16] D. Mirri, G. Iuculano, F. Filicori, G. Pasini, G. Vannini, and G. P. Gualtieri, "A modified Volterra series approach for nonlinear dynamic systems modeling," *IEEE Trans. Circuits Syst. I*, Vol. 49, No. 8, pp. 1118–1128, Aug. 2002.

[17] N. Wiener, *Nonlinear Problems in Random Theory.* Boston, MA: The Technology Press, 1958.

[18] E. G. Lima, T. R. Cunha, H. M. Teixeira, M. Pirola, and J. C. Pedro, "Base-band derived Volterra series for power amplifier modeling," *IEEE MTT-S Int. Microwave Symp. Dig.*, Boston, MA, June 2009, pp. 1361–1364.

[19] T. R. Cunha, J. C. Pedro, and E. G. Lima, "Low-pass equivalent feedback topology for power amplifier modeling," *IEEE MTT-S Int. Microwave Symp. Dig.*, Altlanta, GA, June 2008, pp. 1445–1448.

[20] J. Staudinger, J. C. Nanan, and J. Wood, "Memory fading Volterra series model for high power infrastructure amplifiers," *IEEE Radio & Wireless Symp. Dig.*, New Orleans, LA, Jan. 2010, pp. 184–187.

[21] D. R. Morgan, Z. Ma, J. Kim, M. G. Zierdt, and J. Pastalan, "A generalized memory polynomial model for digital pre distortion of RF power amplifiers," *IEEE Trans. Signal Process.*, Vol. 54, No. 10, pp. 3852–3860, Oct. 2006.

[22] M. J. Korenberg, "Parallel cascade identification and kernel estimation for nonlinear systems," *Annals of Biomedical Eng.*, Vol. 19, pp. 429–455, 1991.

[23] S. J. Kenney, "Device-level behavioral modeling for microwave components," *IEEE MTT-S Workshop on Nonlinear CAD*, Boston, MA, June 2000.

[24] F. Filicori, G. Vannini, and V. A. Monaco, "A nonlinear integral model of electron devices for HB circuit analysis," *IEEE Trans. Microw. Theory Tech.*, Vol. 40, No. 7, pp. 1456–1465, July 1992.

[25] E. Ngoya, N. L. Gallou, J. M. Nebus, H. Buret, and P. Reig, "Accurate RF and microwave system level modeling of wideband nonlinear circuits," *IEEE MTT-S Int. Microwave Symp. Dig.*, vol. 1, Boston, MA, June 2000, pp. 79–82.

[26] A. Soury, E. Ngoya, and J. M. Nebus, "A new behavioral model taking into account nonlinear memory effects and transient behaviors in wideband SSPAs," *IEEE MTT-S Int. Microwave Symp. Dig.*, Seattle, WA, June 2002, pp. 853–856.

[27] A. Z. J. C. Pedro and T. J. Brazil, "Dynamic deviation reduction-based Volterra behavioral modeling of RF power amplifiers," *IEEE Trans. Microw. Theory Tech.*, Vol. 54, No. 12, pp. 4323–4332, Dec. 2006.

[28] A. Neumaier, "Solving ill-conditioned and singular linear systems: a tutorial on regularization," *SIAM Review*, Vol. 40, pp. 636–666, 1998.

[29] G. Golub and C. van Loan, *Matrix Computations*, 3rd ed. London, U.K.: Johns Hopkins University Press, 1996.

[30] C. Moler, *Numerical Computing with MATLAB*, revised ed. SIAM, 2008.

[31] D. Wisell, M. Isaksson, and N. Keskitalo, "A general evaluation criteria for behavioral power amplifier modeling," in *69th ARFTG Conf.. Dig.*, Honolulu, HI, Jun. 2007, pp. 1–5.

[32] P. Aaen, J. Plà, and J. Wood, *Modeling and Characterization of RF and Microwave Power FETs.* Cambridge, U.K.: Cambridge University Press, 2007.

[33] T. R. Cunha, P. M. Lavrador, E. G. Lima, and J. C. Pedro, "Rational function-based model with memory for power amplifier behavioral modeling," *2011 Workshop on Integrated Nonlinear Microwave and Millimetre-Wave Circuits.*, Vienna, Austria, Apr. 2011, pp. 1–4.

[34] M. Isaksson, D. Wisell, and D. Ronnow, "Nonlinear behavioral modeling of power amplifiers using radial basis function neural networks," *IEEE MTT-S Int. Microwave Symp. Dig.*, Long Beach, CA, June 2005.

[35] J. Xu, D. Gunyan, M. Iwamoto, A. Cognata, and D. E. Root, "Measurement-based non-quasi-static large-signal FET model using artificial neural networks," *IEEE MTT-S Int. Microwave Symp. Dig.*, San Francisco, CA, June 2006, pp. 469–472.

[36] Q.-J. Zhang and J. Xu, "Neural networks for behavioral modeling," in *Fundamentals of Nonlinear Behavioral Modeling for RF and Microwave Design*, J. Wood and D. E. Root, (eds.) Norwood, MA: Artech House, 2005, Ch. 6.

[37] J. Wood, D. E. Root, and N. B. Tufillaro, "A behavioral modeling approach to nonlinear model-order reduction for RF/microwave ICs and systems," *IEEE Trans. Microw. Theory Tech.*, Vol. 52, No. 9, pp. 2274–2284, Sept. 2004.

[38] T. Liu, S. Boumaiza, and F. M. Ghannouchi, "Dynamic behavioral modeling of 3G power amplifiers using real-valued time-delay neural networks," *IEEE Trans. Microw. Theory Tech.*, Vol. 52, No. 3, pp. 1025–1033, Mar. 2004.

[39] F. Mkadem and S. Boumaiza, "Physically-inspired neural network model for RF power amplifier behavioral modeling and digital predistortion," *IEEE Trans. Microw. Theory Tech.*, Vol. 59, No. 4, pp. 913–923, Apr. 2011.

[40] S. Haykin, *Neural Networks: a Comprehensive Foundation*, 2nd ed. Upper Saddle River, NJ: Prentice Hall, 1995.

[41] G. Cybenko, "Approximation by superposition of sigmoidal functions," *Math. Control, Signals Systems*, No. 2, pp. 303–314, 1989.

Chapter 7

Digital Pre-Distortion

The design of RF power amplifiers for digital communications systems has several goals, generally mutually incompatible, including: high output power, wide signal bandwidth, high efficiency, high gain, and low distortion. As outlined in the Introduction, Chapter 1, modern digital communications signals are designed for high spectral efficiency, and consequently have large peak-to-average power ratios (PAPR). The high power requirement of the PA derives directly from the power needed for the range of the signal coverage and the PAPR. Even with high spectral efficiency signal modulations, the number of subscribers and their data requirements are pushing up the bandwidth required, from 5–10 MHz for two WCDMA channels to 20–40 MHz LTE bands, and beyond. High-efficiency PA designs are being developed to accommodate these high PAPR and signal bandwidth demands, and high gain is desirable to minimize the power demands on the driver stages, and so improve the overall system efficiency. These modern PA architectures are quite nonlinear.

The nonlinear performance is a significant concern. The levels of distortions that are permitted in the signal channel and as interference in adjacent or nearby channels are clearly specified by international and federal communications agencies, and industry partnerships and standardization bodies, such as International Telecommunications Union Standardization Sector (ITU-T, formerly CCITT), Federal Communications Commission (FCC), and 3rd Generation Partnership Project (3GPP). A linearization or pre-distortion technique is essential to meet these specifications.

A number of linearization methods were outlined briefly in Chapter 1. These included analog feedback techniques at RF and at baseband or intermediate frequency, and analog feedforward methods. In this chapter, we shall focus on pre-distortion techniques carried out on the digital modulation signals at baseband: digital pre-distortion or DPD.

7.1 INTRODUCTION TO PRE-DISTORTION

The basic principle of pre-distortion is that we place a nonlinear component in front of the nonlinear PA, so that the combined action of these two functions on the linear input signal is to produce a scaled (larger!) version of that input: linear gain. This is shown in Figure 7.1. The pre-distorter is a complex gain adjuster placed in front of the PA. Let the power amplifier have a nonlinear gain $G(v)$; then the output voltage for an input signal $v_{in}(t)$ is:

$$v_{out}\left(t\right) = G\left(v_{in}\left(t\right)\right) \tag{7.1}$$

For simplicity here, we will model this nonlinear gain function with an instantaneous (memoryless) polynomial,

$$v_{out}\left(t\right) = a_1 v_{in}\left(t\right) + a_2 v_{in}^2\left(t\right) + ... + a_n v_{in}^n\left(t\right) = \sum_{i=1}^{n} a_i v_{in}^i\left(t\right) \tag{7.2}$$

We can see that the output comprises a linear term and several nonlinear components. The job of the pre-distorter is to add terms to the input signal that, when amplified by the PA, will cancel out these nonlinear components:

$$v_{out}\left(t\right) = G\left(F_{PD}\left(v_{in}\left(t\right)\right)\right) = a_1 v_{in}\left(t\right) \tag{7.3}$$

If the PA can be modeled accurately with a low-degree polynomial, then the pre-distortion function can be the mathematical inverse of the polynomial. This principle can be extended to PAs with memory effects: using a pth-degree Volterra series model for the power amplifier, we can determine the formal p^{th}-order inverse of the series for the pre-distorter function [1].

While this approach has a mathematical elegance, it may be undesirable to use the exact inverse for the pre-distorter. The specified distortion limits may be obtained with a much simpler pre-distortion function, for example. Other mappings of the input-output relationship are also possible, though some may lack a formal

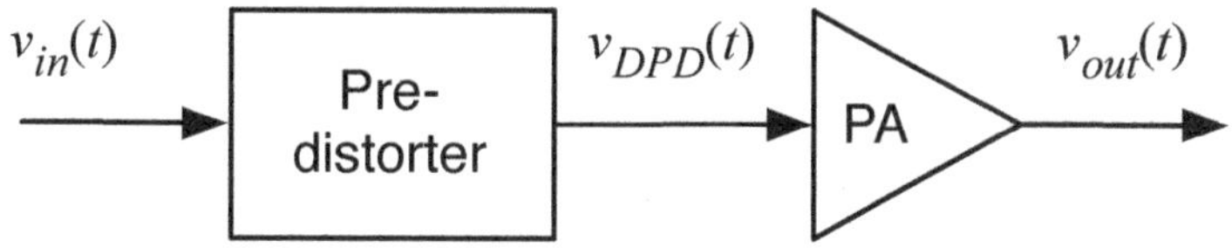

Figure 7.1 A pre-distorter and power amplifier combination

inverse function. In such cases, the PA's output-to-input function is modeled, using nonlinear modeling approaches as have been described earlier. A problem with this approach is that it may not be possible to excite all of the dynamical nonlinear modes in the pre-distorter that are required to be corrected in the PA, as we are dependent on the output signal to excite such modes for identification. Such problems are not new in the fields of system identification [2] or adaptive control [3], though they may be presented as new discoveries in the pre-distortion of RF power amplifiers.

The input-output mapping approach can be used with a static pre-distorter; that is, a function whose coefficients are fixed. In such examples, the PA can be identified *a priori* using a wide set of measurements. Unfortunately, this is rarely the case in a practical transmitter system, where the pre-distorter coefficients' values must be continually adjusted to achieve the optimal linearization performance. This adjustment or adaption is done in order to accommodate variations in the input signal with content, data load, number of channels, or the required average output power of the PA, and variations in the PA behavior with temperature, bias, signal bandwidth and amplitude. For example, in time-division duplex communication, the drive to the PA is switched on and off at the duplex rate, and the power transistors in the PA will cool down during the "off" cycle, and heat up again during the "on" cycle when the large signal is present. The complex nonlinear gain of the PA will therefore have a time dependence and will need to be continuously corrected by the DPD system.

7.1.1 Gain and Bandwidth Concerns

In Figure 7.2 we show the input-to-output power relationship or power drive-up curve for the PA. At low signal levels, the response is close to linear, as indicated by the dotted construction line. At high signal levels, the PA is in compression and generates distortion. At some point, we cannot get any more power out of the device, and the PA is in saturation. As we have seen, we need to operate the PA at the high signal levels for efficiency and power.

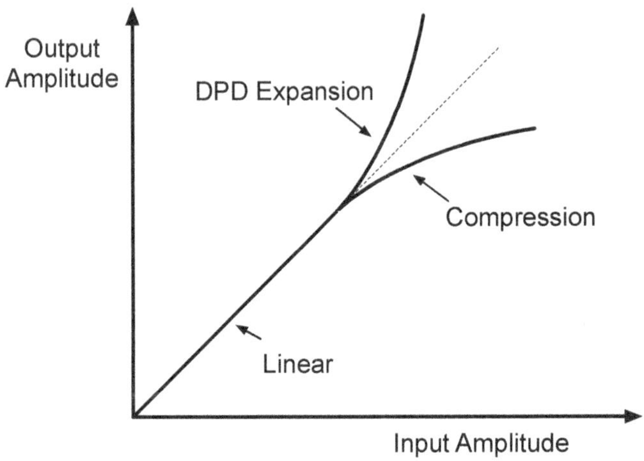

Figure 7.2 The input-output characteristics of the power amplifier and the pre-distorter.

Also in this figure is the input-output response of the pre-distorter. Whereas the PA is compressive, the pre-distorter has an expansive characteristic that is the opposite of the PA. The combination of the two results in a linear input-output relationship overall. There is a similar inverse relationship between the AM-to-PM characteristic of the PA and the pre-distorter's phase behavior. These inverse characteristics only work up to the saturated power level of the PA, so in practice we need to limit the input signal level to prevent the PA from becoming saturated.

The input and output PAPRs are shown on the drive-up curves in Figure 7.3. The input signal PAPR for high-power operation is presented on the input axis, and shows that the output PAPR of the PA alone would be compressed. Applying this signal to the pre-distorter curve shows that the PAPR at the pre-distorter output is higher; but this is the (pre-distorted) input PAPR to the amplifier. To prevent saturation of the PA, the average input power needs to be reduced.

An alternative to this reduction in power capability is to use a technique known as crest factor reduction (CFR), in which the peaks of the digital signal are reduced or removed, to reduce the PAPR of the signal. The CFR function is placed before the pre-distorter, so that the signal seen by the pre-distorter has a lower PAPR, allowing room for the gain expansion characteristic of the pre-distorter. The

CFR can be applied simply but crudely in the analog domain, using a limiter, but much more sophisticated techniques can be applied in the digital signal domain. Crest factor reduction techniques are widely used in digital pre-distortion systems, and several popular techniques are described in more detail in Chapter 8.

Another way of looking at the behavior of the pre-distorter is in the frequency domain. We saw in Chapter 4 that the nonlinear action of the PA is to create harmonic and intermodulation distortion products, and new frequency components, in addition to the compression behavior outlined above. We can use filters to remove the harmonic products, but to remove or reduce the intermodulation components, which are interferers in adjacent channels, we need to use a linearizer. The action of the pre-distorter is to add these distortion components to the signal input to the PA, in such a way as to cancel with the PA's own distortion contributions at the output. These additional components correspond with the PA's higher-order distortion products that can be described by the polynomial model of (7.2). The signal spectrum at the output of the pre-distorter is therefore wider than the original signal input. This is shown schematically in Figure 7.4, where the signal added by the pre-distorter can be seen to act in anti-phase to the PA's spectral spreading.

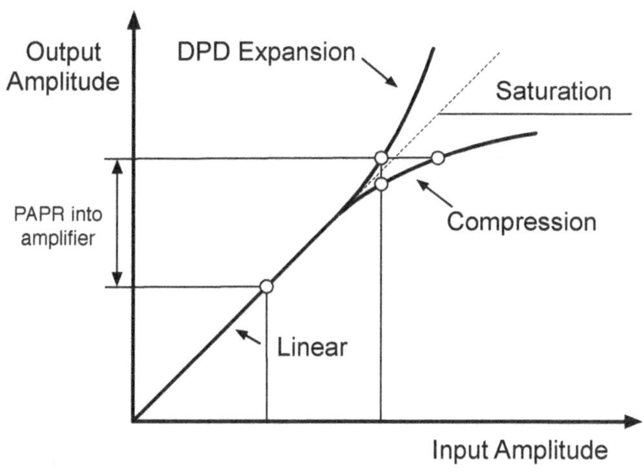

Figure 7.3 The input-output characteristics of the power amplifier and the pre-distorter, indicating the signal peak-to-average power ratios at the input, the output of the pre-distorter, and the PA.

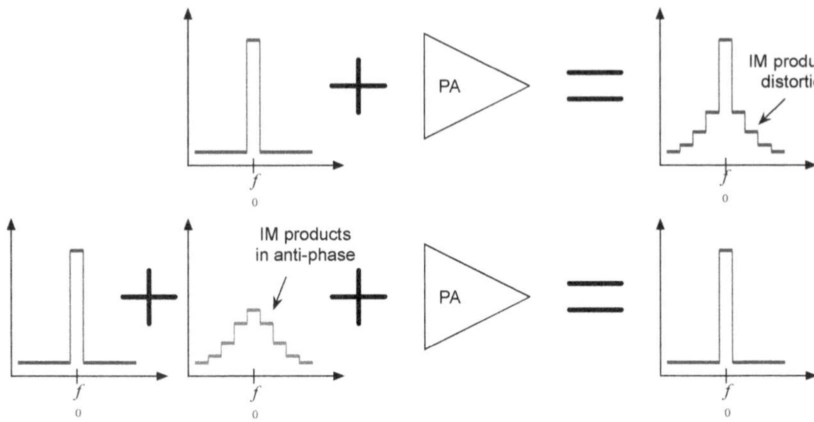

Figure 7.4 The nonlinear action of the PA creates signal products in adjacent channels. The pre-distorter adds these frequency components to the signal, in 'anti-phase' to the PA's distortion, to cancel their effects at the PA output.

While the behavior of the pre-distorter is to improve the linearity of the system,

- it increases the amplitude of the signal presented to the power amplifier;
- it increases the bandwidth of the signal presented to the power amplifier;
- it does not provide any more output power from the PA.

7.1.2 Pre-Distortion of the Digital Signal

We have used the simple pre-distorter and PA schematic of Figure 7.1 to illustrate the basic principles and concerns, but in practice the pre-distortion function is performed in the digital signal domain, directly on the I and Q data streams. This introduces some additional components between the pre-distorter and the power amplifier, in particular, a digital-to-analog converter (DAC) and a frequency up-converter or IQ-modulator, as shown in Figure 7.5. Each of these components will also introduce some distortion that will need to be accounted for by the pre-distorter, although in practice we expect these contributions to be greatly overshadowed by the PA nonlinearities.

The pre-distortion function must now modify the time sequences of $I[k]$ and $Q[k]$ in such a way that the distortion at RF produced at the output of the PA is

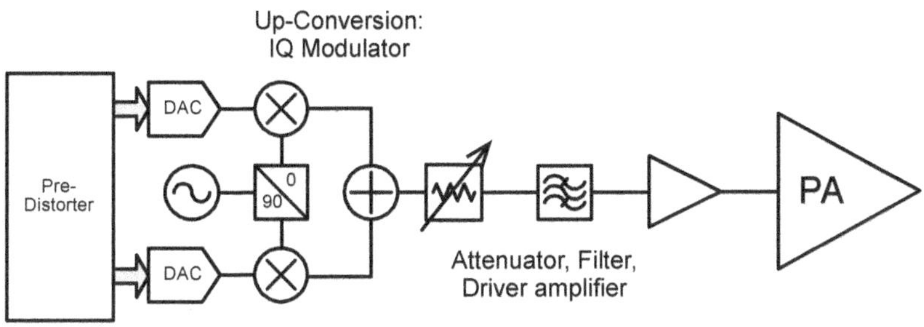

Figure 7.5 A pre-distorter in the digital domain, showing the necessary additional components for up-converting to RF.

minimized. It is less obvious how to do this using a simple input-output mapping technique as outlined earlier, since the input and output signals are now in different domains. But generally, the use of such a static approach to pre-distortion is not used in practice, and adaptive pre-distortion techniques are now virtually ubiquitous.

We add an observation path to the system to enable the adaption. This observation path samples the output from the power amplifier, and feeds the signal back to the pre-distorter so that online identification of the system nonlinearity can be accomplished. As we are performing the pre-distortion in the digital domain, this observation signal must be converted to the baseband frequency, and digitized using an analog-to-digital converter (ADC). This transformation from the RF domain to the digital one must be linear, or at least much more linear than the desired distortion level of the transmitter to avoid introducing distortion at the PA output arising from the observation process. The input and feedback signals must be aligned in time to allow the identification process to take place, the time delay being approximately the same as the group delay of the power amplifier, which tends to dominate the loop.

A high-level block diagram of a transmitter with an adaptive pre-distorter is shown in Figure 7.6. This figure shows the identification block to update the pre-distorter coefficients, and also a block to carry out the crest factor reduction (CFR) function.

We now have the basis of a digital pre-distortion system: the pre-distorter function in the forward path operates continuously on the signal, and the observation path samples the PA output so that we can adapt the pre-distortion function coefficients or parameters as the signal or PA characteristics change. This is, in essence, an adaptive control problem. In the next section of this chapter, we shall

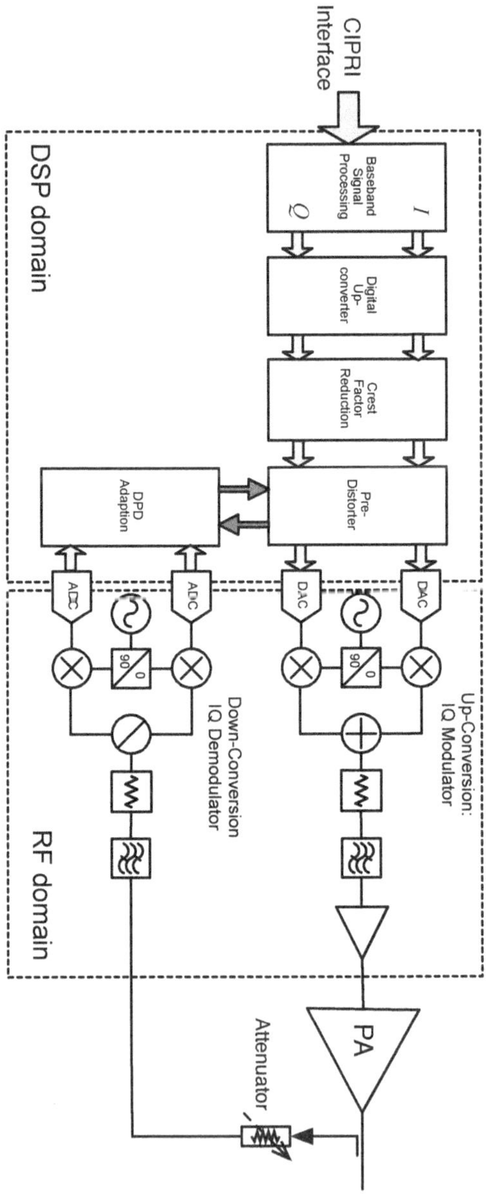

Figure 7.6 Block diagram of a prototype adaptive digital pre-distortion system, showing the observation path and down-converter.

outline the functions of the basic building blocks of this DPD system, in greater or lesser detail, and how they operate on the sampled digital signals. Later, we shall describe the basics of the adaptive digital pre-distortion process, illustrating some typical approaches, and we shall follow this with a description of the practical considerations of implementing an adaptive DPD system.

7.2 THE DIGITAL PRE-DISTORTION SYSTEM

7.2.1 Baseband Signal Processing

The digitally-modulated RF signals that are input to the RF power amplifier need to be managed in time and spectrum. This management is done in the digital domain in the baseband signal processor or modem in the transmitter. The I and Q digital signals representing the information signals for each data or voice channel must be encoded and combined to generate the signal for the transmission channel. For multi-carrier systems, these channels must be further combined and placed in the correct relative positions in the spectrum. These composite digital signals are then up-sampled to an appropriate sample rate or bandwidth for the digital system. This action is probably best illustrated by a couple of simple examples.

For single-carrier GSM and EDGE signals, the users are time-division multiplexed onto the digital signal stream. The data frame comprises a total of eight available time slots, not all of which need be used in any given frame, depending on the number of users and data rate required: the traffic. The baseband processor determines the number of users per slot and number of slots required. There are also synchronizing and other control codes embedded in the signal, in each slot, and these data are also handled by the baseband processor. It also manages the signal power ramping at the beginning and end of each slot, to turn the PA on and off for transmission. For multi-carrier GSM and EDGE signals, these functions must be repeated for each carrier, before the carriers are up-converted and combined in the digital up-converter (DUC).

CDMA and Wideband-CDMA signals are PSK-modulated IQ data, as outlined in Chapter 1, and these modulated signals are then coded or 'spread' across the communications channel. Again, these DSP functions are controlled by the baseband processor, along with the control codes and power ramping functions. For multi-carrier (W)CDMA signals, the individual carriers or channels must be offset in time; for WCDMA this is ⅕ of a slot, where fifteen time slots make up the 10 millisecond radio frame. Each channel must then be offset in frequency, up-sampled, and combined by the DUC.

These input digital signals are up-sampled in the DUC to a high sampling rate, suitable for the subsequent pre-distortion processing. Why do we need to up-sample the signals? Recall from Figure 7.4 that the pre-distorter increases the bandwidth of the signal. To reconstruct this signal in the analog domain, for input into the RF amplifier chain in the transmitter, it needs to be sampled at a higher rate to accommodate this increased bandwidth. If we sampled at a lower rate corresponding to the input signal bandwidth, then we would get aliasing of the pre-distorted signal because it would be under-sampled with respect to its pre-distorted signal bandwidth. This aliasing will corrupt the sidebands that are essential for pre-distortion of the power amplifier. The un-aliased wideband signal cannot then be recovered by re-sampling at the output of the DPD. This means that we must over-sample the input signal at the input to the DPD function. The outcomes from under-sampling – aliasing – and over-sampling are illustrated in Figure 7.7, from Zhu *et al.* [4].

The over-sample rates that we choose depend on the bandwidth of the input signal and the pre-distortion capability we are looking for from the system. The typical rule of thumb is that the input signal sample rate is about five times the signal bandwidth. This enables capture of the fifth-order intermodulation products, which, as we have seen, can include higher-order distortion products as contributions. For a 4-carrier WCDMA signal occupying 20 MHz of bandwidth, this suggest a sampling rate of around 100 MSa/s for the composite digital IQ input signal.

In Table 7.1, we present several wireless signals and their native sample rates, and examples of typical up-sampling rates. The 2G and 3G CDMA signals and LTE all share a common basis, that is, they can have a common over-sampling frequency, albeit they are up-sampled at very different rates. The GSM/EDGE signals do not fall exactly onto this frequency grid. For a multi-protocol signal that includes GSM/EDGE and WCDMA or LTE, this would require resampling of (usually) the GMSK signal. Note that for multi-carrier (multi-channel) signals, the over-sampling rate compared to the native sample rate per carrier is high, but is chosen to obtain a DPD bandwidth that is four to five times the total signal bandwidth. Some of the reasons behind the choices of sample rates and bandwidths shown in Table 7.1 will be discussed in Section 7.4 on the implementation details of a typical DPD system.

As stated, the purpose of the digital up-converter is to take the sampled signals and up-convert them to the sample rate of the DPD digital signal processing system. In the digital domain, this up-conversion is performed by resampling the digital signals. This is done by padding the digital signal with zeros to reach the correct sample rate, and then the signal is interpolated between the zeros to create the up-sampled signal. A digital filter is then used to retrieve the correct frequency and phase response for the output signal. The interpolation and equalization functions

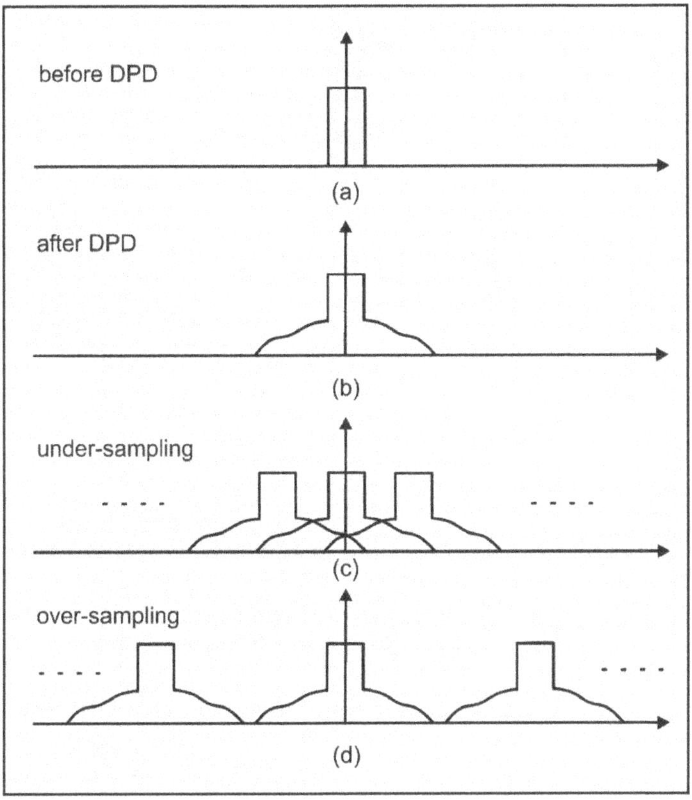

Figure 7.7 The effects of sample rate on the pre-distorted signal. The output signal from the pre-distorter (b) has a much wider bandwidth than the input signal (a). Under-sampling of the input will lead to aliasing of the output signal (c), and the clean DPD signal cannot be recovered by resampling at a higher rate. Over-sampling at the input is required to prevent aliasing (d). From Zhu *et al.* [4]

are carried out using complex filters. The interpolation is often carried out using cascaded differentiator and integrator stages, typically using cascaded integrator-comb (CIC) filters. The differentiator stage provides the delay, and the integrator recovers the signal, but can also introduce wideband noise. Several stages of CIC filter may be used to provide a large over-sampling ratio, though this can lead to instability and possible digital overflow in the chain. Finite impulse response (FIR) digital filters are used for equalization, compensating for the low-pass frequency response of the CIC filters. The FIR filters can also be used for rejection of the

Table 7.1

Sampling Frequencies for Various Wireless Signals, and Typical Over-sample Ratios for DPD.

	Native Rate MSa/s	Common DPD Rate MSa/s	Over-sample Ratio
GSM/EDGE	1.625/6	65	240
CDMA	1.2288	61.44	50
WCDMA	3.84	30.72 or 61.44	8 or 16
4C-WCDMA	3.84	92.16 or 122.88	24 or 32
LTE 20 MHz	30.72	122.88 or 153.6	4 or 5
LTE 2*20 MHz	30.72	245.76	8
LTE 5*20 MHz	30.72	368.64 or 491.52	12 or 16

image signal produced by the interpolation, and perhaps also to provide the correct matched-filter equalization for the modulation, for example, the root-raised cosine filter for WCDMA signals. The filters are programmed digitally by the DSP system to obtain the correct oversampling ratio and equalization for the given signal and bandwidth desired.

The up-sampled signals at the output of the filter chain are independent digital sequences centered at 0 Hz; to create a multi-carrier signal, these signals must be mixed to a complex intermediate frequency (IF) corresponding to the digital IF or clock rate of the DPD system. This frequency up-conversion is carried out by a digital multiplier, driven by a numerically-controlled oscillator (NCO). The frequency spacing for multi-carrier signals is controlled by the NCOs. As with any mixing process, noise in the output is governed by the phase noise of the local oscillator, the NCOs in this case. This noise can be highly correlated as a result of the fixed-precision arithmetic that is often used in the DSP; adding a small amount of digital dither to the NCO signal will whiten this noise and reduce the correlation. The outputs from the multipliers are then summed to create the complex multi-carrier signal waveform at the digital IF or clock rate of the system.

Digital pre-distortion evaluation modules, available from Texas Instruments or Broadcom Corporation (formerly from Optichron, now a part of Broadcom), for example, will generally include a pattern generator in place of the RF transmitter's baseband card. The pattern generator will usually accept digital signal inputs from an arbitrary waveform generator, or as IQ text streams generated from software such as Mathworks MATLAB, or Agilent's ADS or SystemVue. The pattern generator and DUC will then convert this basic IQ data into the up-sampled digital signals for use by the rest of the DPD system.

7.2.2 Crest Factor Reduction

The various techniques of crest factor reduction will be discussed in more detail in Chapter 8, and here we shall provide only a short overview, to place the CFR function in context in the DPD system as a whole.

The objective of crest factor reduction is to reduce the PAPR of the signal, so that the power amplifier can be operated at a higher average power, and hence efficiency, before the signal peaks are clipped. This must be done without any appreciable increase in the in-band distortion or spectral splatter. This would be a worthy goal even without pre-distortion, and, as such, CFR techniques are very appropriate for many wireless communications applications, such as cellular wireless handset PA applications, where Class AB operation is typically used. When CFR is used in conjunction with DPD techniques, the expansive nature of the DPD function operates on the crest factor reduced signal, so that the resulting pre-distorted signal that enters the PA does not have an excessive PAPR, and the PA can still be driven hard to operate at its highest efficiency.

One of the reasons why CFR is successful is that the spectrally-efficient signals such as UMTS and LTE, which have large PAPR of around 10 dB, is that the signal peaks are relatively infrequent. A PAPR of 10 dB means that the signal spends only 0.01% of the time at a power level 10 dB greater than the average power. This means that we can eliminate some of these higher peaks without necessarily causing much loss of information, or increasing the distortion in the signal. A typical performance metric for crest factor reduction is reduction (in dB) of the PAPR. This is usually shown on a CCDF curve, illustrating the hard peak limit introduced by the CFR, as in Figure 7.8. Here the PAPR of the raw digital signal has been reduced from about 10.5 dB to 8 dB.

Crest factor reduction methods range from the simple to the sophisticated. The simplest approach to CFR is a clip-and-filter method. The signal peaks above some defined threshold level are clipped, and then a filter is applied to remove the high frequency components that are associated with the hard clipping. A more sophisticated variation is to take the error signal, that is, the clipped points, and filter this to create a band-limited error that is then subtracted from the original signal.

The threshold limiting can be carried out on the I and Q data independently, known as magnitude clipping, or on the composite envelope, polar clipping. These two approaches are illustrated in Figure 7.9, albeit in a somewhat extreme fashion. It can be seen from this diagram that the clip-and-filter methods will increase

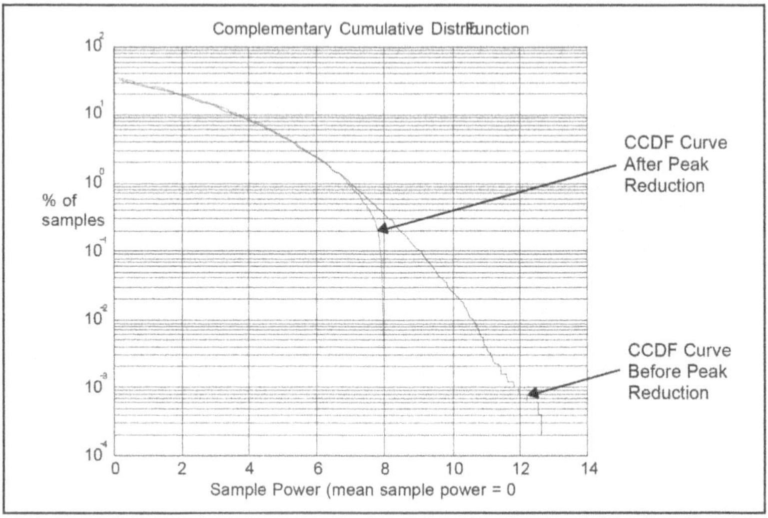

Figure 7.8 The effect of crest factor reduction on the peak-to-average power ratio (PAPR) of the digitally modulated signal, expressed in terms of the CCDF off the signal.

the in-band distortion of the signal, by removing (hopefully only small parts of) constellation points. This is seen as an increase in the EVM of the signal after CFR.

A popular CFR technique is that of *peak cancellation* by pulse injection. In this method, the location of the peak in the data stream is found, and a pulse or tone is added to the signal at that sample instant, the pulse being of a magnitude and polarity as to reduce the signal peak below the allowed peak level. This is shown in Figure 7.10. The pulse shape is controlled by windowing or filtering to cause minimum distortion of the surrounding signal. This technique tends to increase the in-band distortion or EVM of the signal.

The *peak windowing* CFR technique is similar to peak cancellation, in that again the location of the peak in the data signal stream must first be found. The set of samples containing the peak is then windowed in the time domain, or filtered in the frequency domain, with a function to reduce the peak magnitude. In practice, this usually means convolving the subset of data with the impulse response of a suitable lowpass or bandpass digital filter. The filter coefficients can be pre-programmed to yield a given reduction in the peak level. This technique has been applied to OFDM signals such as LTE and WiMAX [5], and tends to increase the sideband energy while maintaining low in-band distortion levels.

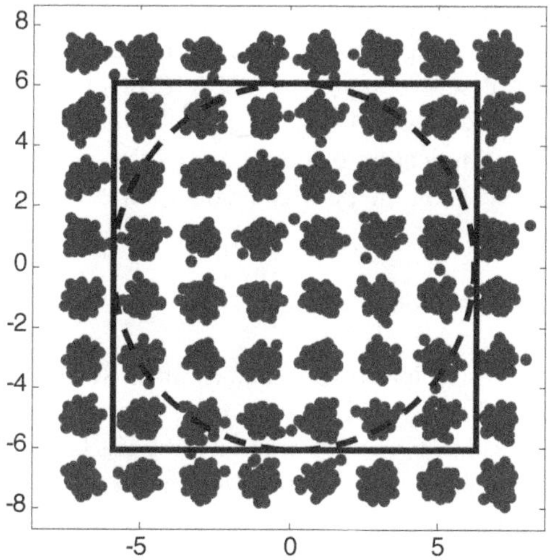

Figure 7.9 Clip-and-filter crest factor reduction techniques, showing the (extreme) effects of clipping of the I and Q signals individually (solid line), and of clipping the magnitude $\sqrt{(I^2+Q^2)}$ (dotted line), resulting in an increase in EVM of the signal.

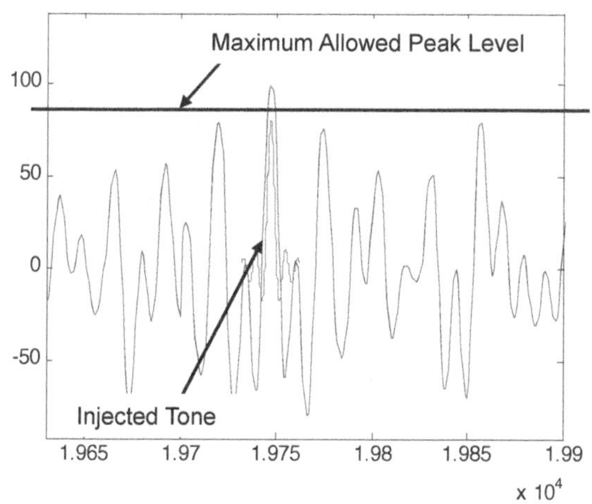

Figure 7.10 Peak cancellation by pulse or tone injection.

The pulse detection and location process operates on a set of signal samples, and closely spaced peaks may not be resolved and detected in a single pass, requiring several passes through the peak detection and cancellation (PD & C) software, or a cascade of several hardware PD & C blocks. This reduces the data throughput for these CFR approaches. The peak windowing technique, being a filtering method, is a multiplication of the filter and the signal, and is more costly in terms of DSP resources than the peak cancellation method, which is an additive technique.

7.2.3 The DPD Function Block

We have seen the basic principle of pre-distortion in Section 7.1. Here we shall describe in more detail the implementation of the nonlinear functions in the pre-distorter block. In particular, two main approaches have been adopted: the use of look-up tables (LUTs) to provide a map relating the desired pre-distorter output to the input voltage, and the use of nonlinear basis functions to describe nonlinear pre-distortion function, requiring generation of the nonlinear functions and the multiplication of the basis functions by the input voltage. This latter can be a calculation-intensive exercise, and techniques for reducing the number of repetitive calculations are used to simplify the process and make the DPD function realizable and reliable.

7.2.3.1 Look-Up Tables

Early pre-distorters were memoryless functions, acting only on the instantaneous input signal to provide correction to the AM-AM compression and AM-PM phase transfer characteristics of the PA. A typical implementation uses either a complex LUT describing the complex gain, or two real LUTs for gain and phase, to provide the output signal for the pre-distorter. The LUT pre-distorter structure is shown in Figure 7.11.

The pre-distorted signal into the PA is given by

$$V_{PD} = V_{in} \times V_{LUT} \tag{7.4}$$

where V_{LUT} is a pre-distortion gain that is read from the table at an index corresponding to the envelope of the input signal V_{in}.

$$V_{LUT} = G_{PD}\left(|V_{in}|\right) \tag{7.5}$$

The indexing into the LUT can be either the signal envelope voltage or power; we have used voltage here, although power is more typically chosen. Each entry in the

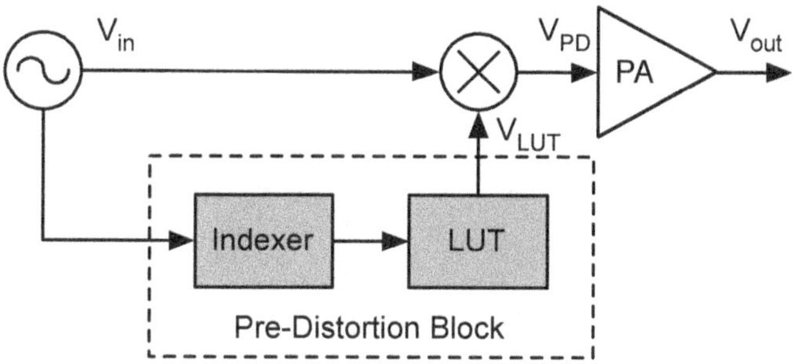

Figure 7.11 Implementation of look-up table (LUT) pre-distorter.

LUT compensates the gain and phase of the PA at the given input power, where the values for the gain and phase are averaged over a small amplitude interval, Δx. This is shown in Figure 7.12, for the PA gain. A similar relation will exist for the phase compensation by the pre-distorter, or the gain and phase can be combined as a complex quantity in a single table.

The number of intervals may be typically 128 or 256, although for some applications this can be much larger. The distribution of the intervals may also be nonuniform over the input power range of the PA, with the intervals becoming much narrower in the more nonlinear region, in compression, where the pre-distortion needs to be most accurate. This was first suggested by Cavers [6], and is known as Cavers' optimal companding.

Such look-up tables for memoryless gain and phase correction are relatively simple and straightforward to implement in software or in silicon on an FPGA or ASIC. Memory effects can also be included in LUT pre-distorter implementations; in simple form, this requires an additional complex LUT for every memory tap. More sophisticated applications of LUTs are associated with nonlinear functional forms, where the LUTs are used to index the coefficient values for the nonlinear basis functions as a function of the input signal power. This will be easier to visualize in the context of the next section.

7.2.3.2 Nonlinear Basis Functions

In this approach, we implement the nonlinear pre-distortion function using nonlinear basis functions, in an identical manner to how we modeled the nonlinear power

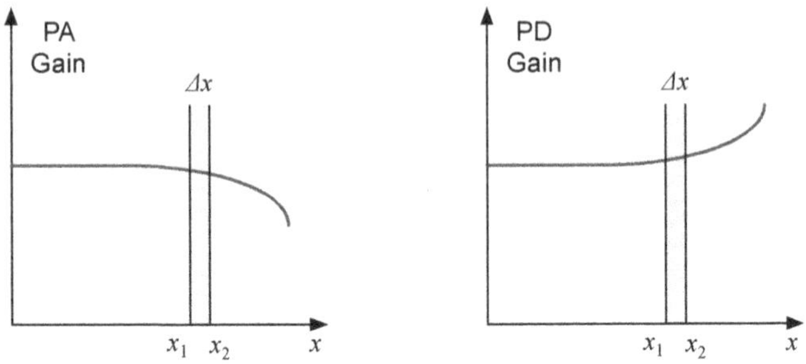

Figure 7.12 The look-up table (LUT) provides a compensating signal for the instantaneous gain and phase characteristics of the PA. The gain compensation is shown in this figure. A similar relationship will hold for the phase. (After Rexberg [7]).

amplifier. The nonlinearities of the pre-distortion function are contained in the nonlinear basis functions, and the pre-distortion function is constructed by combining basis functions in the same linear-in-parameters way:

$$x_{DPD}(t) = \sum_{n=1}^{N} a_n \varphi_n \left(x(t) \right) \tag{7.6}$$

While this is not strictly necessary, it simplifies the identification of the DPD function by reducing it to a least-squares problem of finding the coefficients or parameters a_k. This is especially important in the context of adaptive pre-distortion, described in the next section, where we need to calculate the coefficients on each adaption step. What we can see immediately, though, is that we can use all of the mathematical techniques that we have already studied in Chapter 6 for calculating these coefficients.

Most often, the basis functions will be polynomials in the input signal $u[k]$ and delayed signals $u[k - \tau]$, leading to a Volterra or reduced-Volterra basis. Implementations such as memory polynomials or dynamic deviation reduction (DDR) are commonly used to simplify the Volterra series expression, to enable a straightforward implementation in software, or hardware using FPGA or ASIC.

The calculation of the polynomial components can be a math-intensive exercise, even when pruning and other simplifications are used. There are also some simple algorithmic techniques that can be applied to reduce the number of arithmetic operations that need to be performed. For example, a naïve implementation of the calculation of a seventh-degree polynomial

$$p_7(x) = a_7x^7 + a_6x^6 + a_5x^5 + a_4x^4 + a_3x^3 + a_2x^2 + a_1x + a_0$$

would require 8 + 7 + 6 + 5 + 4 + 3 + 2 = 35 multiplications, and seven additions. A smarter (and fairly obvious) way to avoid this computational burden is to store each product temporarily for use in the calculation of the next higher polynomial degree term:

$$\begin{aligned} x \times x &= x^2 \\ x \times x^2 &= x^3 \\ x \times x^3 &= x^4 \ldots \end{aligned}$$

and so on; this approach requires only six multiplications to calculate the powers, plus a further seven multiplications for the products of the powers and coefficients. We still have the seven additions. This is the basis of Horner's algorithm[1] for calculating polynomials:

$$p_7(x) = ((((((a_7x+a_6)\times x+a_5)\times x+a_4)\times x+a_3)\times x+a_2)\times x+a_1)\times x+a_0 \quad (7.7)$$

which requires seven multiply and add operations. We express the arithmetic calculation in this way because a 'multiply and add' or MAC operation can be accomplished in a single (specific) gate instance in an FPGA, so the operation is performed in a single clock cycle, which makes it very efficient. And arithmetic efficiency is crucial here: the pre-distorter runs continuously and the function is applied to every sample of the transmitted signal. For example, let's consider a memory polynomial pre-distorter with a seventh-degree polynomial and three memory taps. Using Horner's algorithm, that requires seven times four MACs, or 28 MACs per sample. If the signal is being sampled at 245.76 MSa/s, this requires about 6.9 GigaMACs per second. But these are complex data samples that we are using, so these are complex GMACs/s: since each complex multiply is four real multiplies and two real additions, then the processing power that we need to build this pre-distorter is about 28 giga-operations per second. This is quite a lot of processing power required.

An alternative implementation is to use look-up tables to store the pre-calculated values of the six powers of the signal over the range of signal levels expected, each multiplied by the appropriate coefficient. We would then have six

1 sometimes erroneously called "Homer's algorithm," although I think Homer was more famous for his story-telling than his mathematical achievements. Provided that's the Homer I am thinking of...

LUTs for a_7x^7 through a_2x^2, indexed by $|x|$ over the expected input power range. Then at each sample, we use the sampled value $|x|$ to index the six tables, and then perform a single multiply and add operation for $a_1x + a_0$, and seven additions for the instantaneous signal and the three memory taps: four MACs and 28 additions per sample. Given that addition is about four times less expensive to compute than multiplication, the use of LUTs in this fashion saves about 18 MACs per sample, reducing the computing power required in this example to about 10 giga-operations, at the expense of some memory for the LUTs.

7.3 ADAPTIVE DIGITAL PRE-DISTORTION

Reviewing the block diagram of the adaptive digital pre-distortion system illustrated in Figure 7.6, we can see that this is a classic digital adaptive control problem, as shown in Figure 7.13. In the digital control problem, we adjust the signal samples $x[k]$ in the digital domain to control the plant output $y(t)$ in the analog domain; the output from the plant is monitored by an observation path, and converted to the digital domain, where the input and feedback signals are compared and the control algorithm is implemented. We follow the same basic procedure for DPD [7]. The objective in adaptive DPD is to adjust the digital modulating signal in such a way as to minimize the distortion produced by the RF power amplifier.

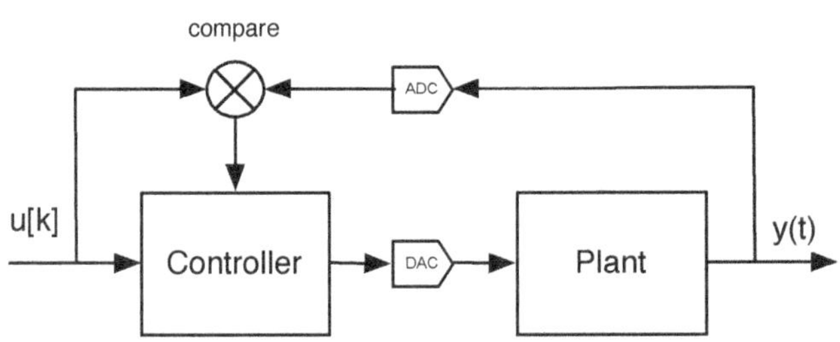

Figure 7.13 Block diagram of a digital adaptive control system.

This digital adaptive control system can be implemented in two simple ways. In the first approach, the controller is inside the feedback loop, as shown in Figure 7.13, and we compare the input signal with the output from the plant. This classic approach is known as model reference adaptive control (MRAC). Alternatively, we can place the controller outside the feedback loop, and we compare the output signal from the controller with the output signal from the plant to generate the control signal. This approach is known as the self-tuning regulator (STR), [3], and is considered to be an open loop system as the controller is not inside the feedback loop. Both of these techniques have been applied to the problem of power amplifier linearization using digital pre-distortion.

7.3.1 'Indirect Learning' DPD System

The DPD system block diagram using the open-loop self-tuning regulator structure is shown in Figure 7.14. In this figure, we have abstracted the RF up- and down-converters and the ADC/DAC data converters into two conversion blocks, between the digital and RF signal domains. In this STR approach, we compare the pre-distorted input signal, $u'[k]$, and the PA observation signal, $y[k]$, for the DPD update. The input side signal, $u'[k]$, is delayed to accommodate the group delay of the PA, and any loop delays in the feedback loop. The alignment of the input and observation signals is crucial to obtaining accurate linearization. The PA output observation signal is attenuated by a factor that is equal to the average power gain of the PA, so that the input and output signal magnitudes are comparable. This is required for stability of the DPD loop.

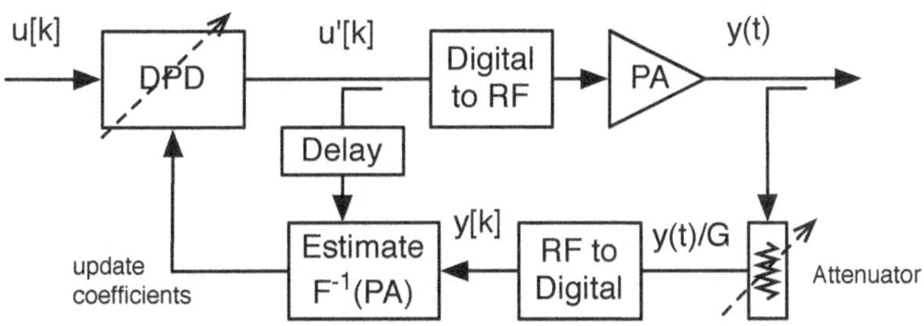

Figure 7.14 Block diagram of an open-loop adaptive DPD system.

The system identification and estimation functions are performed using the pre-distorted input signal, $u'[k]$ and the PA observation signal $y[k]$. These are the digital representations of the input and output signals of the PA, so essentially we are identifying the PA nonlinearity. We must then invert the nonlinear PA function to obtain the pre-distorter coefficients. The inversion calculation can be problematic, especially if the DPD correction includes memory effects. While the Volterra series, for example, has a formal inverse, it requires the calculation of a large number of coefficients, not all of which will be reliably identified by the PA output signal as it has limited dynamics. This can lead to inaccuracies in the inverse, and poor DPD correction.

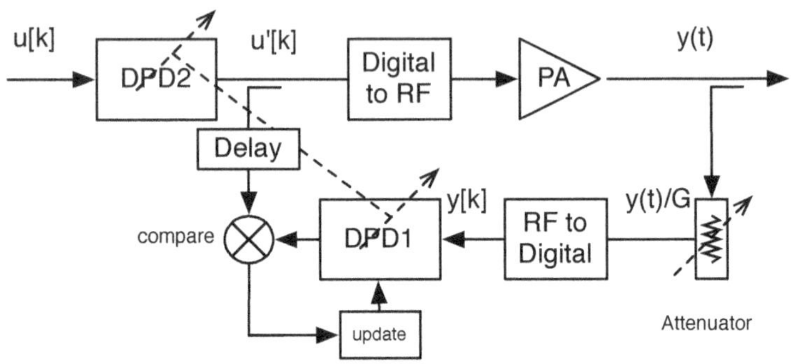

Figure 7.15 A modification of the open-loop pre-distorter uses two pre-distorters to compare similar signals. The error signal is used to minimize this difference by adjusting the pre-distorter parameters.

This problem can be avoided by using the signals $y[k]$ and $u'[k]$ to identify the inverse PA model directly. We recast the STR DPD system as shown in Figure 7.15. Here we use two pre-distorter blocks: the first is in the usual place, in the forward signal path; the second pre-distorter acts on the feedback signal, essentially acting as a post-distorter on the PA output. For some basis functions for describing weakly nonlinear systems with memory, such as polynomial or Volterra Series, it can be shown that pre-distortion of the nonlinear system is functionally the same as post-distortion, provided that the system does not run into limits or saturation. We can then compare the pre-distorted signal with the post-distorted signal to update the DPD coefficients. We can also interpret the operation of this DPD system in the following way: since the DPD block, once converged, is effectively the inverse of the PA, then the block DPD1 is identifying the inverse of the PA directly, as required by the STR. Indirect learning is relatively straightforward to implement, although at the cost of two pre-distorters.

This approach to PA pre-distortion is called indirect learning, and is one of several techniques for controller learning; see, for example, the review by Psaltis *et al.* [8]. Indirect learning was first applied to PA linearization by Eun and Powers [9], who used a Volterra basis for the pre-distorter, and a recursive least-squares adaption method to linearize a traveling-wave tube amplifier with 16-PSK and 16-QAM digital communications signals. An early application of indirect learning to the linearization of solid state power amplifiers is presented by Ding *et al.* [10,11],using a memory polynomial pre-distorter and adaption using least-squares techniques with the Normal equations (see Section 6.5.3).

The pre-distorter coefficient adaption is quite straightforward for indirect learning. It is outlined below, assuming we are using nonlinear basis functions for the pre-distorter. We compare the pre-distorted input signal with the post-distorted output from the PA. The error ϵ at any sample instant k is then

$$\varepsilon = \sum_{n=1}^{N} a_n \varphi_n \left(y[k]\right) - u'[k] \tag{7.8}$$

We then estimate the coefficient errors $\Delta\hat{a}$ by minimizing the difference between the error term and post-distortion update, in a least squares fashion:

$$J = \min \sum_{k} \left| \varepsilon - \sum_{n=1}^{N} \Delta a_n \varphi_n \left(y[k]\right) \right|^2 \tag{7.9}$$

In block matrix form, this can be expressed in the following coefficient update relation, starting from the least-squares (Normal equations) expression describing the coefficient estimation from the output y and the pre-distorted input u'

$$\mathbf{u}_i{}' = \mathbf{Y}.a$$

$$\hat{a}_i = \left(\mathbf{Y}_i^H \mathbf{Y}_i\right)^{-1} \mathbf{Y}_i^H.\mathbf{u}_i{}' \tag{7.10}$$

Then the error term for a given estimation block uses the previously-calculated coefficients, and is written

$$\varepsilon = \mathbf{Y}_{i+1}.\hat{a}_i - \mathbf{u}_{i+1}{}' \tag{7.11}$$

which can be expanded through

$$\mathbf{Y}_{i+1}^H \varepsilon = \mathbf{Y}_{i+1}^H \mathbf{Y}_{i+1}.\hat{a}_i - \mathbf{Y}_{i+1}^H.\mathbf{u}_{i+1}{}' \tag{7.12}$$

$$\left(\mathbf{Y}_{i+1}^H \mathbf{Y}_{i+1}\right)^{-1} \mathbf{Y}_{i+1}^H \varepsilon = \hat{a}_i - \left(\mathbf{Y}_{i+1}^H \mathbf{Y}_{i+1}\right)^{-1} \mathbf{Y}_{i+1}^H \mathbf{u}_{i+1}{}' \tag{7.13}$$

$$\left(\mathbf{Y}_{i+1}^H \mathbf{Y}_{i+1}\right)^{-1} \mathbf{Y}_{i+1}^H \varepsilon = \hat{a}_i - \hat{a}_{i+1} = -\Delta a \tag{7.14}$$

Hence,

$$\varepsilon = -\mathbf{Y}_{i+1}.\Delta a \tag{7.15}$$

yields the error expression for the pre-distorter coefficients. We can now use this coefficient error to update the pre-distorter coefficients through:

$$\hat{a}_{i+1} = \hat{a}_i - \mu.\Delta a \tag{7.16}$$

where the parameter μ is used to speed up or stabilize the convergence of the pre-distorter.

7.3.2 Model Reference DPD System

The closed-loop adaptive control model reference technique is well suited to memory correction in real-time DPD systems. It is shown in Figure 7.16. The control loop is an integrator function that minimizes errors in the DPD function coefficients through the comparison of $u[k]$ and $y[k]$. Again, the input signal $u[k]$ is delayed to allow the input and output vectors to be aligned. The error ϵ in the model reference system is simply the difference between the (scaled) output and the input signals

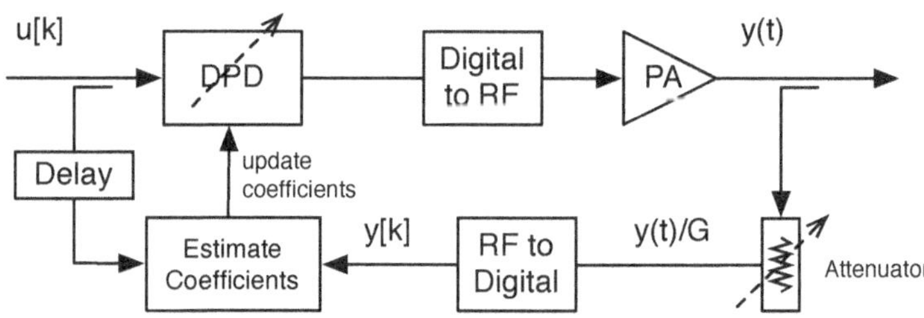

Figure 7.16 Block diagram of a closed-loop adaptive DPD system.

$$\varepsilon = y[k] - u[k] \tag{7.17}$$

and the coefficient errors $\Delta\hat{a}$ are estimated by minimizing the difference between the error and the pre-distortion update, again in a least squares manner:

$$J = \min \sum_{k} \left| \varepsilon - \sum_{n=1}^{N} \Delta a_n \varphi_n \left(x[k] \right) \right|^2 \tag{7.18}$$

For any iteration of the system, the block matrix and vector expression that we are trying to force is

$$\mathbf{y} - \mathbf{u} = \mathbf{U}.\Delta a \tag{7.19}$$

hence, the required change in the pre-distorter coefficients is given by

$$\Delta a = \left(\mathbf{U}^H\mathbf{U}\right)^{-1}\mathbf{U}^H.(\mathbf{y} - \mathbf{u}) \tag{7.20}$$

and the coefficient update relation is then the same as for indirect learning, (7.16):

$$\hat{a}_{i+1} = \hat{a}_i - \mu.\Delta a \tag{7.21}$$

where again μ controls the speed of convergence or stability of the iteration process.

This method needs a starting point for the DPD coefficients, although with care a null set can be used. Alternatively, we can use a model of the PA in a parallel path to provide additional data for the coefficient estimator. This is is illustrated schematically in Figure 7.17.

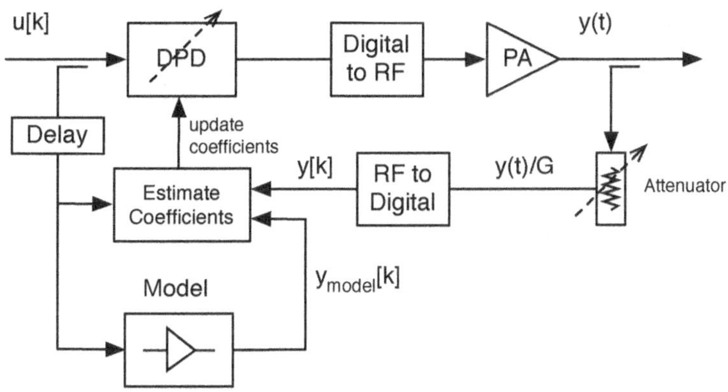

Figure 7.17 Block diagram of model reference control (MRC) system applied to an adaptive DPD system.

7.3.3 DPD Coefficient Adaption Techniques

For both indirect learning and model reference adaptive control, the coefficient identification process is usually carried out by comparing blocks of input and output samples, with block sizes of 2–32k samples being typical, depending on sample rate and signal bandwidth. These data are accumulated in temporary buffer registers. In practical DPD systems, the number of coefficients may range typically from 25 to over 100, depending on factors such as signal bandwidth, the type of PA used, complexity of the modulation signal, and the ultimate correction required: GSM being the most demanding modulation in this respect. The fact that we collect a relatively large number of samples to identify a much smaller number of coefficients for the DPD algorithm means that we have an over-determined problem, and the use of least squares and similar techniques for finding the coefficient values is commonplace.

7.3.3.1 Least Squares Estimation

The least squares algorithm was described briefly in Section 3.4.2. This technique minimizes the absolute distance between the model prediction and the measured data – the error – averaged over the whole data set. In other words, it obtains the best solution for the given data and model. For the regression vector given by the model, $\phi(t)$, and the measured data, y, collected over N samples, the estimate for the model coefficients $\hat{\theta}$ is given by

$$\hat{\theta} = \frac{\sum_{t=1}^{N} \phi(t) y(t)}{\sum_{t=1}^{N} \phi(t) \phi^T(t)} \tag{7.22}$$

Writing the least squares in block matrix form yields the Normal equations that we introduced in Chapter 6, with various forms of their direct solution:

$$\hat{a} = \left(\mathbf{U}^H \mathbf{U}\right)^{-1} \mathbf{U}^H . \mathbf{y} \tag{7.23}$$

where we have made some redefinition of the variables to be consistent with Chapters 6 and 7: $\mathbf{U}$ is the regression matrix, and $\hat{a}$ is the coefficient estimate. Details of the convergence of the coefficient estimate for DPD adaption have been outlined in the previous sections of this chapter.

7.3.3.2 Least Mean Squares

The least mean squares (LMS) algorithm was developed by Widrow and Hoff in 1960 [12], and has since become very popular as an error minimization technique in a wide variety of applications. It is a relatively simple algorithm, easy to implement, and is very robust, delivering consistent and reliable results.

LMS is a stochastic gradient algorithm, which works by finding the gradient of the means square error to update the coefficients at each step. The LMS algorithm does not require the autocorrelation of the DPD regression matrix $\mathbf{U}$, nor the cross-correlation of the regression matrix with the desired signal $\mathbf{y}$, to do this. The expectation operator of stochastic gradient method is not calculated, replaced by an instantaneous estimate. This results in a formulation of low computational complexity, avoiding the matrix inversion of the least squares technique, and the associated errors from ill-conditioning that often arise, which contributes to the robustness of this method.

The LMS algorithm update for the coefficients θ_n is

$$\theta_{n+1} = \theta_n + \mu\varepsilon(n)\varphi_n(t) \tag{7.24}$$

where $\varphi_n(t)$ is the regression vector, and $\varepsilon(n)$ is the error or cost function term at the nth iteration, generally given by the difference between the model output y and the desired output d

$$\varepsilon(n) = d_n(t) - y_n(t) \tag{7.25}$$

In DPD the desired output from the PA is a scaled version of the input signal $u[k]$ at baseband, and so for the model reference DPD controller the error term at the nth iteration is

$$\varepsilon_n = \mathbf{u_n}[k] - \mathbf{y_n}[k] \tag{7.26}$$

and the regression matrix is $\mathbf{U_{n-1}}$; hence, in matrix form

$$\hat{a}_{n+1} = \hat{a}_n + \mu\varepsilon_n\mathbf{U}_{n-1} \tag{7.27}$$

The LMS algorithm can be applied on a sample by sample basis, or in block form as shown above. The parameter μ is a step-size parameter that is used to control the convergence rate and stability of the algorithm. Nevertheless, the coefficient values from step to step may have large variations due to measurement noise. The convergence of the LMS algorithm can be slow, depending on the spread of the eigenvalues of the coefficient correlation matrix, which can be large in DPD systems.

7.3.3.3 Recursive Least Squares

Recursive least squares (RLS) is a least squares formulation that avoids the inversion of the DPD autocorrelation matrix $\mathbf{U}^H\mathbf{U}$, and so like LMS avoids the inaccuracies associated with such an action.

Using indirect learning as an example, the error to be minimized is

$$\varepsilon_n = {u_n}' - y_n.\hat{a}_n \tag{7.28}$$

which is also known as the *innovation*. From the least squares formulation in (7.22), the coefficients are given by $\hat{a}_n$. We would like to calculate the next approximation to the parameters from the information we already have, that is, determine $\hat{a}_n$ from $\hat{a}_{n-1}$. Writing

$$\begin{aligned} \Phi_n &= \sum_{i=1}^{n} u'[i]u'^T[i] = \Phi_{n-1} + u'[n]u'^T[n] \\ \psi_n &= \sum_{i=1}^{n} u'[i]y[i] = \psi_{n-1} + u'[n]y[n] \end{aligned} \tag{7.29}$$

for the autocorrelation of the DPD regression matrix, and the cross-correlation of the regression matrix with the desired signal, respectively, expressed as matrices that are derived from the old value and the newly-determined data, then we can derive the expression for the update of the coefficients:

$$\hat{a}_n = \frac{\psi_n}{\Phi_n} = \frac{\psi_n}{\Phi_{n-1} + u'[n]u'^T[n]} \tag{7.30}$$

The RLS technique avoids the inversion of the autocorrelation matrix by making use of the Woodbury inversion lemma, which states that if a positive-definite matrix $\mathbf{A}$ can be expressed in the form:

$$\mathbf{A} = \mathbf{B}^{-1} + \mathbf{C}\mathbf{D}^{-1}\mathbf{C}^T \tag{7.31}$$

where $\mathbf{B}, \mathbf{C}, \mathbf{D}$ are also positive definite, then the inverse of $\mathbf{A}$ is given by

$$\mathbf{A}^{-1} = \mathbf{B} - \mathbf{B}\mathbf{C}(\mathbf{D} + \mathbf{C}^T\mathbf{B}\mathbf{C})^{-1}\mathbf{C}^T\mathbf{B} \tag{7.32}$$

While this does not immediately look like we have made any simplification, we can now use (7.31) to represent Φ_n in (7.29), where $\mathbf{B}$ is Φ_{n-1}, $\mathbf{C}$ is $u'[n]$, and $\mathbf{D}$ is the appropriately-sized identity matrix. Then the inverse, ${\Phi_n}^{-1}$, is written as

$$\Phi_n^{-1} = \Phi_{n-1}^{-1} - \frac{\Phi_{n-1}^{-1}u'[n]u'^T[n]\Phi_{n-1}^{-1}}{(1 + u'^T[n]\Phi_{n-1}^{-1}u'[n])} \tag{7.33}$$

Usually, this form is simplified by writing

$$\mathbf{P}_n = \Phi_n^{-1}$$

for the coefficient matrix, and introducing a *gain* term $\mathbf{G_n}$

$$\mathbf{G}_n = \frac{\mathbf{P}_{n-1}u'[n]}{(1 + u'^T[n]\mathbf{P}_{n-1}u'[n])} \tag{7.34}$$

Then the expression for the new coefficients $\hat{a}_n$ can be written in terms of the previous coefficient set using $\mathbf{P}_n$, $\mathbf{G}_n$, and equations (7.28) and (7.29)

$$\hat{a}_n = \hat{a}_{n-1} + \varepsilon_n \mathbf{G}_n \tag{7.35}$$

The coefficient matrix $\mathbf{P}$ needs to be initialized, and a typical value is $\delta\mathbf{I}$, a scaled version of the identity matrix, where δ may be in the range from 0.01 to 0.0001.

The RLS technique operates on a sample by sample basis in the data set, is very quick to run, avoiding any explicit matrix inversion, and converges very quickly to the desired target, typically faster than LS or LMS techniques. The RLS method is an example of a Kalman filter, hence the introduction of the gain term in the expression. More sophisticated Kalman filters could also be employed in DPD adaption.

7.4 DPD SYSTEM IMPLEMENTATION DETAILS

The inclusion of linearization by digital pre-distortion in a (base-station) transmitter is not limited to simply the implementation of the pre-distortion algorithm and adaption technique in the digital part of the system. The digital and RF sides of the transmitter have to be designed considering the DPD as an integral part of the system, as illustrated in Figure 7.6, for example. Here we will review some of the architectural and device choices available for both digital and RF domains.

7.4.1 RF Section

The main choice that the designer has to make is what IF architecture to use for the transmit path and for the observation receiver: real or complex IF? In Figure 7.18 we show typical base-station transmitter RF structures: for the transmit path, the usual choice nowadays is a complex IF; and for the observation receiver, we show both real-IF and complex-IF architectures. We shall describe these approaches briefly, below.

As stated, the complex IF architecture is the usual choice for the transmit path. The digital data comes in the form of the *I* (in-phase) and *Q* (quadrature) components of the modulation from the digital baseband system, and the IQ-modulator is a simple and inexpensive RFIC component that enables direct conversion from the digital *IQ* domain to the RF domain as a complex-modulated signal, using an external local oscillator signal to perform the up-conversion to RF. Advantages of

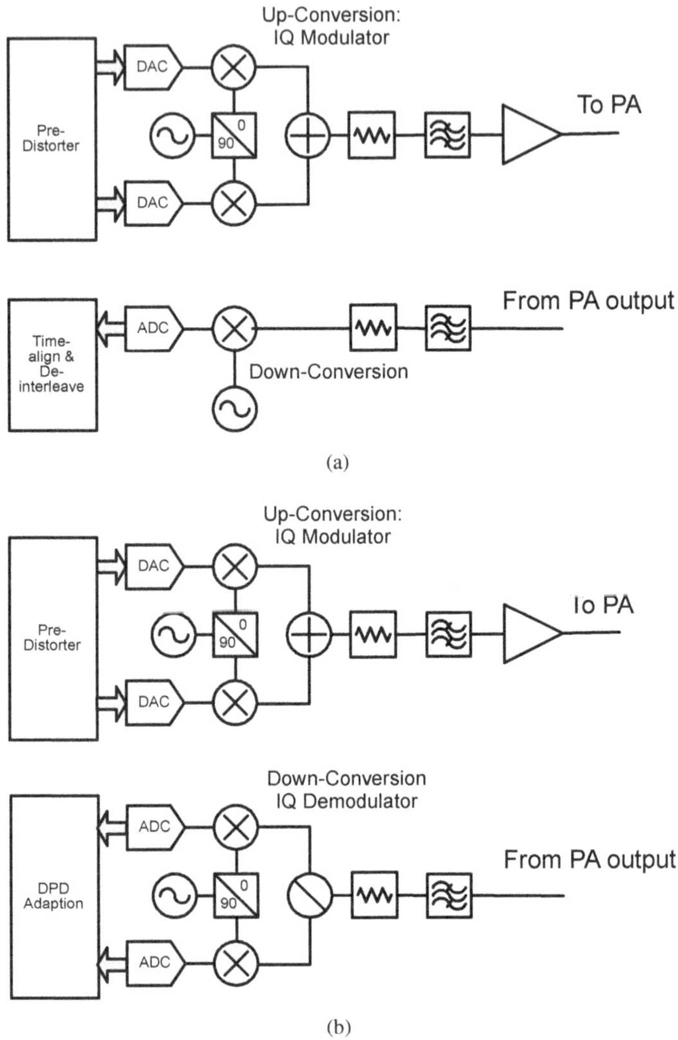

Figure 7.18 Schematics for (a) DPD-enabled transmitter with complex-IF transmit path and real-IF observation path receiver and (b) DPD-enabled transmitter with complex-IF for both transmit path and observation path receiver.

the complex architecture include filtering of the image, as shown in Figure 7.19 [13], and better LO suppression than the real IF architecture. The adoption of a complex transmit path is made easier by the availability of dual DACs in a single package, often at a comparable cost to a single DAC, and availability of the complex mixer or quadrature modulator in a single IC, simplifying the integration of the mixer, filters, and IF amplifier that are required in the real-IF system.

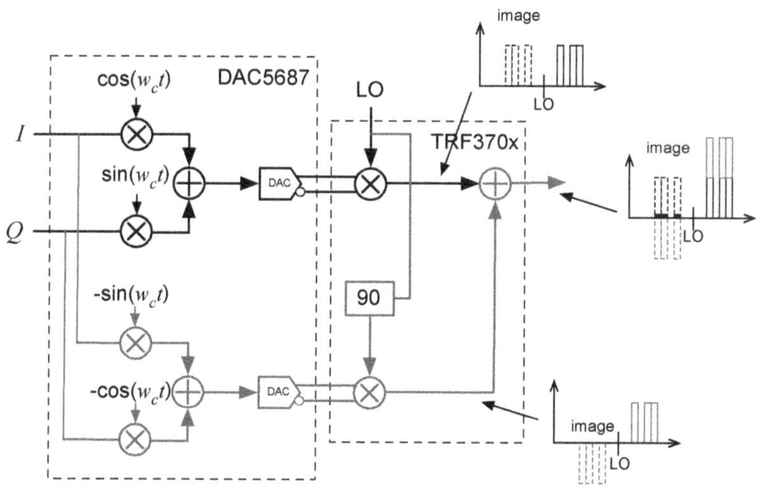

Figure 7.19 Comparison between complex and real (in gray) transmit path architectures, showing the improved image rejection resulting from the addition of the complex components of the signal in the quadrature modulator. (Adapted from R. Sperlich, Texas Instruments [13]; components shown in the dotted rectangles are TI part numbers. The inclusion of these specific TI part numbers does not constitute any specific recommendation.)

For the observation receiver, the real-IF architecture has traditionally been favored, as shown in Figure 7.18(a). This approach requires the use of only a single ADC. The same LO can be used for the transmit path and observation receiver if a high-IF is used for the complex transmitter, simplifying the architecture and resulting in a system that is relatively insensitive to phase noise variations in the LO. The IF must be high enough to ensure a single-conversion radio design; it can be carefully defined with external ceramic filters, and high-IP3 mixers ensure the feedback path for the observation receiver remains linear. A typical choice for the IF is in the 150–300 MHz range, though this precludes the use of commercially-available quadrature demodulators, requiring some system integration. The real-IF approach is a low-risk strategy. The ADC is generally the limiting factor in this architecture, in terms of noise and dynamic range (see Section 7.4.2). Because high-quality ADCs can be very expensive, this is a practical approach, minimizing the number required of this expensive component.

The complex-IF architecture for the observation receiver is also receiving some attention as an attractive alternative to the real-IF approach. In the complex structure, a quadrature mixer or demodulator is used to convert the modulated signal into its in-phase *I* and quadrature *Q* components directly, at the intermediate frequency, so that we have two identical paths for the *I* and *Q* components in the IF section of the receiver. These signals can be converted into digital form using a pair of ADCs. This is shown schematically in Figure 7.20.

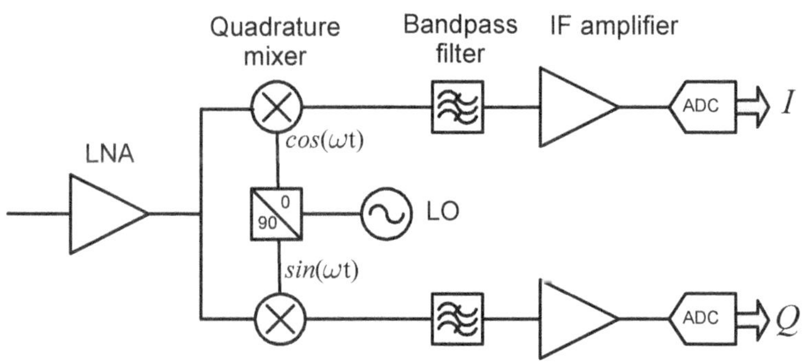

Figure 7.20 Schematic block diagram of a complex-IF receiver, showing the *I* and *Q* paths from the quadrature mixer to the ADCs.

This approach relaxes the demands on the analog-to-digital converters: each of the *I* and *Q* signals is half the bandwidth of the modulated RF signal, and so the sampling rate required at IF is half that of the real-IF receiver. The IF bandwidth can be controlled using ceramic filters. The drawbacks of this approach are that the receiver is a more complex (pun intended) and costly circuit, requiring two mixers and quadrature local oscillator (LO) signal, and two IF paths (Figure 7.20). There is also a potential performance impairment arising from the fact that we now have two independent *I* and *Q* paths in the receiver: *IQ* imbalance.

The *IQ* imbalance is due to differences in the gain and phase response of the two paths: slight differences in the mixers, the IF amplifiers, and the physical path lengths all contribute to this imbalance. The imbalance manifests itself as image signals in the reconstructed baseband modulated signal, that is, an image of the *I* signal appearing in the *Q* signal, and *vice versa*. For narrowband receivers, the *IQ* imbalance can be modeled as a constant gain and phase difference between the two paths, and the correction of this imbalance can be carried out in the analog domain, at the intermediate frequency, or digitally, post-ADC. A simple analog correction, due to Cavers [14], is shown in Figure 7.21. Digital techniques include decorrelation of the *I* and *Q* signals, to remove the image components.

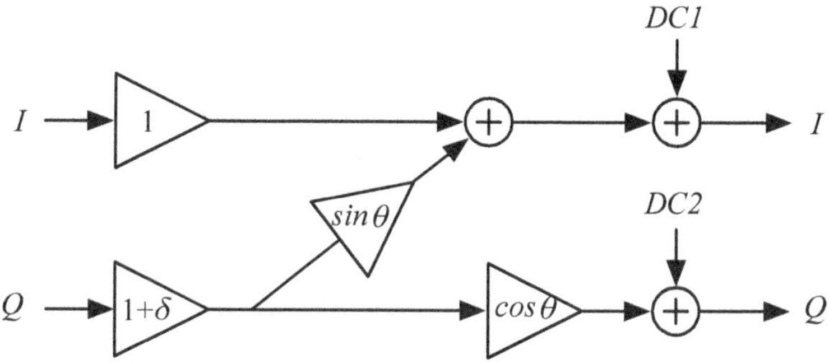

Figure 7.21 A simple analog correction for *IQ* imbalance in a complex-IF receiver. The difference in gains between the two paths is given by δ, and the difference in phase is θ.

For wide bandwidth signals, this simple approach is generally insufficient, as the *IQ*-imbalance is likely to be frequency-dependent. The correction is usually done in the digital domain, either independently or as a part of the DPD algorithm. Care needs to be taken to ensure that the correction of the observation receiver impairments is not included in the DPD correction of the PA impairments, otherwise we will be adding distortion to the PA output.

There is currently quite some interest in the zero-IF or homodyne approach to the complex-IF structure, for both up- and down-converters in the base-station. The attraction of the zero-IF technique is that the bulky ceramic filters used to define the IF bandwidth in the high complex-IF receiver can be replaced by simpler on-chip low-pass filters, enabling the realization of the low-power part of the transmitter, and the receiver, or even complete transceiver, as a monolithic IC, resulting in a "bits-to-RF" component.

The zero-IF approach has the same *IQ* imbalance issues as the complex-IF receiver, and corrections must again be applied. Further, there can also be a DC offset on the *I* and *Q* channels, arising from incomplete cancellation of the RF carrier by the LO, due to LO feedthrough. The DC offset can be removed by analog or digital techniques, as for *IQ* imbalance. The zero-IF structure is also prone to second-order distortions arising from the mixer, and other nonlinear components, resulting in unwanted second-order intermodulation (IM2) products close to DC in the desired signal bandwidth. Complex and real architectures avoid this near-DC problem by using a non-zero IF. The IM2 products arising from the observation receiver itself must be kept very low, to maintain the linearity of this component of the system. This requires very high linearity mixers.

7.4.2 Data Converters

The conversion of the signal between the digital domain and the RF/IF is one of the major components of the transmitter and the observation receiver. The performance of the DACs and ADCs can often be the limiting factor in the performance of the whole transmitter system. The major performance factors to consider are the spurious-free dynamic range (SFDR), which defines the linearity of the converter, and signal-to-noise ratio (SNR), both of which are related to the number of bits in the converter, and the sample rate, which defines the converter bandwidth.

The GSM/EDGE modulation signal has the most stringent requirements on the linearity. The GSM/EDGE transmit mask shown in Figure 7.22 indicates the maximum allowed levels of noise and distortion as a function of the frequency offset from any GSM/EDGE carrier. A maximum channel bandwidth for the (multi-carrier) signal of 10 MHz is considered. Close-in limits below 600 kHz define limits for the distortion of the carrier. For multi-carrier signals, the minimum carrier spacing is 600 kHz, and the limits for frequency offsets beyond 600 kHz indicate the maximum levels allowed for noise and the intermodulation products. The level of −70 dBc is a very tough specification to meet.

For the transmission path, the SFDR for the DAC needs to be better than −80 dBc at frequency offsets greater than 6 MHz; usually a guard of 1 bit (6 dB) is added to this, giving a device specification of at least −86 dBc SFDR. The worst-case SNR is −80 dBc; at offset frequencies greater than 1.8 MHz, the integration bandwidth is 100 kHz, so this yields a required SNR better than −130 dBFS/Hz, to which we may add 6–10 dB as margin. The intermodulation performance needs to be better than −70 dBc, and again a 6-dB margin is usually added. We should also remember that the DAC is operating on the pre-distorted signal, which is expanded in PAPR, and so some headroom should be added to accommodate this gain expansion: in practical DPD systems, this gain expansion is often limited to a maximum of 3 dB, to prevent destructive overdrive of the PA. Additionally, we should accommodate some further headroom to account for thermal effects in the analog up-converter chain: a 3 dB margin is generally sufficient. These two considerations yield a minimum 6 dB margin for the DAC. From the above, we can see that a DAC with 14-bit resolution is marginal, and 16-bit DACs are usually found in base-station transmitters.

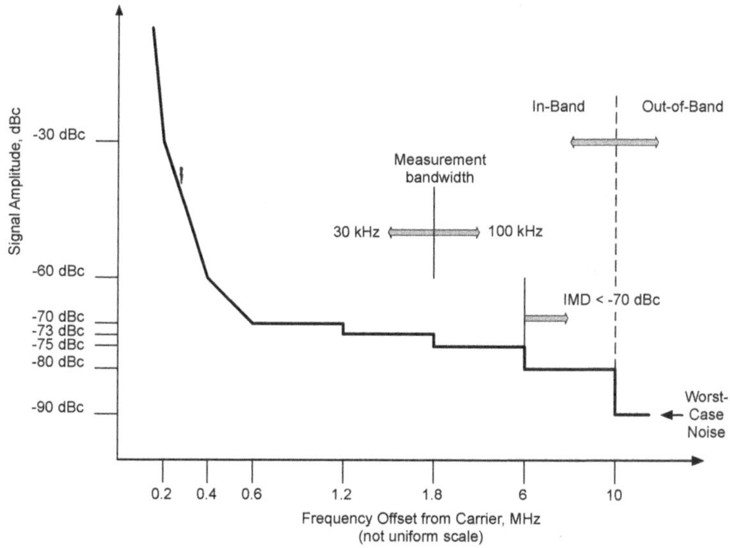

Figure 7.22 The GSM/EDGE transmit mask, indicating the maximum allowable levels of distortion and noise.

These specifications are cast in the definition for a single-carrier GSM/EDGE signal; multi-carrier signals will require further backing off from the full-scale values to accommodate the peak-to-average power ratio of the signals, making the DAC performance requirements more stringent.

The DAC sample rate is determined by the oversampling of the pre-distorted signal that is required to avoid aliasing: a rule of thumb of five times the signal bandwidth is a good guide. For single-carrier LTE, or four-carrier WCDMA, the signal bandwidth is 20 MHz, indicating that the pre-distorter bandwidth should be 100 MHz, and a DAC sampling rate of 200 MHz would be typical for this application.

In the observation path, linearity is the paramount specification for the ADC. The SFDR and SNR requirements are similar to those for the DAC: the corrected IM3 products must be below −70 dBc, and the SFDR and SNR should be at least 10 dB better than this. The linearity and noise performance of the ADC is captured in a specification called effective number of bits or ENOB: this is less than the physical number of bits and accounts for the degradations due to noise and distortion. Because typically the DPD system captures a block of data for the

DPD adaption calculation, we can get some processing gain, and the observation path ADC performance need not be required to meet the instantaneous linearity requirements. To meet a distortion specification of -70 dBc with 6–10 dB margin, we would need a 14-bit ADC performance. Typically, we can use a 12-bit ADC with an ENOB of 10–10.5 dB to obtain successful correction of a GSM/EDGE signal in the transmitter. Again, as we progress to multi-carrier scenarios, the additional PAPR required may require us to demand better performance from the ADC.

The ADC sample rate must be high enough to capture all the required intermodulation products; we usually apply the "five times" rule. The sample rate will depend on the signal bandwidth and order of correction required. For the LTE and WCDMA signals mentioned earlier, the ADC sample rate would be of the order of 200 MHz for a real IF receiver; a complex architecture will permit this to be reduced, and a lower rate of 100 MHz could be used for each of the I and Q ADCs.

7.4.3 Baseband Digital Signal Processing

On the digital side of the data converters is where we carry out the pre-distortion, adaption, and other signal processing activities. These algorithms can be programmed into an FPGA, which is the usual target for development purposes, or a custom ASIC which can offer higher performance, albeit at higher development cost. The pre-distortion correction demands the highest data throughput, and hence the greatest number of numerical operations per second. This is usually done in fixed-point arithmetic in the FPGA or ASIC, from considerations of speed and the size of the digital primitives; 16-bit data path width is typical, and matches the data converters. The DPD adaption requires the calculation of components of the algorithm that may span several orders of magnitude, particularly for polynomial or Volterra-based pre-distorters. This requires the use of floating-point arithmetic, which is usually carried out in a dedicated DSP or CPU chip, though modern FPGAs can include a floating-point CPU, enabling a single-chip solution.

Both FPGA and ASIC approaches enable the use of dedicated hardware components within the chip to accelerate the calculations, including the LUTs for pre-calculation of the polynomial bases, FIR filters for the memory-delay combination of the components of the pre-distorter, and so on.

The data connection between the data converters, which are usually found in the RF domain, and the digital baseband requires the transfer of large amounts of data. For the relatively narrowband signals mentioned in this section, and for a single transmitter and observation receiver, a parallel connection using LVDS is generally used. This provides sufficient bandwidth and resolution. For more integrated transmitter systems and higher bandwidth signals, a serial connection protocol is being implemented: this is the JESD204 interface specification, currently in "B" release. This can provide up to 12 Gsamples per second data transfer rate, and is flexible in terms of data width.

7.4.4 Now You Have Your DPD Evaluation Module...

Perhaps you are not going to build a DPD system from the ground up, developing your own linearization and adaption algorithms, programming the ASIC or FPGA, and building the RF transceiver[2], but you want to use a DPD system to evaluate other parts of the transmitter chain, such as the PA or other transceiver hardware. You are probably going to use an evaluation module from one of the third-party DPD vendors. These vendors currently include Texas Instruments and Broadcom (formerly Optichron), which provide evaluation boards for their ASIC products, and Xilinx and Alterra who can provide evaluation boards for their FPGAs, supplied with their in-house DPD software. These evaluation boards or modules (EVM) will allow you to connect a PA line-up for evaluation of the linearity under DPD correction.

In Figure 7.23 we show a typical bench setup for such an evaluation system. The 'DPD Evaluation Module' shown in the figure is representative of a complete module, such as provided by TI or Broadcom. An FPGA-based system would not include the data converters on the entirely digital evaluation board, and additional DAC and ADC boards are needed: evaluation boards for data converters are available from TI and Analog Devices (ADI). The rest of the hardware is analog and RF.

The output signal from the DPD EVM is of the order of 0 dBm; the PA gain is typically 15–20 dB, and so to achieve the required PA output power levels, a high gain driver chain (40 dB) with an output power of around 10–20 watts is needed. The attenuator and bandpass filter at the output of the EVM are used to control the signal level to the PA, enabling a controlled drive-up and drive-down of power, and to prevent any image signal from being fed to the PA. An isolator is usually included between the driver and PA, to provide a consistent input impedance to the PA to prevent instability, and to prevent any reflected signals from the PA appearing at the driver output, causing additional nonlinearities to be generated. The input and output power at the PA are measured to monitor gain, output power, and compression of the PA. The high output power from the PA needs to be dissipated in a power attenuator, before feeding to the measurement instrumentation, where we would monitor the ACP, EVM, and so forth, to measure how effective the DPD is.

2 but thanks for buying the book anyway.

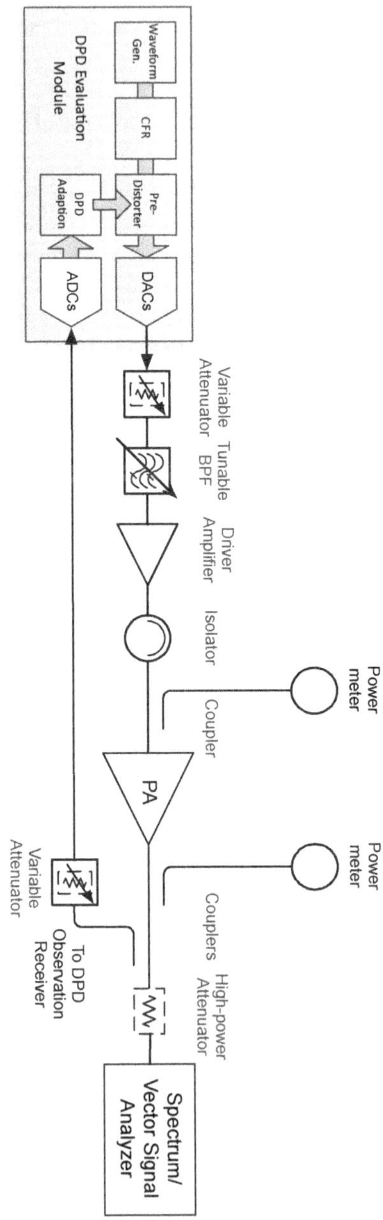

Figure 7.23 Typical DPD bench test system layout.

The receive path is taken via a high-power coupler directly at the output of the PA. This take-off point is recommended over using a point after the high-power attenuator. The modulated energy dissipated in the attenuator can cause memory effect contributions to the observation signal, resulting in incorrect linearization signal being applied to the PA. The variable attenuator in the observation path is used to adjust the signal level to the 'sweet spot' of the receiver, for best linear performance and dynamic range.

There are some practical considerations to think about when building or using such a DPD bench, in order to extract the best performance. The RF section of the DPD receiver is operating on signals that may be around 0 to −10 dBm in power level. The PA will be generating up to 50 dBm of average power, for a macro base-station application, and even though the complete system is connectorized, there will be some radiation from the PA circuit board that can find its way back to the sensitive receiver. Screening of either the PA or the DPD evaluation module, or its receiver at least, is highly recommended. If you are working in a lab where other high-power PAs are being tested, screening of the receiver is essential: a screened room may be a practical option.

The PA needs bias and supply connections. Using simple cables in an RF-rich environment means that these connections behave like antennas for noise and spurious signals, adding unwanted distortion and memory effects into your DPD system that may not be correctable. The use of screened cables for these connections is highly recommended. Power amplifiers, no matter how efficient they are, will generate a lot of heat. Use a good heatsink, perhaps with water cooling, to maintain the PA at a constant ambient temperature, and provide a stable environment for the characterization of the PA's memory effects by the pre-distorter. The driver may also need to be temperature controlled, too.

Modern DPD system implementations are becoming increasingly complex, with memory-capable algorithms such as Volterra basis now becoming more widely adopted. For wide bandwidth signals, this will increase the number of memory components in the expression and result in more complex calculations required for the pre-distorter. Direct adaption methods are being employed for the over-determined solution, for simplicity in implementation, and sample-by-sample adaption techniques such as LMS and RLS are employed to speed up the DPD convergence. While fixed-point arithmetic is still used for speed, larger bit-widths are being developed for improved accuracy.

7.5 ADVANCED TOPICS IN DPD

What we require from the pre-distorter is largely governed by the nature of the communications signal that we are using and the source of the nonlinearity: the PA. At the moment, the PA architecture of choice for macro base-station deployments, of 40–100 W average power output, is the Doherty amplifier. The most demanding signals that this amplifier might have to deal with is the 20 MHz bandwidth single-carrier LTE signal, or 4-carrier WCDMA, as mentioned earlier, in terms of bandwidth, or a multi-carrier EDGE signal for linearity. The pre-distorted signal bandwidth is therefore around 100 MHz, using the"five times" rule-of-thumb, so the Doherty PA needs to be able to handle this signal bandwidth, as does the observation receiver.

With the ratification of LTE Release 10 (also known as LTE-Advanced) in 2012, and the continued development (or evolution?) towards LTE Releases 11 and 12, there are some significant new challenges for linearization using DPD. The most immediate consequence of these new standards is the increased bandwidth permitted. In LTE-10, the system operator is permitted to concatenate up to five contiguous channels of the LTE signal, known as carrier aggregation, effectively to make a single very wide bandwidth signal. The most demanding example is using five 20-MHz LTE signals to give 100 MHz of continuous signal bandwidth. Using the "five times" rule, this would mean that we would need an instantaneous bandwidth of 500 MHz in both PA and DPD observation receiver to accommodate the distortion products. In LTE-11, the signal need not be contiguous in the 100 MHz bandwidth. This does, of course, still require us to keep those unoccupied parts of this spectrum free from distortion products. Examples of the signal spectra are shown in Figure 7.24.

There are a number of possible solutions to this problem that are currently being studied. The most obvious approach is to develop more capable RF and data-converter components to deal with the escalation in bandwidth, along with a move toward complex-IF and zero-IF architectures for the observation receiver, to ameliorate the demands on ADC development. Nevertheless, ADCs with up to 1 GHz sample rates are currently in development. The wide bandwidths and high sample rates lead to higher clock rates in the digital part of the system, increasing the demands on the FPGA and DSP/CPU engines.

Alternative techniques include performing the pre-distortion on a reduced bandwidth signal. This requires that both the pre-distorter and the observation path limit the signal to the same bandwidth. It has been shown that successful linearization can be achieved for a 12-carrier WCDMA signal of 60 MHz bandwidth using only 80 MHz of observed and pre-distorter signal bandwidth [15]. Outside of

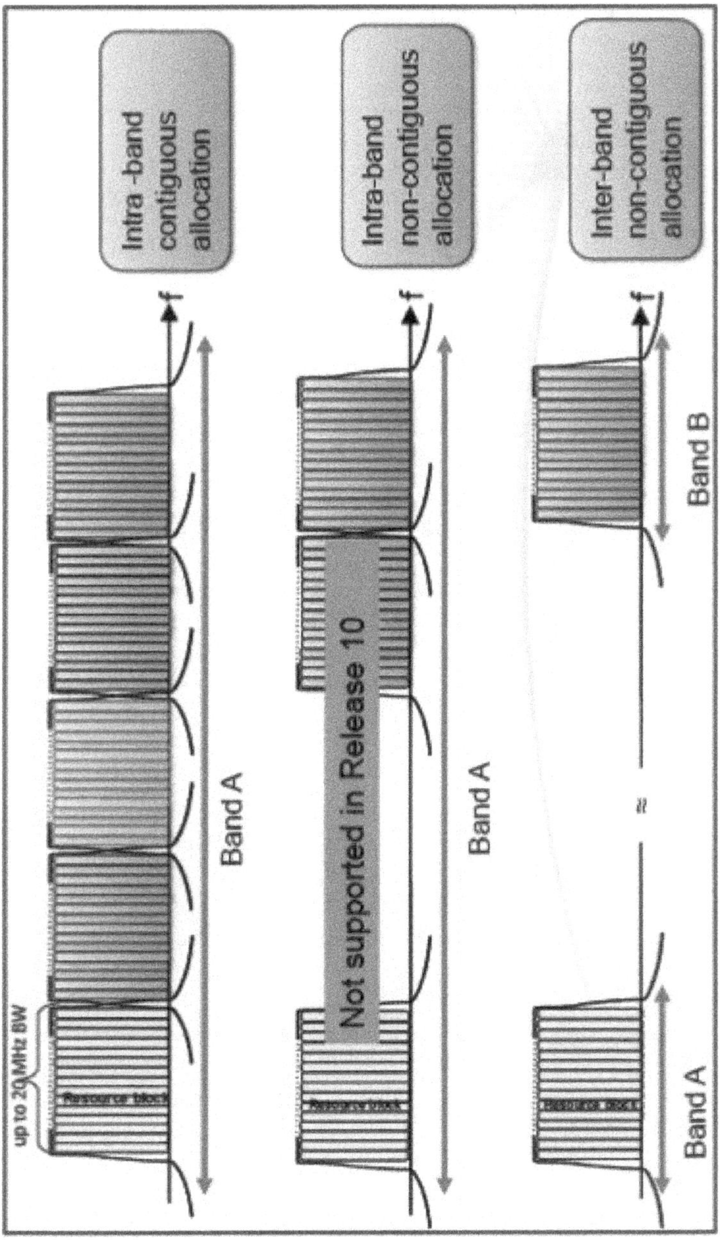

Figure 7.24 Carrier aggregation scenarios permitted in LTE Release 10. (Courtesy Agilent Technologies, Inc.)

the 80 MHz window, the distortion rises significantly, but in a practical base-station transmitter, this distortion signal would be eliminated by the antenna filter at the output of the PA.

Another technique to reduce the bandwidth and sample rate demands, particularly for the observation receiver, is to sub-sample the PA output signal. It has been shown that it is possible to extract the useful information from the signal by sampling at the Nyquist rate for the input signal, not at the Nyquist rate for the bandwidth-expanded pre-distorted signal [16, 17]. This has not received much attention recently, possibly because many of the recently published works on novel DPD approaches use the Indirect Learning architecture, where it is not possible to subsample the PA output and compare with the input signal.

A particularly promising approach to reduced sample rate observation is demonstrated by Hammler *et al.* [18], based on the finite rate of innovation (FRI) principle [17, 19]. Here, a frequency-domain identification of the signal is carried out, using a circular convolution that takes advantage of the cyclic prefix in the LTE signal. The resulting method has shown a potential for a thousandfold reduction in the ADC sample rate. The FRI principle is based on the observation that the data changes only slowly compared to the large oversample rate of the signal, and hence a reduced sample rate, well below the Nyquist rate for the total signal bandwidth, can still extract the useful information from the data.

In addition to the increased signal bandwidth, there is a requirement to be able to linearize multi-protocol signals, by which I mean combinations of LTE signals with multi-carrier WCDMA or EDGE signals, often in a wide total bandwidth but with some parts of the band unoccupied. A typical specification could be LTE with WCDMA or EDGE occupying a total bandwidth of 100 MHz, but with only 40–60 MHz of the total bandwidth occupied by information signal. An example is shown in Figure 7.25, where the signal spectrum comprises an LTE signal at the high end of the spectrum, and multi-carrier CDMA (IS-95) signal at the low end of the allowed spectrum. Other system operators have signals in the intervening space. The two sets of signals must each be linearized to meet their own adjacent channel leakage and EVM specifications, and also any intermodulation between these two sets of signals needs to be reduced.

The Doherty PA is also more nonlinear than a simple Class-AB PA: the main amplifier (Class-AB) is being operated at close to peak power output for most of the time, in its nonlinear regime; the peaking amplifier is usually a Class-C PA, which generates a large number of harmonic and therefore IM products. This combination in the Doherty PA generally requires that the polynomial or Volterra-based predistorter has 11th- or even 13th-degree nonlinearity. This combination of high nonlinear order and wide signal or DPD bandwidth means that a fairly sophisticated

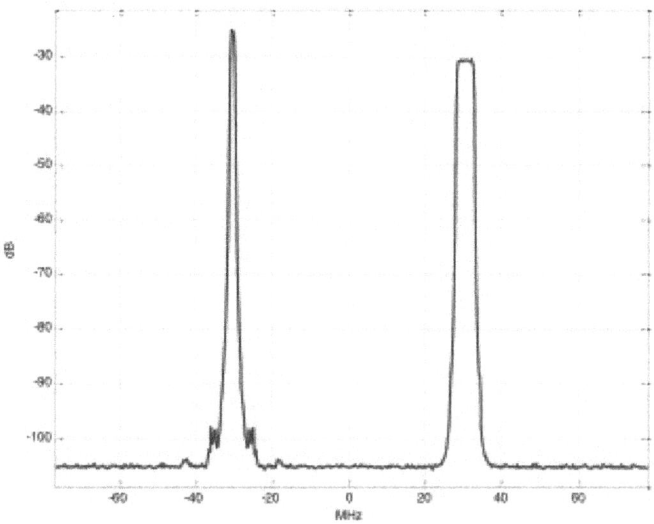

Figure 7.25 Available spectrum showing signals at each end of the allowed bandwidth; they need to be linearized individually and interactively.

linearizer model, often a pruned Volterra series with several memory taps, may be used in the DPD algorithm. Together with the high sample and hence clock rate of the pre-distorter, this requires very large processor power in the FPGA or ASIC. In brute force terms, an 11th-degree memory polynomial with some additional Volterra cross-terms, defined perhaps using DDR algorithm, could easily demand over 100 gigaMACs of processor power: this is quite a significant demand on the hardware.

An approach to accommodating this is to run two copies of the pre-distorter on alternate samples of the signal, and combining at the output or in the analog domain. This halves the clock rate for each path, enabling realization in a lower technology node, which is possibly more cost-effective. Currently, the highest-performing FPGAs and baseband DPD ASICs are built in 28 nm-node CMOS. This will shrink to 22- or 14-nm node to reduce the power dissipation and increase the processor performance, in terms of gigaMACs or gigaFlops.

While the pre-distorter in the transmit path must run all the time, at the DPD sample rate, which places the highest demands on the processor hardware, the adaption often runs intermittently: capturing data, and then running the adaption algorithm to calculate updated DPD coefficients for the forward path pre-distorter. The convergence rate is usually not so much of an issue. This is changing. As multi-protocol and multi-carrier signals become more prevalent, there are a number of system challenges that use be faced. Multi-carrier EDGE signals may change the number of carriers, their frequency position within the available spectrum, and their power, at every slot in the GSM time frame, in other words, every 576 microseconds. For effective linear communications, this requires that the DPD needs to converge to a linearized solution in a significantly shorter time than this. There is much scope for the development of rapid convergence algorithms: LMS is relatively leisurely; RLS can be very rapid, but like LMS is subject to sample-to-sample noise. Other Kalman filter techniques or gradient search methods may be more effective for error estimation in this context.

Looking further to the future, newer high-efficiency PA architectures will bring their own particular issues to the cauldron of linearization. Wider bandwidth signals will require more efficient means of dealing with bandwidth expansion: direct RF-DAC and RF-ADC converters operating at GHz sample rates may be an answer, subsampling techniques may be another. In short, there promises to be much still to do in the development of DPD systems for LTE-10, LTE-11, and 5G communications.

BIBLIOGRAPHY

Reference texts on power amplifier linearization are few and far between. The topic is introduced in a couple of the classic texts on RF power amplifiers

- Peter Kennington, *High Linearity RF Amplifier Design*, Norwood, MA: Artech House (2000), and
- Steve Cripps, *RF Power Amplifiers for Wireless Communications,* 2nd *Ed.*, Norwood, MA: Artech House (2006)

Kennington's book includes analog linearization techniques such as feedforward, and polar and Cartesian feedback; Cripps's more modern text also includes a description of memory DPD.

The multi-author book edited by Fa-Long Luo, *Digital Front End* (Cambridge, 2011), includes several chapters on pre-distortion from different authors, each providing a perspective on the technique.

Much of the important published work on digital pre-distortion methods and applications is to be found in conference papers and journal and transactions articles. The pioneering work of Cavers and colleagues on pre-distortion methods using look-up tables and on IQ error correction, and of Powers and co-workers on the application of control systems approaches and nonlinear basis functions to digital pre-distortion provides a good foundation to the subject. In the early 2000s, the papers by Ding and Morgan *et al.* on the development and application of memory polynomial techniques to the linearization of solid state PAs have provided the introduction to this technique of DPD for many practitioners; the letter by Kim and Konstantinou [20] remains a classic on this subject.

For a more advanced study of digital pre-distortion, I recommend a thorough review of the articles outlined below. The body of work by Neil Braithwaite, and the paper by Rexberg [7] on the application of modern control theory to DPD will repay careful scrutiny; the papers by Hardik Gandhi, by Professor Anding Zhu, and by the team of Professors Bertran, Montoro, and Gilabert, on the development of Volterra techniques for DPD and their applications in ASIC and FPGA implementations provide great insight and a mine of information on this subject. Professors Fadhel Ghannouchi and Slim Boumaiza have both established large university research groups in the area of RF power amplifier research, including the application of various DPD techniques to these advanced PAs, and their approaches and results provide a valuable practical perspective. The application notes and white papers from Texas Instruments, Xilinx, and Alterra provide details on the practical requirements and implementation of DPD systems for cellular wireless applications using custom IC and FPGA.

Specific references to particular DPD systems, applications and results mentioned in the text of this chapter are cited below.

References

[1] M. Schetzen, *The Volterra and Wiener Theories of Nonlinear Systems.* Malabar, FL: Krieger, 2006.

[2] L. Ljung, *System Indentfication: Theory for the User*, 2nd ed., Upper Saddle River, NJ: Prentice Hall, 1999.

[3] R. N. Braithwaite, "General principles and design overview of digital predistortion," in *Digital Front Endn*, F.-L. Luo, (ed.) Cambridge U.K.: Cambridge University Press, 2011, Ch. 6.

[4] A. Zhu, P. J. Draxler, J. J. Yan, T. J. B. and D. F. Kimball, and P. M. Asbeck, "Open-loop digital predistorter for RF power amplifiers using dynamic deviation reduction-based Volterra series," *IEEE Trans. Microw. Theory Tech.*, Vol. 56, No. 7, pp. 1524–1534, July 2008.

[5] M. Pauli and H.-P. Kuchenbecker, "On the reduction of out-of-band radiation of OFDM signals," *IEEE International Conf. on Communications*, Vol. 3, Altlanta, GA, June 1998, pp. 1304–1308.

[6] J. K. Cavers, "Optimum indexing in pre distorting amplifier linearizers," *IEEE 47th Vehicular Technology Conference*, Vol. 2, Phoenix, AZ, May 1997, pp. 676–680.

[7] L. Rexberg, "Digital linearization of power amplifiers - a control problem," *Reglermote 2006*, Stockholm, Sweden, 2006.

[8] D. Psaltis, A. Sideris, and A. A. Yamamura, "A multilayered neural network controller," *IEEE Control Systems Magazine*, Vol. 8, No. 2, pp. 17–21, Apr. 1988.

[9] C. Eun and E. J. Powers, "A new Volterra predistorter based on the indirect learning architecture," *IEEE Trans. Signal Process.*, Vol. 45, No. 1, pp. 223–227, Jan. 1997.

[10] L. Ding, G. T. Zhou, D. R. Morgan, Z. Ma, J. S. Kenney, J. Kim, and C. R. Giardina, "Memory polynomial predistorter based on the indirect learning architecture," *IEEE Global Telecoms Conf. GLOBECOM*, Vol. 1, Taipei, Taiwan, Nov. 2002, pp. 967–971.

[11] L. Ding, G. T. Zhou, D. R. Morgan, Z. Ma, J. S. Kenney, J. Kim, and C. R. Giardina, "A robust digital baseband pre distorter constructed using memory polynomials," *IEEE Trans. Commun.*, Vol. 52, No. 1, pp. 159–164, Jan. 2004.

[12] B. Widrow and J. M. E. Hoff, "Adaptive switching circuits," *IRE WESCON Convention Record*, no. 4, New York, NY, 1960, pp. 96–104.

[13] R. Sperlich, "Implementing practical digital predistorter systems," *IEEE MTT-S Int. Microwave Symp. Workshop* on 'Power Amplifier Linearization,' Boston, MA, June 2009.

[14] J. K. Cavers, "The effect of quadrature modulator and demodulator errors on adaptive digital predistorters for amplifier linearization," *IEEE Trans. Veh. Technol.*, Vol. 46, No. 2, pp. 456–466, May 1997.

[15] C. Yu, L. Guan, E. Zhu, and A. Zhu, "Band-limited Volterra series-based digital predistortion for wideband RF power amplifiers," *IEEE Trans. Microw. Theory Tech.*, Vol. 60, No. 12, pp. 4198–4208, Dec. 2012.

[16] Y.-M. Zhu, "Generalized sampling theorem," *IEEE Trans. Circuits Syst. II*, Vol. 39, No. 8, pp. 587–588, Aug. 1992.

[17] W. A. Frank, "Sampling requirements for Volterra system identification," *IEEE Signal Process. Lett.*, Vol. 3, No. 9, pp. 266–268, Sept. 1996.

[18] N. Hammler, Y. Eldar, and B. Murmann, "Low-rate identification of memory polynomials," *IEEE Int. Symp. on Circuits and Systems (ISCAS)*, Melbourne, Australia, June 2014.

[19] H. Koeppl and P. Singerl, "An efficient scheme for nonlinear modeling and predistortion in mixed-signal systems," *IEEE Trans. Circuits Syst. II*, Vol. 53, No. 12, pp. 1368–1372, Dec. 2006.

[20] J. Kim and K. Konstantinou, "Digital predistortion of wideband signals based on power amplifier models with memory," *Electron. Lett.*, Vol. 37, pp. 1417–1418, Nov. 2001.

Chapter 8

Crest Factor Reduction of Digital Communication Signals

R. Neil Braithwaite

8.1 INTRODUCTION

This chapter discusses crest factor reduction (CFR) techniques suitable for use in a digital transmitter, such as the one shown in Figure 8.1. CFR is a form of digital signal processing applied to the digital input signal $x(n)$. It is often used in combination with digital pre-distortion (DPD) to reduce the requirements on the radio frequency (RF) power amplifier within the transmit chain. When applied properly, CFR allows the power amplifier (PA) to operate at a higher efficiency, thereby improving the performance of the transmitter.

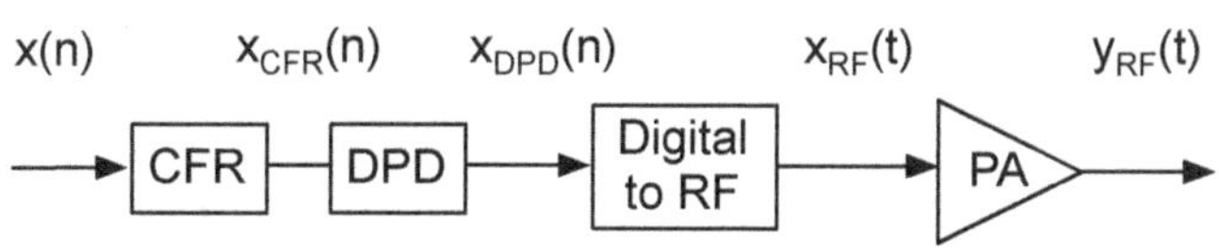

Figure 8.1 Digital transmitter with CFR and DPD.

Wireless communication systems use RF signals to link base-stations and mobile users, allowing for the transfer of digital data. The RF power amplifier is located within the base-station transmitter and is a key component of the downlink connecting the base-station to the mobile. Power amplifiers tend to be most efficient when operated close to saturation. The requirement for linear amplification restricts how close to saturation one can operate the PA.

The nonlinear behavior of the PA can be compensated using linearization techniques, such as digital pre-distortion. As a result, the peak of the input signal can approach the PA saturation point without affecting the linearity of the transmitter. However, the average power level must be backed-off from saturation by an amount equal to the peak to average power ratio (PAPR) of the input signal. This back-off decreases PA efficiency for common amplifier bias conditions, such as class A, AB, B, and C. Reducing the PAPR of the input signal using digital signal processing allows the PA to be operated at a higher efficiency. It also reduces the dynamic range needed to represent the input signal digitally. The digital process of reducing the PAPR is referred to as crest factor reduction.

Crest factor reduction represents a trade-off. As mentioned above, reducing the crest factor allows for increased PA efficiency. In addition, the reduced PAPR will often reduce the complexity of the linearization approach needed to compensate for PA nonlinearities. In fact, CFR is accomplished by degrading the signal quality or reducing the data throughput. For CDMA-based formats, degradations within the bandwidth containing the input signal are considered less significant than degradations that result in an expansion of the signal bandwidth. As a result, CFR of a CDMA signal seeks to limit degradations to the bandwidth spanned by the input signal. For OFDM-based formats, CFR is often achieved using redundant coding or the transmission of auxiliary information, both of which reduce the data throughput of the system.

The outline of this chapter is as follows. Section 8.2 provides background information. The CFR approaches are described in Section 8.3. Finally, Section 8.4 discusses some advanced topics in CFR.

8.2 BACKGROUND

In this section, the background needed to understand crest factor reduction is provided. Section 8.2.1 reviews relevant definitions, concepts, and measurements. The cause of peaks within communication signals is discussed in Section 8.2.2. Section 8.2.3 provides an overview of CFR without discussing specific approaches. The special cases of CFR for downlink LTE signals and multiple antenna systems are introduced in Section 8.2.4.

8.2.1 Review

The peak to average power ratio of a signal $x(n)$ is

$$PAPR = \frac{\max\left(|x|^2\right)}{E\left[|x|^2\right]} \tag{8.1}$$

where $E[]$ indicates expected value. The crest factor (CF) is the square root of the PAPR. The terms PAPR and CF are often used interchangeably, and are generally expressed in dB.

In practice, the PAPR used is based on the statistics of the signal rather than absolute peak. The practical peak is the level (L) at which the signal magnitude has a 10^{-4} probability of exceeding: that is,

$$prob\left\{|x|^2 > \mathrm{L}\right\} = 10^{-4} \tag{8.2}$$

The complementary cumulative density function (CCDF) of $|x|^2$, shown in Figure 8.2, is a useful description of the signal statistics. The CCDF of a signal is often compared to that of a Gaussian waveform because the statistics of multi-carrier signals used in wireless communications tend to approach that of a Gaussian. When referring to a Gaussian waveform, it should be noted that the signal is complex and the Gaussian properties describe the I and Q channels independently. The distribution functions for the magnitude and power are Rayleigh and exponential, respectively.

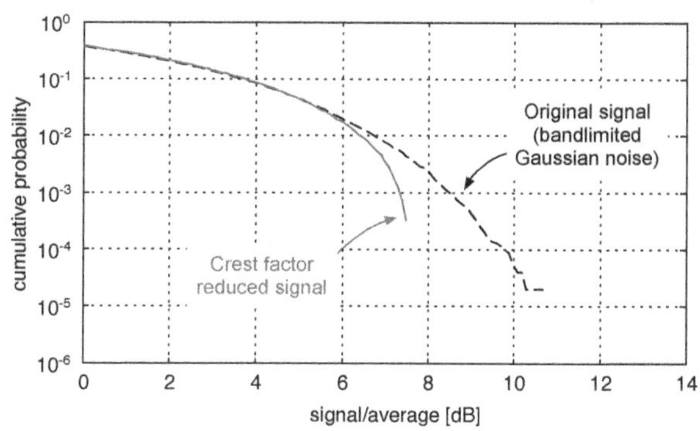

Figure 8.2 Example of a complementary cumulative density function (CCDF).

Let us return to the relationship between PAPR and PA efficiency. The efficiency for class A, AB, B, and C biased amplifiers improves near saturation. Other amplifier configurations such as the Doherty PA [1] have a broader high-efficiency range. For example, the efficiency for an asymmetric Doherty PA whose auxiliary amplifier is twice as large as the main amplifier is shown in Figure 8.3 [2]. Although the efficiency is high for a 9-dB back-off range from saturation, the efficiency shown is based on a constant amplitude input signal. For signals with time-varying envelopes, the probability density function (pdf) is needed to predict the actual efficiency. Gaussian waveforms backed-off by 6 dB and 10 dB from saturation, and crest factor reduced accordingly, are also shown in Figure 8.3. The 6 dB backed-off signal spends more time operating in the high efficiency region of the PA than the 10 dB backed-off signal. As a result, one can expect that the more aggressive CFR operation will allow the PA to operate at a higher efficiency.

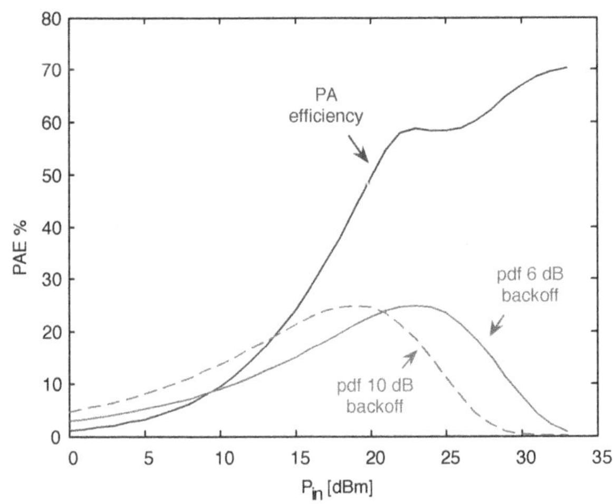

Figure 8.3 PA efficiency and probability density function of the input signal.

A wireless communication system is shown in Figure 8.4. The digital signals in the transmitter and receiver are denoted by $x(n)$ and $z(n)$, respectively. In the RF transmission of digital information, the sampled data sequence $x(n)$ is converted from digital to RF and amplified by a PA. At the receiver, the RF signal is amplified by a low noise amplifier (LNA) and converted back to the digital domain, creating a sampled data sequence $z(n)$, which ideally is the same as that transmitted (that is, $x(n) = z(n)$).

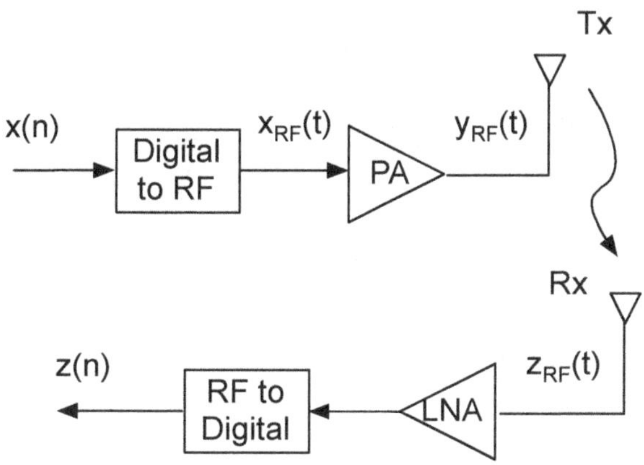

Figure 8.4 Wireless communication system with digital transmitter and receiver.

For a linear system, most of the power transmitted by the base-station is confined to an allocated bandwidth, which is referred to as a channel. Nonlinearities in the transmitter generate a distortion spectrum several times wider than the allocated channel. This broadening of the transmitted spectrum is referred to as spectral regrowth. In-band distortion is also generated, which causes $x(n)$ and $z(n)$ to differ.

The out-of-band and in-band distortions are regulated by specifications associated with the modulation format of the input signal. Out-of-band distortion is measured by the amount of distortion power falling into adjacent channels neighboring the allocated channel. This may be described as the adjacent channel power or the adjacent channel leakage. The ratio of the adjacent and allocated channel powers is referred to as the adjacent channel power ratio (ACPR) or the adjacent channel leakage ratio (ACLR).

To illustrate out-of-band distortion, consider the nonlinear response to a band-limited signal. The output spectrum is the sum of the linear component and intermodulation distortion (IMD), as shown in Figure 8.5. The linear component, also referred to as a carrier signal, remains within the allocated channel whereas the distortion spreads into the adjacent channels. However, a significant portion of the distortion lies within the allocated channel.

The measurements of in-band degradations include error vector magnitude (EVM) and bit error rate (BER), both of which are specified, typically, as a percentage. EVM measurements require a receiver to demodulate the RF signal to a complex digital signal $z(n)$. EVM, illustrated in Figure 8.6 for a QPSK signal,

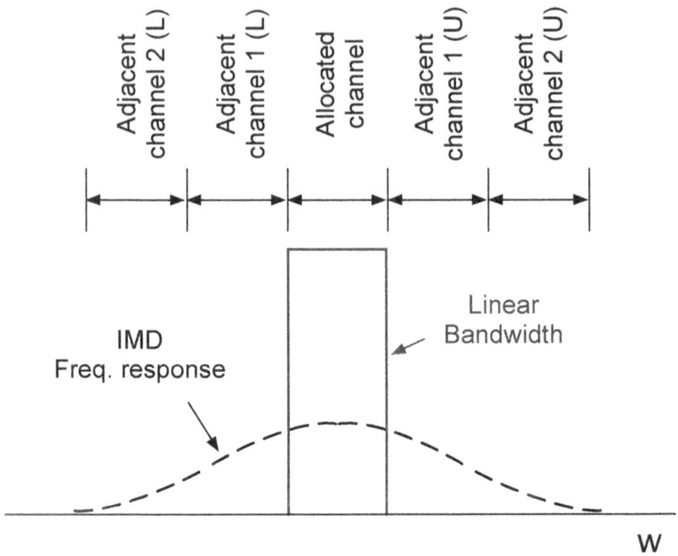

Figure 8.5 Output spectrum comprising the linear signal component, as well as the intermodulation distortion (IMD). Also shown are the allocated and adjacent channels.

is defined as

$$EVM = \sqrt{\frac{E\left[|z(n) - x(n)|^2\right]}{E\left[|x(n)|^2\right]}} \tag{8.3}$$

The EVM measurement often assumes that scaling and rotation have been applied to the received signal $z(n)$ to remove any offsets common to all received samples. In most cases, the received signal is also equalized before measuring EVM.

Bit error rate measurements also require a receiver; however, the signal must be demodulated beyond the complex digital signal to a bit sequence. Bits originating at the input of the transmitter are mapped onto an IQ constellation to create the complex digital signal $x(n)$. At the receiver, the complex digital signal $z(n)$ is quantized along the I and Q axes, which maps $z(n)$ to the nearest point in the constellation. A bit error occurs when the error vector magnitude is large enough that the constellation points associated with $x(n)$ and the quantized $z(n)$ differ. The BER can be specified before or after forward error correction (FEC) if redundant coding is used.

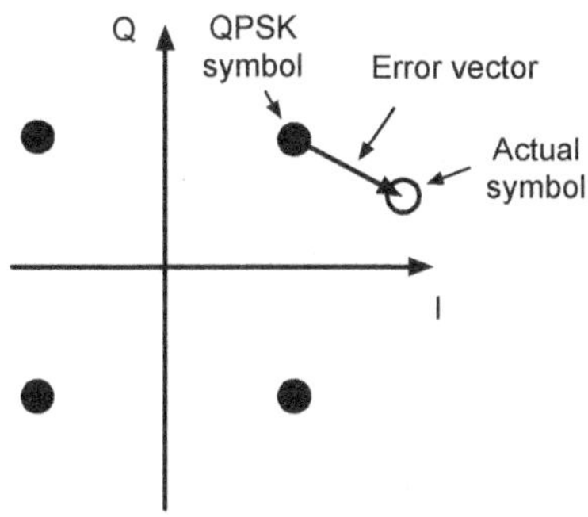

Figure 8.6 Error vector magnitude (EVM) for a QPSK signal.

Most IQ constellations represent several bits. For example, QPSK, 16-QAM, and 64-QAM symbols represent 2, 4, and 6 bits, respectively. The higher modulation rate associated with 64-QAM is obtained by reducing the distance between constellation points, compared to QPSK or 16-QAM; however, the BER is more sensitive to EVM as a result.

It is interesting to note that the BER sensitivity to error vectors is not uniform throughout the constellation. Consider a 16-QAM constellation, shown in Figure 8.7. The signal $z(n)$ can expand outward from the outer constellation points without causing a bit error. For the outer regions marked in gray within Figure 8.7, the distance to the nearest incorrect constellation point is greater than or equal to the quantization interval, denoted by Δ. Thus, there is a low probability that $z(n)$ in these gray regions will experience a bit error despite having a large outward pointing error vector. Some of the CFR approaches [3] exploit this property of the IQ constellation to reduce the peak by degrading EVM, without increasing the BER.

The bit mapping to the constellation is often performed using a Gray code so that neighboring points along the I- or Q-axis differ by one bit only. The bit mapping for the 16-QAM constellation is shown in Figure 8.8. The interesting property of the Gray code is that it wraps around the I and Q axes. That is, the outer constellation points at $z(n) = 1.5\Delta + j1.5\Delta$ and $1.5\Delta - j1.5\Delta$ differ by one bit (1100 and 1110, respectively), where Δ is the quantization interval. The number of bit differences between two constellation points is referred to as the Hamming distance.

In some cases, it is acceptable to generate the occasional bit error. If a bit error is necessary, it can be beneficial from a CFR viewpoint to choose the constellation point that has the largest EVM from the set of points having a Hamming distance of

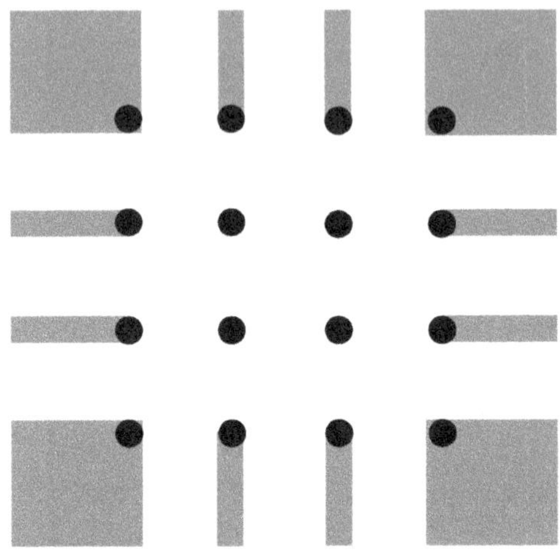

Figure 8.7 A 16-QAM constellation. The outer regions marked in gray have a large error vector magnitude without creating a bit error.

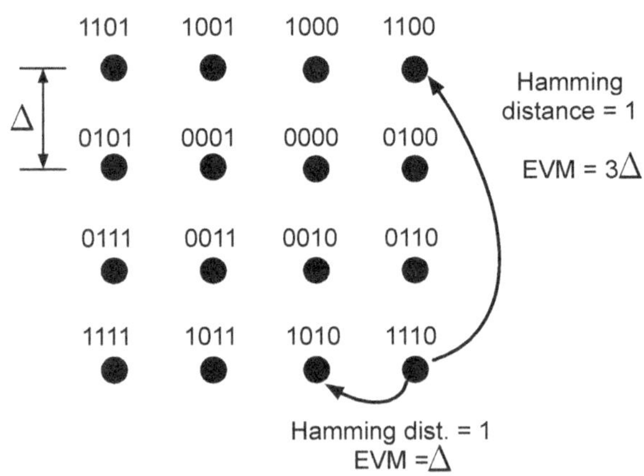

Figure 8.8 Bit mapping for a 16-QAM constellation. The bit mapping is based on Gray codes so that neighboring points along the I- or Q-axis differ by one bit (Hamming distance = 1). The Gray code wraps around the I and Q axes so that outer constellation points also have a Hamming distance of unity.

one. This is shown in Figure 8.8 for the 16-QAM constellation. Negating the I- or Q-axis for an outer point results in a Hamming distance of one (one bit error), but an EVM of 3Δ. Most constellation points with a Hamming distance of one will have an EVM of Δ only. That is, the CFR has three times the in-band degradation available for peak reduction, for a given BER. This is exploited in some CFR methods [4].

8.2.2 Cause of Signal Peaks

A complex symbol within a communication signal can often be described as a zero mean, random variable with a known probability density function (pdf) determined by the underlying modulation (BPSK, QPSK, 16-QAM, or 64-QAM). When several independent symbols are added together, the probably density function usually changes, where the central limit theorem [5] suggests that the *pdf* of the combined signal will appear more Gaussian than that of the original signal. In most cases, this results in a higher PAPR.

Digital signals start out as binary values with a PAPR of 0 dB. OFDM and CDMA modulation increase the variation in the signal magnitude and the PAPR. In addition, the PAPR is affected by the filter used for band-limiting the channel. Multi-access modulation (OFDM and CDMA) and filtering are discussed below.

OFDM uses an inverse fast Fourier transform (IFFT) to convert complex modulation symbols (BPSK, QPSK, 16-QAM, or 64-QAM) into a time-domain sequence $x(n)$, as shown in Figure 8.9. Groups of symbols are loaded into a buffer of length M, performing a serial-to-parallel conversion, and assigned to sub-carriers of the IFFT. Thus, each symbol will have two indices, k and l, which indicate the sub-carrier frequency and the time position of the buffered symbol ensemble. The time-domain representation of the OFDM signal becomes

$$x(n) = \sum_{k} X(k,l) \exp\left(jk\Delta\omega n_l\right) \tag{8.4}$$

where $\Delta\omega$ is the sub-carrier spacing,

$$n_l = n - l \cdot T_{symbol} \tag{8.5}$$

$X(k,l)$ are the complex-valued modulation symbols, $x(n)$ is the complex-valued time-domain signal, and T_{symbol} is the time duration of a symbol. A cyclic prefix (CP) is added before transmission.

The size of the buffer (M) is usually less than the size of the IFFT (N) because guard bands are used on the outer sub-carriers. However, from a CFR perspective, the key point to note is that the time sequence $x(n)$ is derived from the

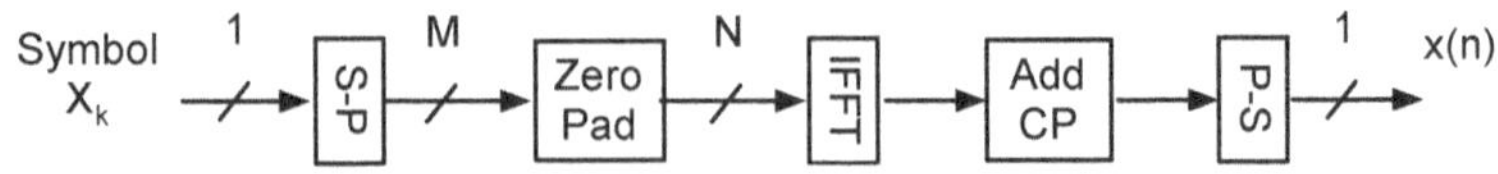

Figure 8.9 OFDM signal generation.

sum of several weighted sub-carriers. The peak occurs when all of the sub-carriers are aligned in phase. The PAPR based on the absolute peak (rather than the 10^{-4} probability) is M. The number of active sub-carriers can be as high as $M = 1200$ for LTE, which has the potential for an enormous PAPR. Thus, CFR is needed to ensure that such phase alignments between sub-carriers can not happen.

OFDM formats tend to specify a low value of EVM. For example, WiMAX (IEEE 802.16) requires the relative constellation error (RCE) to be less than -31 dB for 64-QAM-3/4. RCE and EVM measure the same in-band degradations. A RCE of -31 dB corresponds to an EVM of 2.8%, which is not generous. Thus, CFR methods that trade off data throughput are preferred for OFDM signals over methods that degrade the in-band signal quality.

CDMA-based signals are formed in a manner similar to OFDM except that the sub-carriers are replaced by Walsh codes. Walsh codes are binary sequences (± 1) that are orthogonal to each other. They are defined in part by their length. For example, there are four Walsh codes W_i for a code length of 4 samples, which are defined by

$$H_4 = \begin{bmatrix} W_0 \\ W_1 \\ W_2 \\ W_3 \end{bmatrix} = \begin{bmatrix} 1 & 1 & 1 & 1 \\ 1 & -1 & 1 & -1 \\ 1 & 1 & -1 & -1 \\ 1 & -1 & -1 & 1 \end{bmatrix} \tag{8.6}$$

where H_4 is the Hadamard matrix of size 4. Larger Walsh code sets are defined recursively, using the Hadamard matrix, as

$$H_{2n} = \begin{bmatrix} H_n & H_n \\ H_n & -H_n \end{bmatrix} \tag{8.7}$$

For the original IS-95 version of CDMA, the Walsh code set is based on H_{64}. That is, the set comprises 64 orthogonal codes, which are 64 samples in length. The coded signal amplitude is

$$|x(n)| = \sum_{i=0}^{63} \rho_i \cdot d_i(l) \cdot W_i(k) \tag{8.8}$$

where ρ_i are scalar weights, $d_i(l)$ are binary data (± 1), whose duration equals the length of the Walsh code ($N = 64$), $k = 1 : 64$, and

$$n = l \cdot 64 + k \tag{8.9}$$

From a CFR perspective, the key point is that the time sequence $x(n)$ is derived from a sum of weighted Walsh codes. If all of the Walsh codes within the set are active, the PAPR is $N = 64$, assuming that the codes are weighted equally.

In practice, most of the Walsh codes are inactive at a given time (that is, $\rho_i = 0$) for IS-95. The upper bound of the resulting PAPR is equal to the number of active codes. Thus, the PAPR is determined primarily by the Walsh code utilization. However, the PAPR can be reduced below this bound by the proper selection of the active Walsh codes for a partially utilized code set [6].

Newer CDMA formats allow the length of the Walsh codes to be variable. This is referred to as an orthogonal variable spreading factor (OVSF) where the Walsh code length is adjusted based on the data rate. In IS-95, it is assumed that each code carries a voice signal and therefore has the same data rate. With data services, the data rates can vary substantially. A Walsh code from the SF = 4 set (defined by H_4) carries as much data as four codes from the SF = 64 set (defined by H_{64}). However, it is beneficial from a PAPR point of view to assign shorter Walsh codes to higher data rate connections rather than send the data in parallel on Walsh codes from the SF = 64 set [7]. For example, the PAPRs for fully utilized SF = 4 and SF = 64 sets are 6 dB and 18 dB, respectively, yet the data throughput is the same. A PAPR improvement is also obtained by using a mixture of code lengths (SF $\leq$ 64) to create $x(n)$, compared to the use of parallel SF = 64 codes.

The CFR used for CDMA signals typically reduces the PAPR of the input signal by degrading the in-band signal quality. It is shown in Section 8.4 that a WCDMA HSDPA+ 64-QAM signal [8] can be crest factor reduced to a PAPR as low as 5.7 dB and it will still pass the in-band specification. The WCDMA signal is very robust with respect to degradations in the in-band signal quality, and the CFR approach selected should exploit it. The same is true for most CDMA-based formats.

A filtered signal will often have a peak that is larger than the original input signal. A peak occurs when a segment of a digital sequence is correlated with the filter kernel. This should not be surprising given that the classic method of generating Gaussian noise is to start with a binary random sequence as the input and band-limit it using a lowpass filter [9]. This can cause the PAPR to increase from 0 dB to 9.6 dB.

Consider an interpolation filter used in the digital to analog conversion. The filtered signal is

$$x_h(t) = \sum_{n=-M}^{M} h(t - nT)x(n) \tag{8.10}$$

where T is the sampling period, $h(t)$ is the filter kernel, and $x(n)$ is the digital input signal. The peak output for a binary input ($x(n) = \pm 1$) is

$$\max\left(|x_h(t)|\right) = \sum_{n=-M}^{M} |h(t - nT)| \tag{8.11}$$

The kernel for a raised cosine filter with a roll-off factor of zero is

$$h(t) = \frac{\sin\left(\pi t \cdot T^{-1}\right)}{\pi t \cdot T^{-1}} \tag{8.12}$$

This kernel is also referred to as a *sinc* function. If it is assumed that the filter kernel size $(2M + 1)$ approaches infinity, we get

$$\max\left(|x_h(t)|\right) = \sum_{n=-\infty}^{\infty} \left| \frac{\sin\left(\pi(t \cdot T^{-1} - n)\right)}{\pi(t \cdot T^{-1} - n)} \right| = \infty \tag{8.13}$$

where the maximum occurs mid-sample at $(t \bmod T) = 0.5$. That is, an unbounded peak can appear between digital samples when the input signal is converted to analog or interpolated to obtain a higher sampling rate.

In practice, a digital implementation of the filter would be truncated to a finite M, thereby constraining the peak. For example, if $h\{(n + 0.5)T\}$ is truncated to 14 samples, filtering will increase the PAPR for a BPSK input signal from 0 dB to 8 dB. This 8-dB level is approximately the same as the 10^{-4} probability peak. It should be apparent that this high-order (steep roll-off) filter kernel can cause the PAPR of the signal to increase significantly.

Consider a different filter kernel that is a Gaussian function (not related to the Gaussian *pdf* mentioned earlier):

$$g(t) = \exp\left(-\alpha t^2\right) \tag{8.14}$$

If it is assumed that the filter kernel size $(2M + 1)$ approaches infinity, we get

$$\max\left(|x_h(t)|\right) = \sum_{n=-\infty}^{\infty} |g(t - nT)| = \sqrt{\frac{\pi}{\alpha}} \tag{8.15}$$

In this case, the peak is bounded even for large values of M. From the above two examples, it is apparent that using a filter kernel with a more relaxed roll-off in the frequency domain results in a lower peak for a binary input. Thus, one should use the most relaxed filter kernel allowable by the specifications of the modulation format to avoid reconstructing peaks after the CFR.

Let us revisit the mid-sample peak associated with the first kernel. Although the peak is unbounded, it can occur only when the binary sequence has a resonant pattern, that is,

$$x(n) = \begin{cases} (-1)^n & \text{for} \quad n \geq n_0 \\ (-1)^{n+1} & \text{for} \quad n < n_0 \end{cases} \tag{8.16}$$

will create a peak at $t = (n_0 - 0.5) * T$. The probability of this exact sequence (or its negative) occurring is 2^{-2M}, which is small for large values of M. The CFR approach needs to alter the binary sequence to ensure that this pattern never occurs.

8.2.3 CFR Overview

The above text can be summarized as follows. A weighted summation of independent signals creates a new signal with a complex Gaussian distribution. Crest factor reduction attempts to truncate the tails of the Gaussian distribution by disrupting the randomness of the summation. For example, we can introduce correction samples into the summation that are derived from, and therefore correlated to, the original samples to reduce the peak. These are viewed as redundant or parity samples containing no new information. Alternatively, we can buffer the input signal and apply a transformation $T[x(n)]$ that does not increase the signal bandwidth. Possible transformations for time signals include interleaving and modulation by pseudo-random sequences. For OFDM signals, different phase shifts can be applied to the individual sub-carriers. Several transformations from a known set are tested and the signal $T_i[x(n)]$ with the lowest PAPR is selected. Information identifying the transformation used is sent separately to the receiver as overhead data to allow the original samples to be restored later. Overhead data reduces the throughput of system. However, many modulation formats contain overhead data such as pilot, and if such auxiliary information is encoded into the existing overhead data, the original throughput is maintained.

Other methods of peak reduction are available that degrade aspects of the system performance. Small errors can be added to the individual samples that add coherently to reduce a peak, when needed. If the individual errors are less than the quantization interval within the receiver, the original samples can be restored without error. Alternatively, a bit error can be added to reduce the peak. The original samples are demodulated successfully if the number of bit errors occurring due to

CFR does not consume the excess redundancy associated with FEC coding. The above-mentioned modifications reduce the error margin of system in order to reduce the PAPR.

Another approach to limiting signal peaks is to reduce the number of independent samples added together. An extreme example would be to use time-domain duplexing (TDD), transmitting only one sequence of BPSK, QPSK, 16-QAM, or 64-QAM modulated data at any given time. Unfortunately, this results in a loss of the desirable characteristics of a multi-access modulation format including the use of channel-dependent scheduling to limit the effects of fading. As a result, this is not always a viable option for limiting the PAPR in a multi-access system.

Within many of the CFR methods described in Section 8.3, several waveforms are created: the clipped input waveform x_{clip}, the error waveform x_{peaks}, the CFR correction waveform Δx, and the crest factor reduced signal x_{CFR}. The input signal clipped to the desired level L is

$$x_{clip}(n) = \begin{cases} \mathrm{L}\frac{x(n)}{|x(n)|} & \text{for} \quad |x(n)| > \mathrm{L} \\ x(n) & \text{otherwise} \end{cases} \tag{8.17}$$

The error waveform, which corresponds to the amount that a peak exceeds the clipping threshold L, is

$$x_{peaks}(n) = x(n) - x_{clip}(n) \tag{8.18}$$

The error waveform is filtered to create the correction signal used to obtain the CFR: that is,

$$\Delta x(n) = \sum_{\tau} x_{peaks}(n - \tau)h(\tau) \tag{8.19}$$

where $h(t)$ is the filter kernel used typically to limit the bandwidth to the allocated channel. The crest factor reduced signal is

$$x_{CFR}(n) = x(n) - \Delta x(n) \tag{8.20}$$

For transform-based CFR methods, the crest factor reduced signal is

$$x_{CFR}(n) = \min_{PAPR(i)} \left(T_i\left[x(n)\right]\right) \tag{8.21}$$

The preferred transformation for an OFDM signal is to phase-shift individual sub-carriers where the permitted phase shifts are often quantized to 0 or π radians. In general, the optimal vector representing the phase shifts of the active sub-carriers is discovered by searching the space of possible vectors. Searches for optimal vectors (or other transformations) are often exhaustive; however, several sub-optimal search methods are described in Section 8.4 which terminate early if a sufficient vector is found that achieves the desired PAPR. Thus, much of the research in CFR involves the development of efficient search methods.

Let us look at the effect of CFR on the signal, without concern for the specific approach used. CFR reduces the peaks exceeding the selected PAPR threshold. Consider the magnitude of a noise waveform with Gaussian statistics, as shown in Figure 8.10. Two PAPR thresholds are selected: 6.5 dB and 8.5 dB. It is apparent from Figure 8.10 that it is more difficult to achieve a PAPR of 6.5 dB than 8.5 dB because the former has more peaks to reduce than the latter. For the Gaussian waveform, reducing the PAPR to 6.5 dB affects 1% of the samples, whereas a PAPR of 8.5 dB affects only 0.1%. The problem associated with a higher rate of peak occurrence is that attempting to cancel several peaks in close proximity has the potential to create a new higher peak elsewhere in the signal. Thus, it is important to use CFR in moderation.

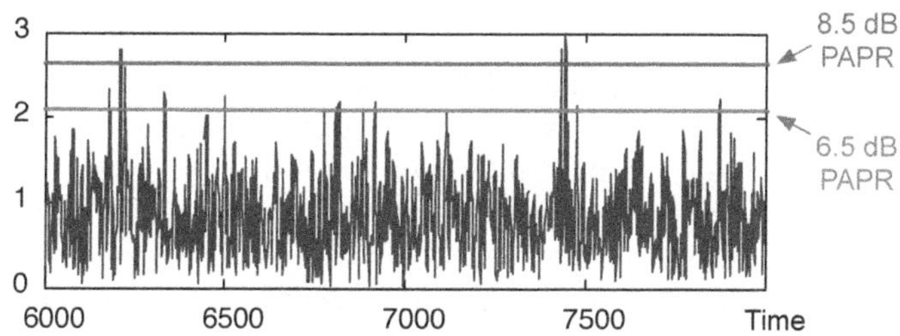

Figure 8.10 Magnitude of a Gaussian waveform. PAPR levels of 6.5 dB and 8.5 dB are indicated.

Another important aspect of CFR is to know when not to use it. If the average power level of the PA is backed-off because of low traffic load or reduced cell coverage, then the CFR target should be relaxed accordingly. This will result in a lower error vector magnitude for the backed-off signal and will allow the modulation coding rate for LTE (and other modulation formats) to be increased.

Crest factor reduction should not be applied to an individual carrier prior to combining it with other carriers. The combining process, assuming that the inter-carrier information is uncorrelated, will tend to increase the PAPR of the combined signal towards that of a Gaussian signal. Thus, unless the PAPR of the individual carrier is worse than Gaussian, we would not recommend CFR of the individual carriers. A better approach would be to buffer the individual carrier signals into time blocks (if possible) and adjust the relative phase between the carriers to minimize the PAPR of the combined signal. The preferred phase shift vector will vary for each time block. This is similar to the PTS approaches described later, except that it can be applied to any modulation format or combination of formats instead of being restricted to OFDM. The challenges are to determine the best phase shift vector using the fewest mathematical operations and to get the phase-shift information to the receiver.

In summary, crest factor reduction has drawbacks. CFR degrades the system performance, usually by increasing the in-band degradation or reducing the data throughput. In some cases, CFR requires modifications to the receiver to extract auxiliary information or redundant data from the in-coming signal. CFR will also consume specification margins related to ACLR, EVM, or BER. However, if the CFR approach is selected to best utilize the margin available for the modulation format used, the overall system can be optimized to meet the specification at higher PA efficiencies.

8.2.4 Special Cases

In addition to the general application of CFR, two special cases are considered in section 8.3: the downlink LTE waveform and multiple antenna transmissions. In downlink LTE signals, the minimum allocation for data is a resource block, which is 12 sub-carriers wide in frequency and one slot (7 OFDM symbols, typically) in time duration (see Section 8.4 for details). In some of the CFR approaches, the PAPR for the seven OFDM symbol intervals spanning a slot must be optimized jointly. It is advantageous to use CFR methods where the PAPR can be optimized for each OFDM symbol interval individually because they outperform jointly optimized CFR methods.

Multiple antenna transmissions include MIMO (multiple input, multiple output) and beam forming. MIMO exploits the diversity of the RF channel connecting the base-station and mobile user to allow spatial multiplexing where multiple data streams are transmitted over a common set of sub-carriers. The data streams are pre-coded and mapped to the available antennas. The pre-coding matrix performs a weighted summation of the data streams to create the signals presented to the antennas. In general, the signal at each antenna is different. As a result, the position of peaks in MIMO will not be aligned between antenna signals.

In beam-forming, the relative phase between signals at each antenna is selected to create a directional beam pattern. Consider beam-forming for the case of downlink OFDMA (which includes LTE). Since OFDMA is a multi-access system, the signals are generated for multiple users where each user may be transmitting a different radiation pattern. In such cases, the position of peaks will not be aligned between antenna signals, as in the MIMO case.

In both cases, CFR approaches applied independently to each signal entering an antenna can degrade the directionality of the beam forming or reduce the capacity of the MIMO link. Thus, it is necessary for some CFR approaches to optimize the multiple antenna signals jointly so that each antenna radiates the same CFR correction waveform or experiences the same phase shift vector. As mentioned earlier, jointly optimized CFR methods are less effective than CFR applied to the antenna signals individually.

8.3 CFR APPROACHES

Several CFR approaches exist that reduce the peaks in a signal without generating distortion outside of the allocated bandwidth. The approaches discussed include the clip and filter, peak windowing, partial transmit sequence, tone reservation, constellation extension, and coding.

8.3.1 Clip-and-Filter Approaches

A method of CFR that is applicable to any modulation format is to clip peaks exceeding a specified level and filter the waveform to remove out-of-band distortion. Three such approaches are discussed: direct clipping, peak cancellation, and error waveform subtraction. The error waveform subtraction is also referred to as noise shaping.

The direct clipping approach [18] is shown in Figure 8.11. It has the advantage of simplicity. The digital signal is up-sampled (over-sampled) by N_{ov} times the Nyquist rate. The signal is clipped when the magnitude exceeds the desired PAPR level L. The clipped signal $x_{clip}(n)$ is then band-limited by the channel filter.

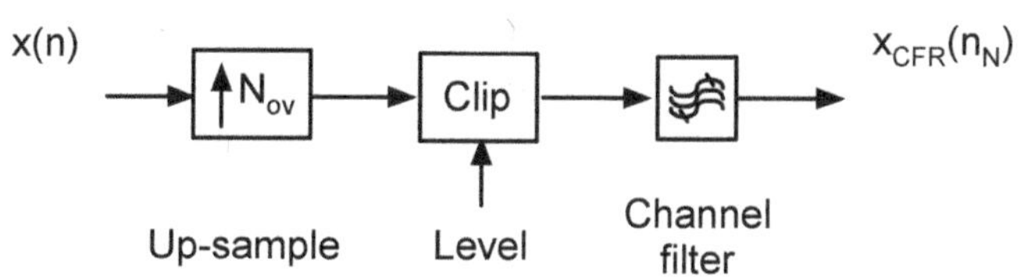

Figure 8.11 Direct clipping approach to CFR.

For OFDM signals, clipping moves the complex-valued symbols from their assigned constellation positions distributing the constellation error over all of the available sub-carriers. This can be problematic if the synchronization signals or reference (pilot) signals are degraded. In addition, high-order constellations, such as 64-QAM, are especially sensitive to EVM and therefore the clipping should be applied sparingly.

There are also some potential problems related to using the channel filter for CFR. The channel filter is designed to produce a pulse shape for the linear component of the signal that matches the specification associated with the modulation format. It is not necessarily optimized for the error correction portion of the CFR waveform. For example, if two clipped peaks occur in close proximity, interaction during filtering can generate a peak that is larger than the clipped level. Some clip-and-filter-based approaches use multiple CFR passes to limit the final PAPR. It is not advisable to apply the direct clipping method shown in Figure 8.11 repeatedly because the pulse shape of the linear signal component may be altered. The signal processing for the linear signal must be separated from that of the clipped error signal to allow for multiple CFR passes, as will be discussed later.

The effect of filtering on the PAPR of the signal is also influenced by the over-sampling rate N_{ov} prior to the clipping. The over-sampling rate prior to CFR should be $N_{ov} \geq 4$ to avoid recreating peaks due to interpolation [10]. If N_{ov} is too low, any interpolation filtering applied after the CFR, such as that done by the reconstruction filter within the DAC, will cause the PAPR to rise discernibly.

Consider a two-carrier WCDMA signal, for example. The individual WCDMA signals have a symbol rate of 3.84 Msamples/second (Ms/s) and are assigned to 5 MHz channels. For the two-carrier signal, the individual carriers can be separated by 5 MHz, 10 MHz, or 15 MHz. These two-carrier signals are referred to as 11, 101, and 1001 configurations, respectively, where the 0 indicates an idle channel between carriers. Table 8.1 shows the effects of interpolation on the PAPR. The post-CFR signal is interpolated to a sample rate of 61.4 Ms/s from a pre-CFR sampling rate of 46.08 Ms/s. The PAPR increases by 0.03 dB, 0.4 dB, and 0.73 dB for the 11, 101, and 1001 cases, respectively [11]. Table 8.1 also illustrates why the over-sample rate must be based on the bandwidth of the composite signal. A sample rate of 46.08 Ms/s is not sufficient ($N_{ov} < 4$) for the latter two cases. The sample rate selected should be greater than 40 Msps, 60 Msps, and 80 Msps for the 11, 101, and 1001 cases to avoid PAPR expansion due to interpolation filtering.

A variation on the clip-and-filter approach [12] is peak cancellation. It is shown in Figure 8.12. The linear and CFR correction signals are separated. The peaks in the linear signal, as measured at the output of the channel filter, are detected. A band-limited pulse is generated at each peak location. The magnitude

Table 8.1

Effect of Interpolation after CFR for Two-Carrier WCDMA Signals Where the PAPR Before Interpolation is 6.5 dB.

Sample Rate	Carrier Configuration		
	11	101	1001
Pre-interpolation 46.02 MS/s	6.51 dB	6.50 dB	6.50 dB
Post-interpolation 61.44 MS/s	6.54 dB	6.90 dB	7.23 dB

of the pulse is set equal to the excess peak above the clipping threshold. The phase of the pulse is the same as the phase of the linear signal at the peak. The frequency response of the pulse is determined by a filter. The sequence of filtered pulses is subtracted from the linear signal to produce a band-limited signal with the desired PAPR.

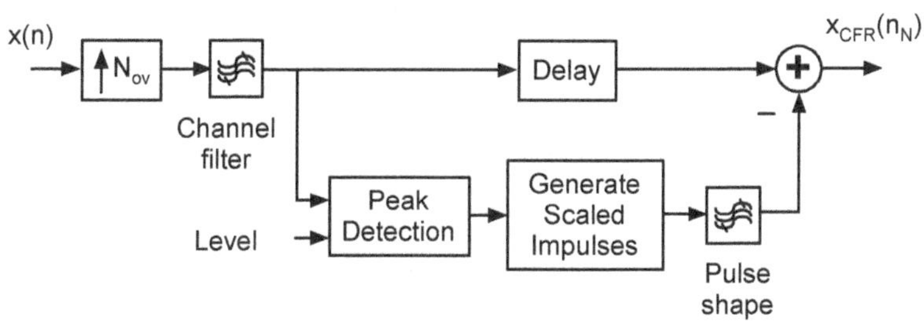

Figure 8.12 Peak cancellation approach to CFR.

The signals involved in the peak cancellation approach are shown in Figure 8.13. The top sub-figure depicts the magnitudes of the original and crest factor reduced signals for a Gaussian waveform. The middle sub-figure shows the pulses from the peak detection circuit. The magnitude of the correction signal, after filtering, is shown in the bottom sub-figure. Note that the filter kernel spreads the pulses in time and that some overlap between the filtered pulses is apparent. In general, the peak cancellation approach works well when the detected pulses are well separated. Temporal overlap of neighboring pulses after filtering can create new peaks, which is not desirable.

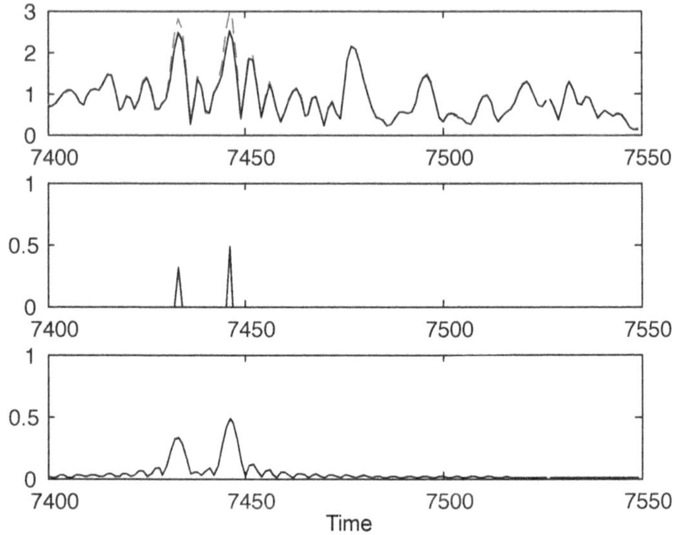

Figure 8.13 Signal involved in peak cancellation. (Top) Magnitudes of the original and crest factor reduced signals. (Middle) Pulses from the peak detector. (Bottom) Magnitude of the correction signal.

An advantage of the peak cancellation approach is that the pulse-shaping filter does not have to be the same as the channel filter. This can be exploited in many ways. For example, it is possible to generate several correction waveforms using different pulse-shaping filters. The best of the waveforms, in terms of avoiding new peaks, can be selected.

The peak cancellation approach can also be implemented using multiple CFR passes. Since the linear component of the signal is processed separately until the desire correction signal is found, the channel filter response is preserved.

Another variation of the clip-and-filter approach is to find peaks above a threshold and create a clipped error waveform [13]. The error waveform is filtered to create a correction waveform that is subtracted from the original signal, as shown in Figure 8.14. The EVM is constrained by adjusting the scaling of the correction waveform subtraction. The scale factor is also used to compensate for any overlap associated with the filtering of neighboring peaks. As in the peak cancellation approach, the filter used to band-limit the error signal can be different from the channel filter. Several candidate CFR correction signals can be generated in parallel or the correction can be derived from multiple CFR passes.

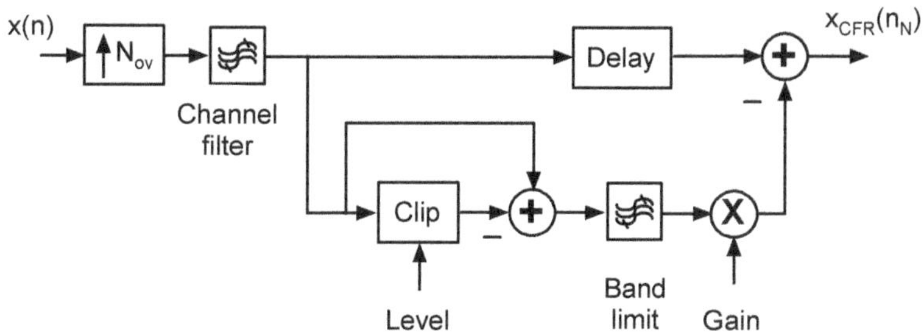

Figure 8.14 Error waveform subtraction.

The above-mentioned CFR approaches are not optimized for OFDM signals. As mentioned earlier, OFDM has tighter specifications on the in-band degradation. An OFDM suitable method [14] for controlling the EVM degradation on a per sub-carrier basis is shown in Figure 8.15. The OFDM signal is converted to the time domain and an error waveform is created. The error waveform is converted back to the frequency domain and the error present on each sub-carrier is measured. If for any sub-carrier the error magnitude exceeds the threshold allowed by the modulation type (BPSK, QPSK, 16-QAM, or 64-QAM), the error magnitude is reduced. Band-limiting of the error waveform in the frequency domain is achieved by reducing the error magnitude for the guard band sub-carriers to the level allowed by the spectral mask specification.

One of the advantages of this approach is that different error magnitude limits can be assigned to each sub-carrier. For example, we recommend that non-data sub-carriers associated with pilots, reference signals, and synchronization signals be transmitted free of CFR degradations. The EVM thresholds for the data sub-carriers should be selected to constrain the constellation errors with the goal of avoiding bit errors. For example, a QPSK sub-carrier can be degraded more than a 64-QAM sub-carrier. In addition, certain error vectors do not affect the bit decision during quantization. The outer constellation points can expand outward without causing a bit error, as shown earlier in Figure 8.8. Error vectors in such regions do not have to be limited in magnitude.

The clip-and-filter methods are well-suited for downlink LTE signals, assuming that the EVM increase can be tolerated. In some of the other CFR approaches, the PAPR for the seven OFDM symbol intervals spanning a slot must be optimized jointly. This is not the case for the clip-and-filter approach, which is advantageous.

We recommend exercising moderation when using clip-and-filter methods for multiple antenna transmissions, such as MIMO and beam-forming. In general,

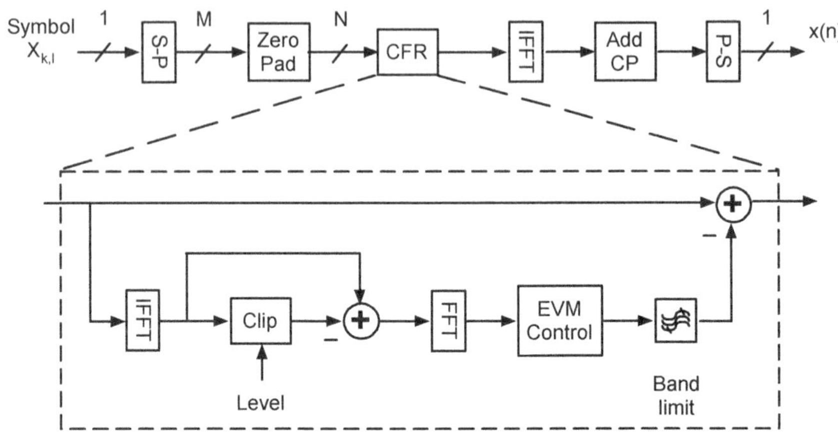

Figure 8.15 OFDM version of error waveform subtraction. The EVMs of individual sub-carriers are controlled to avoid bit errors.

the CFR correction waveform applied will be different for each antenna signal. This may lead to problems beyond the increased EVM. For the OFDMA beam forming case, nulls in the radiation pattern may be steered to minimize interference in a multi-access environment. The CFR correction waveform will be radiated in bursts from a single antenna in a broad pattern, possibly omni-directional. This interference may negate some of the benefits sought from the beam forming. For the MIMO case, the CFR correction waveform from a single antenna may increase the EVM unevenly among the MIMO data streams when pre-coding is used. While it is possible to optimize multiple signals jointly so that each antenna radiates the same CFR correction waveform, this is not recommended because the EVM generated would increase. More research is needed to determine whether these problems are significant enough to discourage the use of the clip-and-filter approach for multiple antenna transmissions.

8.3.2 Peak Windowing

The peak windowing approach [15] to CFR reduces the PAPR by attenuating the waveform over a short interval of time surrounding a peak. The block diagram for peak windowing is shown in Figure 8.16. Once a peak in the waveform is detected, a tapered window function centered on the peak multiplies the waveform attenuating the peak and its neighboring samples. The window function will have typically a

Gaussian shape that is truncated in the time domain. Other suitable functions include the Hann, Hamming, and Kaiser windows.

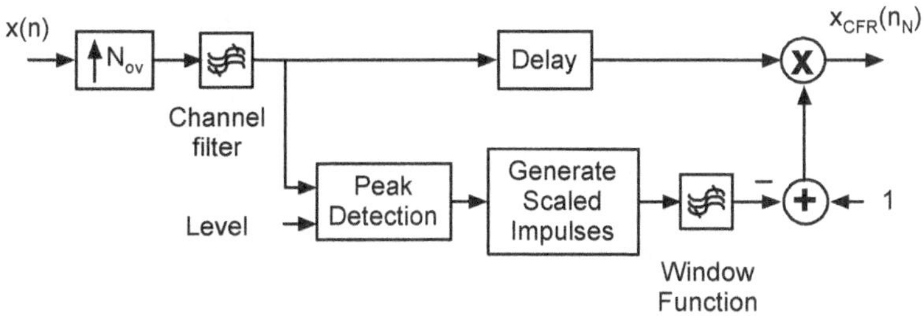

Figure 8.16 Peak windowing.

Let a peak be defined by the sample having the largest magnitude within an interval of time spanned by the window function that also exceeds a threshold L. The location in time of the detected peak is denoted by $\delta_{peak}(n)$. The peak windowed waveform becomes

$$x_{pw}(n) = (1 - c(n))\, x(n) \tag{8.22}$$

where

$$c(n) = \sum_m \gamma_m \sum_\tau g(\tau)\delta_{peaks}\left(n - \lambda_m - \tau\right) \tag{8.23}$$

$$\gamma_m = \max\left(1 - \frac{\mathrm{L}}{|x|}, 0\right) \tag{8.24}$$

λ_m is the location of peak m and $g(n)$ is a Gaussian window.

Multiplying the time sequence by a window is the same as convolving the spectrum by the Fourier transform of the window. Thus, a signal that is band-limited to an allocated channel will leak energy into the adjacent channel when peak windowing is applied. The width of the window in time is chosen to control the leakage. Long windows reduce leakage, which is desirable, but also reduce the signal energy transmitted, which is not. In the extreme case, a window spanning the entire time sequence is no different than backing-off the amplifier, which would do little to improve PA efficiency.

To illustrate the trade-off between adjacent channel leakage and transmitted power, consider the Gaussian window. The Gaussian window in the time domain is also a Gaussian function in the frequency domain. Let the standard deviations of Gaussians define the width in time (Δt) and the bandwidth in frequency (Δf). The product $\Delta t * \Delta f$ is constant for the Gaussian, which means that the narrower its Fourier transform is in frequency, the wider the Gaussian window must be made in time. The amount of margin available in the $ACLR_1$ and $ACLR_2$ measurements will determine in part the selection of the window width Δt.

The ACLR degradation is also determined by how aggressive the peak reduction is applied. If the reduction of the peak is modest, the spectral leakage will be low in power regardless of the window chosen. That is, it is possible to select the window based on the amount of peak reduction required where a wider window (in time) is used when the higher reductions are required.

If the ACLR does not have sufficient margin, the peak windowing can be combined with the error waveform subtraction method, shown earlier in Figure 8.14. The error waveform used in place of the clipped signal is $-c(n)x(n)$, which is the difference between the original signal $x(n)$ and the peak windowed signal $x_{pw}(n)$. This error signal is filtered as needed to meet the ACLR specifications [16]. This is one of many examples where different CFR approaches can be combined. In terms of suitability for use in LTE downlink signals and multiple antenna transmissions, the peak windowing approach is similar to the clip-and-filter approaches.

8.3.3 Partial Transmit Sequence

Other methods for CFR adjust the phase of the sub-carriers to prevent peak forming alignments. These include partial transmit sequence (PTS) [17–19] and selective mapping (SLM) [20]. The sub-carriers are multiplied by different phase vectors producing a set of potential time sequences. The time sequence with the lowest PAPR is transmitted. The disadvantage of such approaches, typically, is that side information regarding the phase shift vector used must be sent to the receiver to allow demodulation. The benefit of the PTS approach is that there is no EVM degradation.

Recall that the OFDM signal is

$$x(n) = \sum_{k} X(k,l) \exp\left(jk\Delta\omega n_l\right) \tag{8.25}$$

The crest factor reduced signal for PTS is

$$x_{CFR}(n) = \sum_{k} X(k,l) \exp\left(jk\Delta\omega n_l + \theta_k(l)\right) \tag{8.26}$$

where $\theta_k(l)$ is a phase shift for sub-carrier k and time interval l. The phase shift vector $\bar{\theta}$ for the OFDM time interval l is

$$\bar{\theta} = \begin{bmatrix} \theta_1 & \cdots & \theta_N \end{bmatrix}^T \tag{8.27}$$

It is selected to reduce the peaks of the OFDM signal and changes with each OFDM time interval.

Assume that the OFDM signal has M active sub-carriers. To reduce the complexity of the PTS approach, the sub-carriers are grouped into disjoint blocks of size B, where $B < M$, as shown in Figure 8.17. The block size is often selected to be $B = 2^{\beta}$ where β is an integer because of computational efficiencies inherent within radix-2 FFT/IFFT operations. The number of sub-carrier blocks, denoted by N_{blks}, is M/B. The sub-carrier blocks are zero padded to a size N and then inverse Fourier transformed to create N_{blks} partial transmit sequences, each of which is phase shifted by θ_k with the goal of reducing the PAPR.

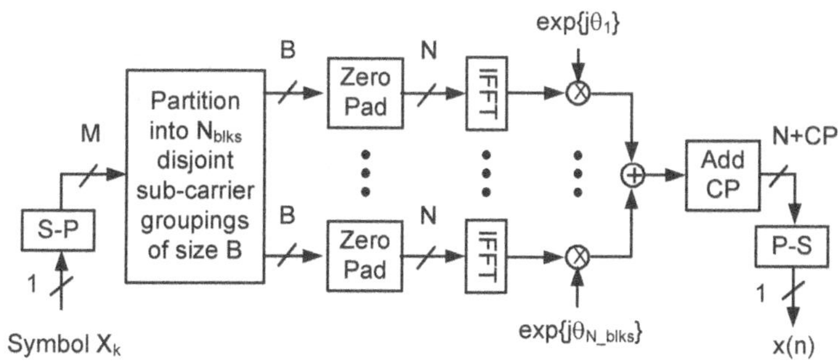

Figure 8.17 PTS approach for CFR of an OFDM signal.

An alternative implementation of the PTS approach is shown in Figure 8.18. The phase shift vector is applied to the sub-carriers before the IFFT. A common phase is assigned to each sub-carrier within a given block B.

The phase shift is often quantized into a finite number of levels, N_{θ}. The number of sub-carrier blocks N_{blks} and discrete phase shifts N_{θ} affects the complexity of the phase vector search, the amount of side information needed by the receiver, and the ability to reduce the peak. The number of possible phase shift vectors is $[N_{\theta}]^{N_{blks}}$. A common phase shift to all blocks does not affect the CFR, so the first phase shift is often set to zero ($\theta_1 = 0$). Thus, the number of phase shift vectors in the search space for the best CFR is $[N_{\theta}]^{N_{blks}-1}$.

A variation of the PTS is the selective mapping (SLM) approach. SLM uses the implementation shown in Figure 8.18. It creates and tests a finite number of pre-selected phase shift vectors, without requiring specific sub-carrier groupings. By assigning each vector an identifying tag, the side information is reduced and is

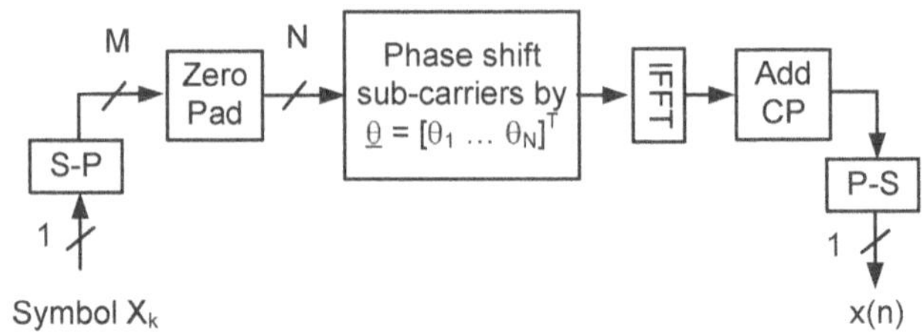

Figure 8.18 An alternative PTS implementation for CFR.

independent of sub-carrier groupings. The search for the lowest PAPR is exhaustive; nevertheless, the total number of pre-selected vectors is known, making the search complexity fixed and bounded. The drawback of the SLM approach is that the number of vectors considered is typically small, making the CFR performance less than that of PTS.

The PTS approaches are often divided into two classes: optimal and sub-optimal. In our opinion, the concept of an optimal solution is misguided because it is dependent on the size of the sub-carrier blocks and the number of discrete phase shifts allowed. That is, it is the optimum of a reduced-size subset made manageable by enforcing arbitrary sub-carrier groupings and coarse quantization of the phase shifts. The true optimum would phase shift each sub-carrier independently; however, the transmission of such a large phase shift vector to the receiver as side information would consume almost as much data throughput capacity as the original signal. This is undesirable.

A more sensible goal for PTS is to reduce the PAPR of the signal to a target level using as few computations as possible and sending the least amount of side information regarding the phase shift vector to the receiver. The fact that several possible phase shift vectors could achieve the desired PAPR target should not be a concern, even if the selected phase shift vector is sub-optimal. Several sub-optimal searches are described in Section 8.4; these terminate early if the desired PAPR is achieved.

The PTS approach requires modification when used in downlink LTE waveforms or multiple antenna transmissions. An alternative PTS approach for downlink LTE waveforms is described in Section 8.4. It requires that the same phase shift vector be used for an entire slot, which spans typically seven OFDM symbols in time, to avoid degrading the pilot-based (reference signal-based) channel estimation. Similarly, the same phase shift vector must be applied to signals for each antenna for MIMO and beam forming applications. The requirement for joint optimization makes the PTS approaches less effective for multi-antenna CFR.

8.3.4 Tone Reservation

A class of CFR methods that uses some of the OFDM sub-carriers as peak reducers is referred to as tone reservation [21]. The drawback of this approach is that throughput is lost because fewer sub-carriers are available for transmitting data. The benefit of tone reservation is that there is no EVM degradation on the data sub-carriers.

Two waveforms are created within the tone reservation approach: the data waveform $x(n)$ and the correction waveform $\Delta x(n)$, as shown in Figure 8.19. Unlike the peak cancellation method from the clip-and-filter section, the correction waveform does not use any of the data sub-carriers. Although this results in no EVM, there are fewer sub-carriers available for creating the correction waveform. This means that the filter kernel used in creating the correction waveform will have a wider temporal extent, compared to those used in the clip-and-filter approaches. This, in turn, will limit the performance of the CFR process because unwanted interaction between peak canceling sub-carriers is more likely.

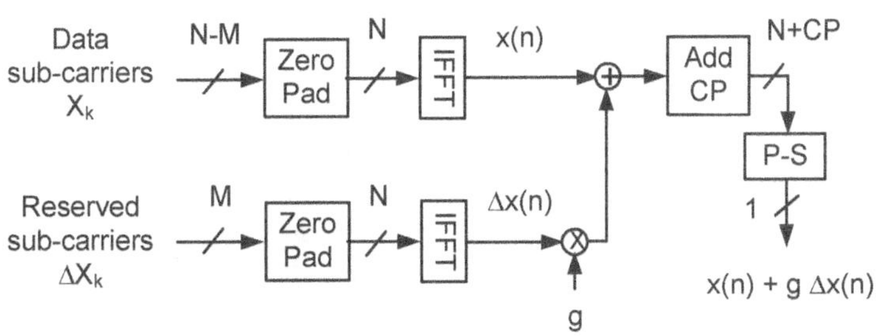

Figure 8.19 Tone reservation.

Consider two methods of selecting the sub-carrier frequencies reserved for CFR: localized and distributed. Localized spacing represents a contiguous block of sub-carriers, as shown in Figure 8.20. Compared to the entire signal, the correction waveform is severely band-limited. That is, the correction filter kernel has a wide main lobe width when represented in the time domain. A compromise is often required when subtracting a wide main lobe from a narrow peak, where the gain of the correction waveform may need to be reduced to avoid creating new peaks.

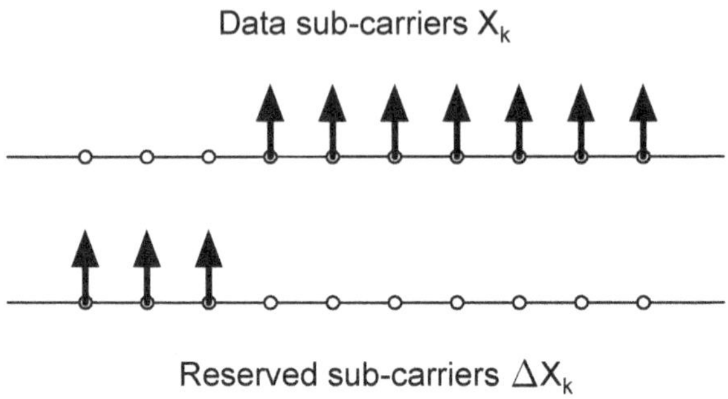

Figure 8.20 Localized reserved sub-carriers occupying a contiguous block of sub-carriers.

Now consider the distributed sub-carrier spacing. The reserved sub-carriers are selected to be equally spaced in frequency, as shown in Figure 8.21. Assume that the spacing between neighboring reserved sub-carriers is S. This sub-sampling in the frequency domain results in repetition within the time domain. That is, the correction waveform created will be a unique sequence for the first N/S samples in the time domain, and this sequence will be repeated S times over the OFDM time interval. As in the localized case, the gain of the correction waveform may need to be reduced to avoid creating new peaks.

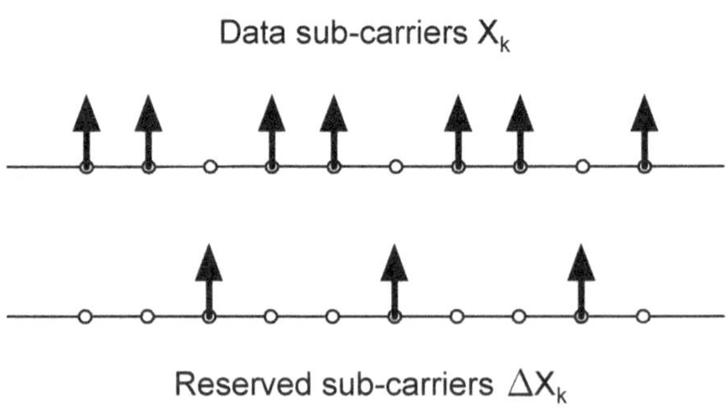

Figure 8.21 Distributed reserved sub-carriers that are spaced equally in frequency.

Regardless of how the reserved sub-carrier frequencies are selected, there are two basic methods to finding suitable correction coefficients. The first method constrains the magnitudes of all of the reserved sub-carriers to be a constant value. The correction waveform is obtained by searching for the best phase shift vector from the set available (search methods are described in Section 8.4). Assume that the permitted phase shifts are quantized to N_θ levels. The number of phase shift vectors in the search space is $[N_\theta]^{N_\Delta}$, where N_Δ is the number of sub-carriers reserved for the correction waveform. The search space expands if more than one gain level is considered as the scale factor for the correction waveform.

The second method for selecting the correction coefficients uses narrow bandwidth or sub-sampled (in the frequency domain) versions of the clip and filter approach. If one prefers the peak cancellation method, the signal peaks are detected, pulses are generated, and the pulse shaping is applied. The pulse shape, however, must be derived from the reserved sub-carriers only. In this case it would be wise to test several pulse shapes and incorporate a correction scale factor to avoid creating new peaks. A Gaussian pulse is used in [22] for peak cancellation where the relaxed roll-off of the Gaussian is exploited to reduce the likelihood of new peak creation. Alternatively, if error waveform subtraction is favored, the OFDM version shown in Figure 8.15 is used where the EVM control module sets all of the correction coefficients associated with data sub-carriers to zero. A correction scale factor is applied before the EVM control to compensate for the sparseness of the reserved sub-carriers within the frequency domain.

The temporal repetition of the distributed sub-carrier spacing provides a new (untested) method of generating the correction waveform. First, compute the target error waveform from the difference between the original and clipped signals. Split the error waveform into S segments. Create two time sequences of length N/S. The first sequence is the sum of all S segments. The second sequence comprises the maximum absolute values of the S segments from each sample. For each sample of the sum sequence that is larger in magnitude than the corresponding sample in the max sequence, normalize the former to match the magnitude of the latter. Repeat the segment S times and perform a FFT of length N to obtain the coefficients for the reserved sub-carriers. Perform filtering by setting the sub-carrier coefficients in the guard band to a sufficiently small magnitude to pass the spectral mask specification. Although this approach still has the potential to create peaks that are larger than the clipping threshold, it should be possible to obtain useful results with a single CFR pass.

A variation of tone reservation is to use the pilot signals to reduce peaks. Pilots are generally constant amplitude, BPSK-modulated sub-carriers used for estimating the RF channel response. If the binary sequence used to modulate the pilots can be selected from a set, then CFR can be achieved by selecting the sequence resulting in the lowest PAPR. A blind estimation is used at the receiver to determine the most likely binary sequence used.

Pilot-based CFR has also been proposed [23] by shifting the frequency indices of the pilot sub-carriers by an offset. That is, the distributed pilot sub-carriers are

$$k_{pilot}(i) = k_0 + i \cdot S \tag{8.28}$$

where k_o is the sub-carrier offset, which can have a value between 0 and $S - 1$. The offset k_o resulting in the lowest PAPR is selected. Although the offset k_o is considered side information needed to demodulate the signal at the receiver, the estimation of k_o can be performed blind because the pilot magnitude is often larger than the data sub-carriers and the pilot phase is restricted to 0 or π radians.

Pilot approaches will not work for downlink LTE because the reference signals (pilots) are not sent every OFDM symbol. In addition, the phase sequence and locations of reference signals are defined by the LTE specification.

Tone reservation is well-suited to multiple antenna transmissions because it does not affect the data sub-carriers. Only the reserved sub-carriers are phase-shifted or altered within the antenna signals. Since the information is discarded at the receiver, the signals at each antenna can be CFR optimized individually. For downlink LTE, tone reservation can be applied to unused resource blocks for CFR, which would be a form of contiguous spacing of reserved carriers (see Figure 8.20).

8.3.5 Constellation Extension (CE)

CFR can also be achieved by altering the IQ constellation so that the mapping between data bits and complex-valued modulation symbols is not unique. This is referred to as constellation extension [24]. Data bits are mapped to one of two opposing constellation positions so that c_{IQ} and $-c_{IQ}$ represent the same data, where c_{IQ} is the complex-valued modulation symbol, as shown in Figure 8.22. The downside of this approach is that one bit is lost in the constellation mapping which reduces the throughput for the QPSK, 16-QAM, and 64-QAM to ½, ¾, and ⅚ of the original value, respectively. The benefit is that there is no EVM degradation.

Constellation extension is used as part of an OFDM signal. Each sub-carrier is phase shifted by 0 or π radians independently to minimize the PAPR. The challenge is to search for the phase shift vector efficiently. A search for the optimum solution is often exhaustive. Sub-optimal solutions (see Section 8.4) are used typically, which involve an early termination of the search if a solution with a sufficiently low PAPR is found. Thus, the more relaxed the PAPR target, the faster the sub-optimal search will be completed, on average.

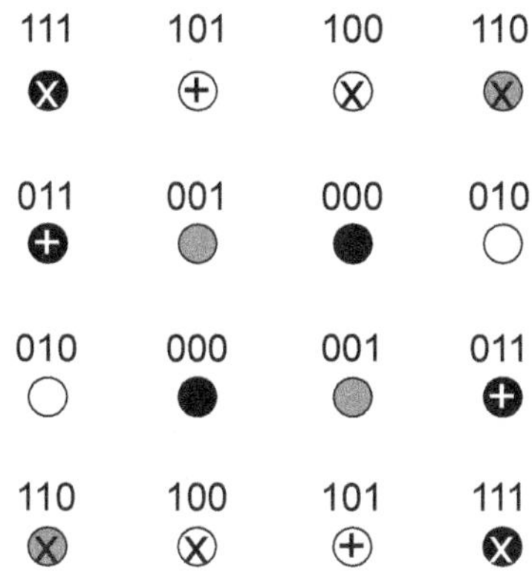

Figure 8.22 Constellation extension for the 16-QAM case. Data bit mapping is not unique because c_{IQ} and $-c_{IQ}$ represent the same data. The PAPR is reduced by selecting the sign for each sub-carrier in a favorable manner. One bit is lost in the mapping, reducing the data throughput by ¾ for the 16-QAM case.

In most cases, it is not necessary to apply constellation extension to every sub-carrier. For example, there is no point changing the sign of more than half of the sub-carriers because $x_{CFR}(n)$ has the same PAPR as $-x_{CFR}(n)$. Thus, it is the recommendation of the author that CE be applied to only a sub-set of the sub-carriers to limit the data throughput loss. The number of CE sub-carriers needed would depend on the PAPR target where a higher PAPR target requires fewer CE sub-carriers for CFR.

A variation of the constellation extension is selective sub-carrier degradation [4]. Bit errors are introduced intentionally, so that c_{IQ} becomes either $(c_{IQ})^*$ or $-(c_{IQ})^*$, where $()^*$ indicates complex conjugate. Redundancy coding is required to compensate for the bit errors, which reduces the data throughput.

It can also be useful to combine tone reservation and constellation extension. That is, the reserved sub-carriers are allowed to carry some information, losing only the single bit associated with CE. In doing this, the throughput loss associated with reserved sub-carriers is reduced.

The CE approach is well-suited for downlink LTE signals because the bit lost to CFR is ignored by the receiver. As a result, joint optimization of symbols within a slot is not required. On the other hand, CE is not well-suited for multiple antenna transmissions because the same phase shift vector must be applied to signals for each antenna. In this case, joint optimization is required, making CE less effective.

8.3.6 Coding

There are several coding techniques that reduce the PAPR of a signal. Coding trades off data throughput for a lower PAPR, but does not increase EVM. Block coding is discussed below for OFDM signals; however, it is not recommended for LTE. Two other coding methods that reduce the PAPR are described below, which are used by or compatible with existing standards, namely downlink CDMA IS-95 and uplink LTE.

Block coding can be used to reduce the PAPR for OFDM systems when the sub-carriers are modulated using constant modulus constellations such as BPSK and QPSK. Block coding maps a data set of size M to N sub-carriers where $M < N$ with the goal of preventing phase shift combinations of sub-carriers that cause peaks [25]. That is, $N - M$ sub-carriers become parity signals, which provide redundancy to the transmit waveform. This provides the potential for error correction. Block coding works well for a small number of sub-carriers. Unfortunately, it becomes intractable for downlink LTE waveforms, which have 600 sub-carriers for a 10 MHz bandwidth. In addition, LTE sub-carriers can be modulated using 16-QAM or 64-QAM, neither of which have constant modulus constellations.

An example of coding designed for downlink CDMA IS-95, but extendable to newer CDMA formats, can be found in [26]. Walsh codes are grouped into sets of four of which some of the codes are used for peak reduction. This encoding is applied recursively to create a PAPR of 0 dB, prior to filtering, for a fully utilized Walsh code set of $SF = 4^{\beta}$, where β is an integer. For $SF = 4$, there are three data carrying codes and one encoding code. As the set size increases, a larger proportion of the code set is assigned to encoding. For $SF = 16$ and $SF = 64$, encoding consumes 7 and 37 Walsh codes, respectively.

This approach illustrates several practical problems in CFR that are often overlooked. For example, the goal of obtaining a PAPR of 0 dB pre-filter is not reasonable. Since the resulting signal is BPSK, the interpolation filter for IS-95 will increase the PAPR of the final signal to 5.7 dB. In addition, it is rare that the IS-95 code set is utilized fully because of the interference it causes for neighboring cells. Nevertheless, the approach is interesting and could be applied in moderation to provide useful results for newer CDMA-based formats with far less throughput loss or neighboring cell interference.

The uplink for LTE uses a pre-coded OFDM signal that is referred to as single carrier, frequency division multiple access (SC-FDMA) and is shown in Figure 8.23. SC-FDMA starts as time symbols, which are buffered and converted to the frequency domain using a FFT. The frequency symbols are mapped to the OFDM sub-carriers and transformed back to the time domain using the IFFT. The size of the FFT (M) is smaller than the IFFT size (N). The mapping shifts the sub-carrier index k by a frequency offset k_o and fills the unused sub-carriers with zeros. This operation is the same as an up-conversion to IF and an interpolation by a factor of N/M with filtering. Thus, the PAPR will be the same as the original time sequence of complex data, filtered by a sinc function. The PAPR using the 10^{-4} probability level varies typically from 6.5 dB to 7.4 dB, depending on the modulation of the data (QPSK or 16QAM) [27]. This is an improvement over the downlink LTE signal whose 10^{-4} PAPR is 9.5 dB typically.

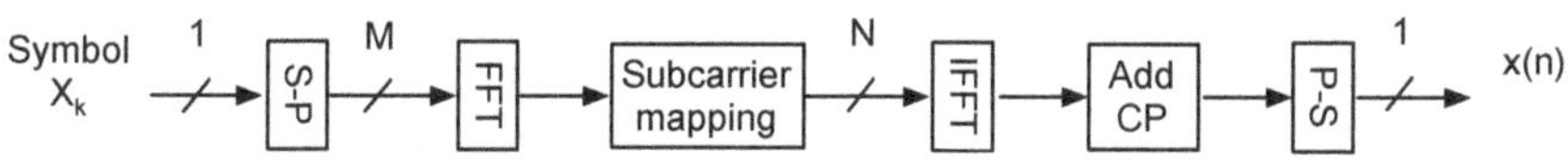

Figure 8.23 SC-FDMA signal generation.

8.4 ADVANCED TOPICS IN CFR

The previous section described many basic aspects of CFR. In the following, advanced topics are discussed. These include search techniques for phase shift vectors, CFR for HSDPA+ signals, and CFR for downlink LTE signals.

8.4.1 Search Techniques

Many of the above-mentioned CFR approaches are based on phase-shifting sub-carriers by quantized values ($\theta = m\pi/(N_\theta - 1)$ radians where $m = 0 \cdots N_\theta - 1$). They rely on searches to find phase shift vectors that provide the desired PAPR. Searches for the optimal phase shift vector are often exhaustive; several sub-optimal search methods have been developed which terminate early if a sufficient vector is found.

The iterative flipping algorithm [18] is a sub-optimal alternative to an exhaustive search. For the PTS implementation, the sub-carriers are divided into N_{blks} blocks. The phase shift vector is set initially to zero. The first block is not phase shifted because a common phase shift between blocks has no effect on the CFR of the signal. The phase of the second block is adjusted to test each of the quantized phases and the phase shift resulting in the lowest PAPR is retained. The third and successive blocks are phase shifted, in a serial manner, retaining the lowest PAPR of those tested. Earlier shifted blocks are not retested so once a block is phase shifted, that value is retained as part of the final vector. The search stops after all of the blocks (except the first) have been tested or the PAPR target is reached. The complexity of the search is $N_{blks} * (N_\theta - 1)$, assuming there is no early termination. In contrast, the complexity of an exhaustive search is $N_\theta^{(N_{blks}-1)}$.

Another PTS-compatible technique is the neighborhood gradient search (NGS) [17]. The NGS tests exhaustively all phase shift vectors within a neighborhood surrounding the present vector $\bar{\theta}_0$. The separation between two phase shift vectors is defined by the Hamming distance measure, which has two possible definitions. The first Hamming distance measure is

$$d_{Ham,1}(i,0) = \sum_n f_H\left(\bar{\theta}_i(n) - \bar{\theta}_0(n)\right) \tag{8.29}$$

where

$$f_H(a) = \begin{cases} 1 & \text{for} \quad |a| > 0.5\Delta\theta \\ 0 & \quad\text{otherwise} \end{cases} \tag{8.30}$$

$\bar{\theta}_i$ is the phase shift vector i, and $\Delta\theta$ is the quantization of the phase, $(\Delta\theta = \pi/(N_\theta - 1))$. This represents a count of the number of phase elements that are different between two vectors θ_i and θ_0. The second Hamming distance measure is defined as

$$d_{Ham,2}(i,0) = \sum_n \frac{\left|\bar{\theta}_i(n) - \bar{\theta}_0(n)\right|}{\Delta\theta} \tag{8.31}$$

and represents the total number of phase quantizations by which vectors $\bar{\theta}_i$ and $\bar{\theta}_0$ differ. It should be apparent that $d_{Ham,2} \geq d_{Ham,1}$. The two Hamming distance measures are equivalent when the phase shift is quantized to 0 and π radians.

Consider the search of the neighborhood comprising all phase shift vectors that are within a Hamming distance of one from the present vector $\bar{\theta}_0$, for the case of $\Delta\theta = \pi$ radians. A phase shift of π radians is applied to one sub-carrier grouping each pass. All possible positions are tested to determine the lowest PAPR possible for a single phase shift applied. Multiple passes are used to introduce additional phase shifts to the vector. The complexity of the search is $N_{blks} * N_{iter}$ where N_{iter} is the number of passes (iterations) made.

The CFR techniques discussed so far minimize the L_{inf} norm. That is, we wish to minimize (2), which is based on a max function. Search techniques that minimize other norms, such as L_p where $p < \inf$, have more favorable convergence properties. That is, L_p norms with finite values of p have computable derivatives that allow for gradient-based searches. One such approach, [28], optimizes the phase shift vector based on a L_4 norm initially, and then re-optimizes the vector based on increasingly larger values of p. This approach tends to avoid local minima that would be present if one started with a high value of p. The steady-state vector will be a local minimum of the L_{inf} norm that is close to the global minimum of the L_4 norm. Although there is no guarantee that the steady-state phase shift vector will be the global minimum for the L_{inf} norm, it often is in practice. Fortunately, the only requirement is that the steady-state performance meets the PAPR target.

An optimal search technique with lower complexity than exhaustive is sphere decoding [19]. Sphere decoding generates only solutions for which the L_2 norm is constrained below a specified threshold. The computation of the viable solutions uses a Cholesky decomposition to create a set of simultaneous equations with an upper triangular matrix structure. Specifying some of the elements of the phase shift vector constrains the search space for the remainder of the elements (which are quantized phase shifts). This is shown in [19] to reduce the search time by as much as 50 % compared to an exhaustive search.

8.4.2 CFR and WCDMA HSDPA+ 64-QAM Signals

In addition to EVM and BER, CDMA-based signals measure in-band degradations using the relative code domain error (RCDE). RCDE is related to the amount of power detected by the receiver within inactive Walsh codes. Walsh codes are orthogonal under normal conditions; however, leakage into unused Walsh codes occurs due to nonlinearities in the transmitter, or impairments in the receiver such as LO phase noise or synchronization problems. For the case of CFR, the effect of nonlinearities on the RCDE is the most relevant.

Consider the effect of PA nonlinearities on HSDPA+ 64QAM signals, in terms of RCDE and ACLR. Assume that the input is a 3GPP WCDMA test model 6 (TM6) signal [8]. The ACLR and RCDE for a single carrier TM6 signal and a class A biased amplifier driven at various power levels is shown in Table 8.2 [29]. Both the ACLR and RCDE pass the 3GPP specifications at low input power levels. As the input power increases, the ACLR fails first. An additional 4 dB of input power is required before the RCDE approaches within 1 dB of the 3GPP specification limit. Thus, the single-carrier TM6 signal has significantly more margin for the RCDE specification compared to the ACLR.

Let us now look at the effect of CFR on the measured RCDE for the single carrier TM6 signal. CFR based on a clip-and-filter approach reduces the PAPR of the input signal by degrading the in-band signal quality. The TM6 signal can be crest factor reduced to a PAPR of 5.7 dB and still pass the RCDE specification (RCDE < -20 dB), as shown in Table 8.3. In general, the CFR should not consume the entire RCDE specification margin because other sources of signal degradation are present in the transmitter and receiver. However, the key point being made is that the WCDMA signal is very robust with respect to degradations of the in-band signal quality, and the CFR approach selected should exploit it.

Table 8.2

ACLR and RCDE Measurements for a Single Carrier 3GPP WCDMA Test Model 6 Signal after Nonlinear Amplification by an Over-driven Class-A Amplifier.

Power dBm	ACLR$_1$(L) dB	ACLR$_1$(U) dB	ACLR$_2$L) dB	ACLR$_2$(U) dB	RCDE dB
-15	-55.3 Pass	-56.7 Pass	-64.1 Pass	-64.2 Pass	-40.8 Pass
-10	-42.4 Fail	-42.5 Fail	-62.3 Pass	-60.2 Pass	-33.6 Pass
-6	-30.4 Fail	-29.8 Fail	-53.3 Pass	-49.1 Fail	-20.9 Pass
3GPP	-45.0	-45.0	-50.0	-50.0	-20.0

Table 8.3

RCDE Measurements for a Single Carrier 3GPP WCDMA Test Model 6 Signal for Different PAPR Levels after CFR.

PAPR	10.7 dB	8.2 dB	7.2 dB	5.7 dB
RCDE	-41.6 dB	-36.5 dB	-24.3 dB	-20.7 dB

8.4.3 CFR for Downlink LTE Using Phase-Shifted Resource Blocks

A CFR approach specific to downlink LTE is described in [30]. It is a PTS-based method that exploits characteristics of LTE to transmit low PAPR waveforms, without requiring auxiliary information to be sent to the receiver. The downlink LTE waveform comprises resource blocks, which are groups of time-frequency elements that span one slot (typically seven OFDM symbols) and 12 sub-carriers, as shown in Figure 8.24. In addition to data, resource blocks are also assigned several reference signals, or pilots, which the demodulator within the receiver uses to equalize the propagation effects of the downlink channel.

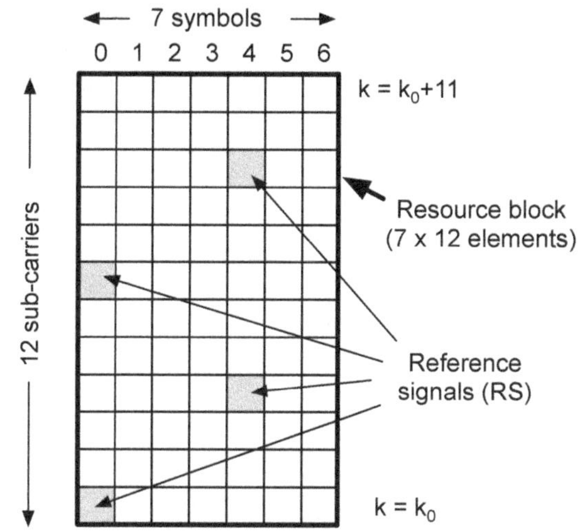

Figure 8.24 Resource block and reference signals for LTE downlink.

In this PTS approach, a common phase shift is applied at the transmitter to all elements within the resource block so that

$$\theta_k(l) = \theta_b \tag{8.32}$$

where $b = floor[k/12]$ and $l = 1:7$. A different phase shift is assigned to each resource block to achieve the CFR. This is an example of a joint optimization of the phase shift vector, which in this case spans seven symbols in time.

At the LTE receiver, the demodulation process equalizes the transmit chain by comparing the received reference signals with their intended values and interpolating the computed frequency responses to nearby sub-carriers that carry data. For the proposed approach, interpolation is restricted to each resource block. The phases of the four reference signals within the resource block are used to compute a mean phase shift $\theta_{ave}(b)$. The CFR phase shift vector θ_b becomes part of $\theta_{ave}(b)$ and is thus removed by the equalization process. This eliminates the need to transmit side information regarding θ_b. Thus, resource block phase shifts will not affect the demodulation process if the equalization at the receiver is performed on individual blocks separately.

At this point any PTS method can be used to determine an adequate phase shift vector θ_b. The problem faced by most PTS algorithms is that the computational complexity increases with the size of the phase shift vector, which equals the number of resource blocks for the downlink LTE case. The 50 resource blocks used in a 10 MHz bandwidth create a phase shift vector that is considered large for standard PTS algorithms.

It is preferable to use a search method that requires fewer IFFT/FFT computations to find the phase shift vector θ_b. The proposed method extracts the peaks from the original time-domain signal (after the IFFT) and correlates the peaks with the complex-valued modulation symbols within the resource block grid. The phase shifts in the proposed CFR are smaller than NGS ($|\Delta\theta| = 0.05\pi$ radians instead of π) but are applied to all of the resource blocks each iteration. Recall that the crest factor reduced signal for PTS is

$$\mathrm{x_{CFR}(n)} = \sum_k X(k,l)\exp\left(jk\Delta\omega n_l + \theta_b\right) \tag{8.33}$$

For small angles, $|\Delta\theta_b|$, the correction signal is approximated by

$$\Delta x(n) = j \cdot \sum_k X(k,l)\theta_b \exp\left(jk\Delta\omega n_l\right) \tag{8.34}$$

Thus, we get

$$\theta_b \approx \frac{j \cdot \sum\limits_{l=0}^{0} \sum\limits_{k=12b}^{12b+11} X(k,l)\Delta X^*(k,l)}{\sum\limits_{l=0}^{6} \sum\limits_{k=12b}^{12b+11} |X(k,l)|^2} \tag{8.35}$$

where $\Delta X(k,l) = FFT\{\Delta x(n)\}$. The correction signal $\Delta x(n)$ is equal to the band-limited version of the error signal, $x_{peaks}(n)$. A FFT is applied to obtain the complex-valued modulation symbols corresponding to $x_{peaks}(n)$:

$$X_{peaks}(k,l) = FFT\left\{x_{peaks}(n)\right\} \tag{8.36}$$

From (8.35), it can be seen that the cross-correlation of the modulation symbols of the original and error signals for the sub-carriers within resource block b,

$$C_{RB}(b) = \sum_{l=0}^{6} \sum_{k=12b}^{12b+11} X(k,l) \cdot X^*_{peaks}(k,l) \tag{8.37}$$

defines the phase shift direction needed for peak reduction. A phase shift θ_b is introduced such that

$$\theta_b = \begin{cases} |\Delta\theta| \operatorname{sgn}\left(\operatorname{Im}\left[C_{RB}(b)\right]\right) & \text{for} \quad \operatorname{Re}\left[C_{RB}(b)\right] > 0 \\ 0 & \qquad\qquad \text{otherwise} \end{cases} \tag{8.38}$$

where $|\Delta\theta|$ is a small angle.

The proposed search method starts by making the signal-under-test equal to the original signal $x(n)$. The PAPR is computed, and if the PAPR is below the target value, the signal-under-test is transmitted without modification. If CFR is required, the following steps are performed. The phase increment is selected initially to be a small value, such as $\Delta\theta = 0.05\pi$ radians. The phase shift vector θ_b, obtained by applying equations (8.36) through (8.38), is used to phase shift the resource blocks. The phase-shifted signal is generated and the PAPR is computed. If the PAPR is worse, the phase increment $\Delta\theta$ is reduced by half. If the PAPR is better, the phase-shifted signal becomes the new signal-under-test. The process is repeated until the PAPR is below the target value or until the maximum number of iterations is reached. The phase shift vector producing the lowest PAPR of those tested is transmitted. The complexity of the search for the phase shift vector, in terms of IFFT/FFTs, is $2 * N_{iter} + 1$ per symbol where N_{iter} is the number of iterations ($N_{iter} = 3$ is used below).

Consider a downlink LTE signal with a 10 MHz bandwidth, which has 50 resource blocks. The CCDFs of the downlink LTE waveform, with and without CFR, are listed in Table 8.4. Also listed is the CCDF of a Gaussian waveform for comparison. When the maximum number of iterations and target CFR are set to $N_{iter} = 3$ and $\lambda = 8$ dB, respectively, the proposed CFR reduces the peak from 11.3 dB to 8.8 dB. The CFR is improved further by increasing N_{iter} to 5 and lowering λ to 7 dB, as shown in Table 8.4.

The NGS CFR method was applied as well and used as a baseline. The number of resource blocks phase shifted by π radians (per slot) using NGS varied from 2 to 7, with a mean of 4.2 (the early termination threshold was set to 8 dB, maximum of 7 passes). The average complexity for 50 resource blocks is 210 (4.2*50) IFFT computations per symbol, which is too large to be practical. It is possible to reduce the complexity of NGS by computing a bank of 50 IFFTs that correspond to each resource block, zero padded to the original IFFT size; however, this would increase memory usage by a factor of 50. In either case, the complexity of the proposed CFR in terms of IFFT/FFT computations (7 per symbol for $N_{iter} = 3$) is significantly less than NGS (210 per symbol or 50 per symbol plus memory usage).

A comparison of the entries within Table 8.4 shows only a modest difference in the PAPR performance between the proposed CFR and the baseline NGS CFR. This is due to the fact that the selected phase shift vector must optimize seven OFDM symbols jointly, reducing the peak responses in each time interval while avoiding the creation of new larger peaks. This becomes increasingly difficult as the clipping threshold λ is reduced, as seen by the fact that the PAPR (0.000326%

probability in Table 8.4) increases when λ is reduced from 7 dB to 6.5 dB. Increasing the maximum number of iterations from 3 to 5 reduces the PAPR in each case; however, it is not clear that the incremental CFR improvement (< 0.3 dB) is enough to justify the additional computational expense. In summary, $N_{iter} = 3$ and $\lambda = 7$ dB is recommended because it represents a better trade-off between PAPR performance and computational complexity. Note that this method is an example of using a L_2 norm, equation (8.37), to direct a gradient-based search for a suitable crest factor reduced waveform.

Table 8.4

CCDF for CFR Applied to Downlink LTE

CFR Approach:	IFFT/FFTs	CCDF (peak-to-average)		
Proposed CFR	per symbol	0.1%	0.01%	0.000326%
$N_{iter} = 3, \lambda = 8dB$	7	8.0 dB	8.5 dB	8.76 dB
$N_{iter} = 3, \lambda = 7dB$	7	7.8 dB	8.4 dB	8.59 dB
$N_{iter} = 3, \lambda = 6.5dB$	7	7.7 dB	8.3 dB	8.76 dB
$N_{iter} = 5, \lambda = 8dB$	11	7.9 dB	8.2 dB	8.51 dB
$N_{iter} = 5, \lambda = 7dB$	11	7.7 dB	8.1 dB	8.34 dB
$N_{iter} = 5, \lambda = 6.5dB$	11	7.5 dB	8.1 dB	8.51 dB
NGS CFR (baseline)	210	8.1 dB	8.6 dB	8.81 dB
DL-LTE (no CFR)	1	8.4 dB	9.5 dB	11.33 dB
Gaussian Noise	-	8.4 dB	9.6 dB	11.02 dB

8.5 CONCLUDING COMMENTS

Within this concluding section, existing CFR approaches are summarized and a list of companies offering commercially available CFR products is provided. Our thoughts regarding future research directions and a summary of the chapter are also provided.

8.5.1 Summary of Existing Work

The approaches discussed in this chapter reduce the crest factor of the signal in various ways. In-band degradations are applied to the signal using addition or multiplication. Phase shift vectors are applied to sub-carriers of OFDM signals. Correction waveforms are created using reserved (non-data) sub-carriers. Non-unique mappings of bits to constellation symbols and coding are used to avoid symbol sequences that create peaks. Existing methods of limiting peaks not discussed include introducing time offsets between carriers for multi-carrier CDMA signals and prewhitening the input signal for OFDM. The tone injection method of CFR was not discussed; however, it is similar to the constellation extension approach in that a non-unique mapping of bits to constellation symbols is used.

In some cases it is simply better to allow others to do the work. Commercially available CFR systems have been developed by companies such as Texas Instruments, Optichron (now Broadcom), and PMC Sierra. Many FPGA vendors, such as Xilinx and Altera, have developed CFR algorithms optimized for their products as a form of sales support. However, it is still important to understand the CFR algorithm used to assess its limitations for the application required.

8.5.2 Future Work

Many comments were made earlier about the effect of CFR on MIMO and beam forming. Tone reservation appears to be the best CFR method for multiple antenna OFDM-based signals. Clip-and-filter and peak windowing are the easiest CFR methods to implement, making them the most likely to be used. Thus, actual experiments are needed to quantify the degradation associated with using clip-and-filter and peak windowing in multiple antenna applications.

New CFR approaches are of interest. This may be motivated by the introduction of a new modulation format or the desire for a proprietary algorithm that can be patented. The easiest paths to creating new CFR algorithms are to combine the basic approaches in novel ways or to draw upon well-studied theories and techniques from different fields. Peak windowing can be combined with error waveform subtraction. Tone reservation can be combined with constellation extension. In addition, searches based on lattice reduction techniques, similar to the sphere decoding method mentioned earlier, can be incorporated in PTS-type CFR to create something new.

The CFR module often precedes the DPD module within a digital transmitter as shown in Fig. 8.1. It has been shown in [31, 32] that it is possible to combine the CFR function into the DPD module. The error waveform for CFR is created using a weighted sum of basis functions present within the DPD module. The basis function selection and coefficient estimation are altered to consider the optimal trade-off between linearity and crest factor. But this may result in a better transmitter.

Crest factor reduction is based on the assumption that reducing the peak of the signal is beneficial to the performance of the digital transmitter. This is true, in general. However, it is important to remember that the goal of CFR in this chapter is to allow the linearized power amplifier to operate at a higher efficiency while meeting the linearity requirements of the modulation format used. It is our belief that CFR should consider the PA characteristics, as well as the characteristics of the linearization method used. For example, it may be more useful to suppress any part of the signal that is stimulating nonlinear modes that are difficult to linearize rather than just clip the peaks to a predefined level. For example, the slew rate of the signal envelope may also be important in the CFR design. Research is required to determine if this belief is true and how it can be used to improve the performance of the transmitter.

An example of a signal modification not related to the peak response, but used to improve the transmitter, is zero elimination [33,34]. The signal is modified to prevent the IQ trajectory from passing too close to the origin. This technique has utility for polar-type transmitters that vary the drain voltage as a part of the amplitude modulation (for example, envelope tracking). Zero elimination reduces the bandwidth of the envelope signal and avoids problems associated with trying to bring the PA output to zero using the drain voltage only.

8.5.3 Conclusion

As mentioned at the beginning of this chapter, crest factor reduction represents a trade-off. Reducing the crest factor of a signal allows amplifiers to run more efficiently at rated power. However, this is achieved at the expense of signal quality or data throughput. Fortunately, each modulation format has excess margin available with respect to some system performance metric. If the CFR approach is selected to best utilize the margins available, the overall system can be optimized to meet the specification at higher PA efficiencies.

A key piece of advice is to be reasonable in the selection of the PAPR target when working with band-limited signals. A 0-dB PAPR signal grows easily to 6 dB when pulse-shaped by a channel filter. If a 6 dB PAPR is considered too large, then one should reconsider the use of OFDM or CDMA.

References

[1] S. C. Cripps, *RF Power Amplifiers for Wireless Communications*, 2nd ed. Norwood, MA: Artech House, 2006.

[2] N. Maslennikov, 2010, personal communication.

[3] B. S. Krongold and D. L. Jones, "PAR reduction in OFDM via active constellation extension," *IEEE Trans. Broadcast.*, Vol. 49, No. 3, pp. 258–268, Sept. 2003.

[4] R. N. Braithwaite, "Crest factor reduction system and method for OFDM transmission systems using selective sub-carrier degradation," U.S. Patent 7 742 535, June, 2010.

[5] A. Papoulis, *Probability, Random Variables, and Stochastic Processes*. New York: McGraw-Hill, 1984.

[6] R. N. Braithwaite, "Using Walsh code selection to reduce the power variance of band-limited forward-link CDMA waveforms," *IEEE J. Sel. Areas Commun.*, Vol. 18, No. 11, pp. 2260–2269, Nov. 2000.

[7] R. N. Braithwaite, "Exploiting data and code interactions to reduce the power variance for CDMA sequences," *IEEE J. Sel. Areas Commun.*, Vol. 19, No. 6, pp. 1061–1069, June 2001.

[8] *3GPP: technical specification group radio access network; base station BS: conformance testing (FDD)*, 3GPP Std. TS 25.141, Rev. V9.5.0, Sep. 2010.

[9] G. C. Anderson, B. W. Finne, and G. T. Roberts, "Pseudo-random and random test signals," *Hewlett-Packard Journal*, Vol. 19, No. 1, pp. 2–20, Sept. 1967.

[10] C. Luke, 2007, personal communication.

[11] J. Wood, 2009, personal communication.

[12] G. A. Awater, R. D. deWild, and A. Hendrik, "Transmission system and method employing peak cancellation to reduce the peak-to-average power ratio," U.S. Patent 6 175 551, Jan. 2001.

[13] M. J. Hunton, "System and method for peak power reduction in spread spectrum communications systems," U.S. Patent 6 449 362, Sept., 2002.

[14] M. J. Hunton, "OFDM communications system employing vest factor reduction with ISI control," U.S. Patent 20 070 058 743, Mar. 2007.

[15] "Crest factor reduction," Application Note, Altera, Tech. Rep. 1.0, June 2007.

[16] R. G. Batruni, "Crest factor reduction," U.S. Patent 8 369 809, Feb., 2013.

[17] S. H. Han and J. H. Lee, "PAPR reduction of OFDM signals using a reduced-complexity PTS technique," *IEEE Signal Process. Lett.*, Vol. 10, No. 11, pp. 887–890, Nov. 2004.

[18] L. J. Cimini and N. R. Sollenberger, "OFDM communication system and method having a reduced peak-to-average power ratio," U.S. Patent 6 928 084, Aug., 2005.

[19] A. Alavi and C. Tellambura, "PAPR reduction of OFDM signals using partial transmit sequence: an optimal approach using sphere decoding," *IEEE Communications Letters*, Vol. 9, No. 11, pp. 982–984, Nov. 2005.

[20] C.-L. Wang and Y. Ouyang, "Low-complexity selected mapping schemes for peak-to-average power ratio reduction in OFDM systems," *IEEE Trans. Signal Process.*, Vol. 53, No. 12, pp. 4652–4660, Dec. 2004.

[21] J. Tellado and J. M. Cioffi, "Peak power reduction for multi carrier transmission," *IEEE GLOBECOMM*, Rio de Janeiro, Brazil, Dec. 1999.

[22] L. Guan and A. Zhu, "Gaussian pulse-based two-threshold parallel scaling tone reservation for PAPR reduction of OFDM signals," *Int. J. Digital Multimedia Broadcasting*, Vol. 2011, Apr. 2011, [Online]. Available: http://www.hindawi.com/journals/ijdmb/2011/470310/

[23] S. Hosokawa, S. Ohno, K. D. Teo, and T. Hinamoto, "Pilot-tone design for peak-to-average power reduction in OFDM," *IEEE Int. Symp. Circuits and Systems ISCAS*, Kobe, Japan, May 2005, pp. 6014–17.

[24] Y. J. Kou, W.-S. Lu, and A. Antoniou, "Peak-to-average power ratio reduction algorithms for OFDM systems via constellation extension," *IEEE Int. Symp. Circuits and Systems ISCAS*, Kobe, Japan, May 2005, pp. 2615–2618.

[25] T. Jiang and G. Zhu, "Complement block coding for reduction in peak-to-average power ratio of OFDM signals," *IEEE Commun. Mag.*, Vol. 43, No. 9, pp. S17–S22, Sep. 2005.

[26] T. Wada, T. Yamazoto, M. Katayama, and A. Ogawa, "A constant amplitude coding for orthogonal multi-code CDMA systems," *IEICE Trans. Fundamentals*, Vol. E80-A, No. 12, pp. 2477–2484, Dec. 1997.

[27] Y. Li, D. Prikhodko, Y. Tkachenko, and R. Zhu, "Compact power amplifier for LTE mobile terminals using coupling variation reduction technique," *MPD Microwave Product Digest*, p. 24, Apr. 2011.

[28] P. Guillaume, J. Schoukens, R. Pintelon, and I. Kollar, "Crest-factor minimization using nonlinear Chebyshev approximation methods," *IEEE Trans. Instrum. Meas.*, Vol. 40, No. 6, pp. 982–989, Dec. 1991.

[29] R. N. Braithwaite, "The effects of power amplifier nonlinearities and CFR on 64-QAM HSDPA waveforms," 81st *ARFTG Conference*, Seattle, WA, June 2013, pp. 1–3.

[30] R. N. Braithwaite, "Crest factor reduction for down-link LTE by transmitting phase-shifted resource blocks without side information," *European Wireless Conf. EuWIT*, Rome, Italy, Sept. 2009, pp. 13–16.

[31] R. N. Braithwaite, "Implementing crest factor reduction (CFR) by offsetting digital predistortion (DPD) coefficients," *Int Workshop on Integrated Nonlinear Microwave and Millimetre-wave Circuits (INMMIC)*, Dublin, Ireland, Sept. 2012, pp. 1–3.

[32] R. N. Braithwaite, "A combined approach to digital predistortion and crest factor reduction for the linearization of an RF power amplifier," *IEEE Trans. Microw. Theory Tech.*, Vol. 61, No. 1, pp. 291–302, Jan. 2013.

[33] M. J. Hunton, "System and method for eliminating signal zero crossings in single and multiple channel communications systems," U.S. Patent 6 931 240, Aug. 2005.

[34] J. Wang, A. Zhu, X. Zhu, and T. J. Brazil, "Vector hole punching technique using circle-tangent shift and unused tones," *IEEE Trans. Microw. Theory Tech.*, Vol. 57, No. 11, pp. 2682–2691, Nov. 2009.

Chapter 9

RF Pre-Distortion

Frédéric Roger

9.1 INTRODUCTION

Traditional RF PA linearization solutions relying solely on digital signal processing are readily available today. System OEM and PA manufacturers alike can choose from a variety of commercial solutions to increase the performance of their PA modules, or decide to implement their own custom DPD system. These solutions are mature, well understood, and are the primary choice when it comes to high-power PA linearization. Nevertheless, while offering very good performance, DPD has a significant cost.

First, implementing a DPD system requires significant engineering resources and system understanding to assemble their different parts. DPD requires, on one hand, the use of a high performance digital signal processing unit and, on the other hand, high-performance RF and analog components such as mixers or high-speed analog-to-digital converters (ADC) and digital-to-analog converters (DAC).

Second, emerging applications are creating new challenges that state-of-the-art DPD systems are not always able to solve, such as the linearization of low-power PAs in multi-antenna arrays or picocell applications for instance. For those 5W or less PAs, DPD is simply not an option: it is too large and power hungry.

Finally, new standards and markets are calling for even larger linearization bandwidth that traditional DPD cannot support today.

With the need for an alternative and simpler linearization method, we, at Scintera, decided to take an orthogonal approach based on an analog implementation called RF pre-distortion (RFPD). RF pre-distortion faces unusual design challenges both at the system and implementation level. First and foremost, little can be found in the literature. Few integrated and discrete implementations of analog pre-distorters (APD) have been reported but these pre-distorters were either not integrated [1, 2], did not integrate the RF and IF on the same die [3, 4], and did not reach the same complexity. Traditional RF systems such as transceivers are very well described and the impact of their building blocks' specifications on the final performance has been thoroughly analyzed and reported. Analysis methods and simulation platforms are readily available and can be used by system and circuit engineers alike.

RFPD, because of its novelty, does not offer the same kind of resources. Simulation methods for PA/linearizer combinations are not very accurate and it is extremely difficult to predict final performances accurately. We were able to improve these methods over time, but at the time of the genesis of SC1887, the first commercial RFPD solution, we had to make many decisions relying on analysis rather than simulation.

In this chapter, we will give an overview of the RF pre-distortion system that we have been developing [5]. We shall start by explaining the fundamentals of RF pre-distortion and emphasize the main differences with DPD. We will then present the architecture of our RFPD solution, before taking a look at the technical issues specific to the implementation of such a system. Indeed, while an analog implementation offers many advantages, it also suffers from imperfections that can ultimately limit the performance of the system. Finally, we will take a look at a few implemented solutions, and finish with some measurement results comparing Scintera's commercial RFPD linearizers.

9.2 FUNDAMENTALS OF RF PRE-DISTORTION

9.2.1 RFPD vs. DPD

First things first! Before going into the details of RFPD, we need to understand what a traditional DPD system looks like. In Figure 9.1, we recognize two signal paths: the pre-distortion path (TX) and the feedback or observation path (FB). The transmit path (TX) is composed of the digital baseband immediately followed by a DPD processor. These pre-distorted in-phase and quadrature digital signals are then converted into the analog domain. The two resulting analog baseband signals,

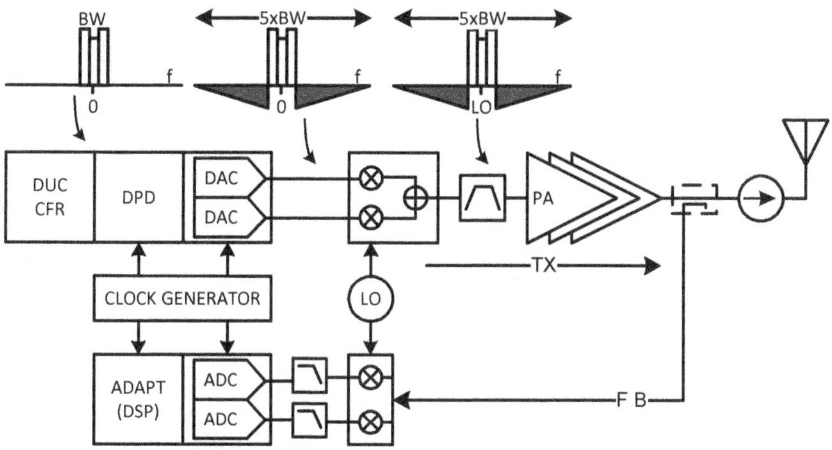

Figure 9.1 DPD application schematic.

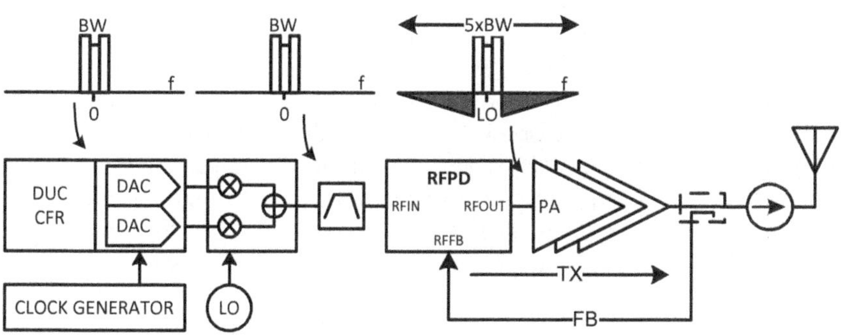

Figure 9.2 RFPD application schematic.

I and Q, are then up-converted to the RF frequency and combined using an analog quadrature modulator. At this stage, we should like to mention a fundamental drawback of DPD: as the linearization BW is typically about five times larger than the signal BW, all the devices following the DPD processor must be upgraded to also support the larger BW, resulting in higher component cost and power consumption. Finally, in order for the system to adapt the predistortion coefficients, the loop is closed by connecting the PA output to a monitoring path. This so-called feedback signal is down-converted, filtered, digitized, and fed to a DSP for processing and computation of the DPD coefficients.

RFPD is based on a different, yet very similar approach as can be seen in Figure 9.2. Because the predistortion function is implemented and applied to the transmit signal using only analog domain signal processing circuitry, the predistorter can be connected in front of the PA line-up, right after the reconstruction filter: no change to the existing baseband or up-conversion circuitry is required. This results in substantial savings compared with DPD, as neither the DACs nor the up-converter need to support the full linearization BW. The adaptation path implementation is very similar to DPD but it benefits by being fully integrated on the same die as the TX path

Compared with DPD, RFPD offers significant benefits. It is a cost-effective and easy-to-integrate technology because of the achievable form factor (currently available as a dual-channel $11x11$ mm IC). Further, because none of the external baseband components needs to run at the full linearization BW, the DAC, TX modulator, and clock generator can run at about one-fifth or lower the speed of their counterparts in a DPD system. Given that the power consumption and cost of these devices scale with their frequency of operation, any BW reduction leads to substantial savings (Table 9.1). Because RFPD is a true $RFIN \rightarrow RFOUT$ solution, no complex digital or algorithm development is required on the customer side, allowing PA designers with no predistortion knowledge to linearize their PA. This also means that the linearized PA function can be abstracted and developed entirely in the RF domain, independently of the baseband part of the system, as no access to the baseband processor is required. Finally, the high integration of all components necessary to the predistortion in a single IC allows for substantial bill of material (BOM) reduction and power savings that no DPD system can currently achieve.

9.2.2 Implementation

The integration of an RFPD pre-distorter is straightforward and can be added to any existing PA line-up in a fraction of the time required by conventional DPD

Table 9.1
RFPD vs. DPD

Component		DPD		RFPD
System speed	☹	X 5 (BW expansion)	☺	X 1
Complexity, cost	☹	Large device count, expensive data converters	☺	Single chip, integrated feedback, relaxed DAC speed
Reliability	☹	Multiple high performance chips	☺	Single chip
Performance	☺	Excellent	☹	Good

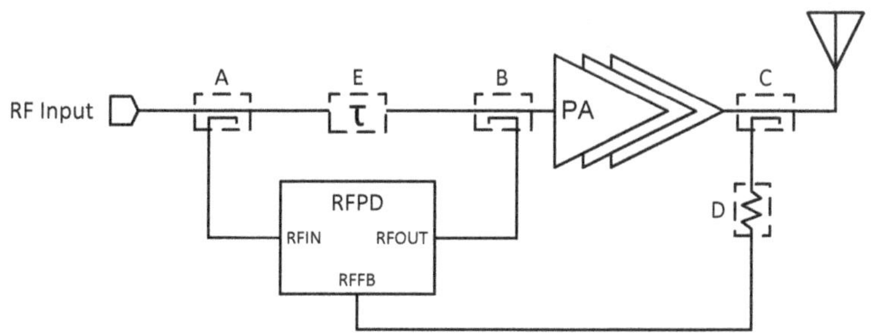

Figure 9.3 Application schematic of RFPD.

systems, as it is fully integrated and plug-and-play. Figure 9.3 illustrates RFPD's simplicity of integration. The RF input signal is connected to an input called *RFIN* through the directional coupler, *A*. The pre-distorted signal, *RFOUT*, is added back to a delayed version of *RFIN* and the delay *E*, which compensates for the group delay introduced by the RFPD IC, is implemented at the input of the PA (coupler *B*). Lastly, the output of the PA is connected to the observation port *RFFB* through the coupler *C* and attenuator *D*.

9.2.3 From RF to Baseband... and Back to RF

Because our goal is to process the transmit signal in the analog domain, we need to understand how the high-order terms needed for the pre-distortion signal are generated.

We can consider the simple case of a two-tone signal being distorted by an amplifier: $X = \cos(\omega_1) + \cos(\omega_2)$, where ω_1 and ω_2 are two frequencies within the signal bandwidth. Taking the cube of this signal generates harmonics and intermodulation products:

$$\begin{aligned}(X)^3 = &\frac{1}{4}(\cos(3\omega_1) + \cos(3\omega_2) + 3(\cos(2\omega_1 + \omega_2) + \cos(\omega_1 + 2\omega_2)) \\ &+ 9\cos(\omega_2) + 9\cos(\omega_1) \\ &+ \mathbf{3\cos(2\omega_2 - \omega_1) + 3\cos(2\omega_1 - \omega_2)})\end{aligned} \tag{9.1}$$

The first two lines of (9.1) show the-high order products that will fall far outside the desired frequency band. They will be filtered by the *TX* filter and can be ignored. The frequencies $2\omega_1 - \omega_2$ and $2\omega_2 - \omega_1$ are what are commonly referred to as *IMD3*. They are the distortion products that we aim to cancel.

How do we create these terms in the analog domain? It is basically a two-step approach. We start by squaring the RF signal (9.2).

$$\begin{aligned}(X)^2 = &\frac{1}{2}(\cos(2\omega_1) + \cos(2\omega_2)) + \cos(\omega_1 + \omega_2) + 1 \\ &+ \mathbf{\cos(\omega_1 - \omega_2)}\end{aligned} \tag{9.2}$$

This equation contains three different terms: a high-frequency term, a DC term, and the signal we are interested in: $\cos(\omega_1 - \omega_2)$.

The high-frequency term is not a concern because it is far out of band and can be filtered easily. The appearance of a DC component, on the other hand, is an issue that will need to be taken care of, as will be seen later in Section 9.4.1.

If we ignore the unwanted terms for the time being, we notice a very interesting property of this *RF squarer*: it demodulates the RF signal and outputs a baseband signal. Having a baseband signal at our disposal will allow us to make very efficient analog signal processing by creating all the necessary high-order terms at baseband rather than RF. This means that power and silicon area can be greatly reduced. The same correction engine can then be used for any RF carrier frequency or transmission protocol.

In our example, we can now multiply the baseband signal by a variable coefficient c_3 and generate the correction signal Y:

$$Y = c_3 \cos(\omega_1 - \omega_2) \tag{9.3}$$

The last step consists in multiplying the PD signal by the original RF signal as shown in the following:

$$
\begin{aligned}
X.Y =& (\cos(\omega_1) + \cos(\omega_2))c_3 cos(\omega_1 - \omega_2) && (9.4)\\
=& \frac{c_3}{2} \underbrace{(\cos(\omega_1) + \cos(\omega_2))}_{Original Signal} + \frac{c_3}{2} \underbrace{(\cos(2\omega_2 - \omega_1) + \cos(2\omega_1 - \omega_2))}_{Pre-distortion signal} && (9.5)
\end{aligned}
$$

Et Voila! We have just created a new RF signal consisting of the original tones ω_1, ω_2 as well as their third-order distortion products, $2\omega_1 - \omega_2$ and $2\omega_2 - \omega_1$ whose amplitude can be varied by the coefficient c_3. This process is shown in spectral terms in Figure 9.4.

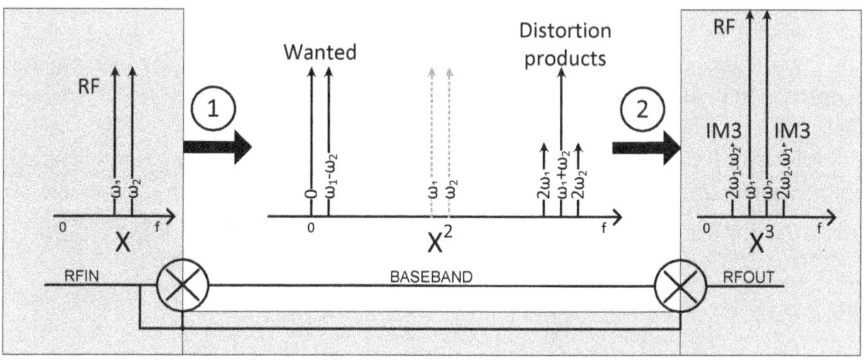

Figure 9.4 X, X^2, and X^3.

9.2.4 Volterra Series Simplification

In the following, it is assumed here that the reader is familiar with the general concept of pre-distortion using a Volterra series, as outlined in earlier chapters. Ideally, we would like to generate a Volterra series of the form given by (9.6).

$$
H(X) = \sum_{p=0}^{\infty} \sum_{n=0}^{\infty} c_{p,2n+1} X^{2n+1}(\omega t - \tau_p) \tag{9.6}
$$

where τ is a fixed delay.

We will start by constraining this equation to a practical example using 4 memory kernels and 7th degree polynomials ($p = 4, n = 3$):

$$H(X) =$$

$$\begin{aligned} &c_{0,1}X(\omega t-\tau_0)+c_{0,3}X^3(\omega t-\tau_0)+c_{0,5}X^5(\omega t-\tau_0)+c_{0,7}X^7(\omega t-\tau_0)\\ &+c_{1,1}X(\omega t-\tau_1)+c_{1,3}X^3(\omega t-\tau_1)+c_{1,5}X^5(\omega t-\tau_1)+c_{1,7}X^7(\omega t-\tau_1)\\ &+c_{2,1}X(\omega t-\tau_2)+c_{2,3}X^3(\omega t-\tau_2)+c_{2,5}X^5(\omega t-\tau_2)+c_{2,7}X^7(\omega t-\tau_2)\\ &+c_{3,1}X(\omega t-\tau_3)+c_{3,3}X^3(\omega t-\tau_3)+c_{3,5}X^5(\omega t-\tau_3)+c_{3,7}X^7(\omega t-\tau_3) \end{aligned} \quad (9.7)$$

Equation (9.7) is not practical for hardware implementation, mostly because it would require 4 delayed versions of the input RF signal $X(\tau_0, \tau_1, \tau_2, \tau_3)$. On-chip RF delays tend to be narrowband, lossy, and large. A straightforward implementation of this equation would also necessitate nonlinear RF operators to generate the high order terms $X^{(3,5,7)}$. Multipliers or diodes can be used but they also tend to be power hungry, lossy, and have a limited dynamic range. Luckily, only a subset of the information contained in (9.7) is actually needed. As a matter of fact, we are only interested in the intermodulation terms and not in the harmonics of the original signal.

As we have seen in Section 9.2.3, we chose a two-step approach where all the signal processing is realized at baseband rather than RF. By taking advantage of the fact that half of the energy of the even powers of RF signal are falling down to baseband, we can rewrite (9.6) and separate BB and RF signals:

$$H(X) = \sum_{p=0}^{\infty}\sum_{n=0}^{\infty} c_{p,2n} \underbrace{X(\omega t-\tau_p)}_{RFsignal}\underbrace{X^{2n}(\omega t-\tau_p)}_{BBsignal} \quad (9.8)$$

We then approximate (9.8) and replace the sum of RF signals by a single RF term:

$$H(X) = \underbrace{X(\omega t-\tau_{RF})}_{RFsignal}\underbrace{\sum_{p=0}^{\infty}\sum_{n=0}^{\infty} c_{p,2n}X^{2n}(\omega t-\tau_p)}_{BBsignal} \quad (9.9)$$

Simulations and experience have shown that this approximation was good enough to get the targeted system performance. Finally, we can expand (9.9) and get the transfer function:

$$H(X) = X(\omega t-\tau_{RF})(\alpha_{RF}+\sum_{p=0}^{3}\sum_{n=1}^{3} c_{p,2n}X^{2n}(\omega t-\tau_p)) \quad (9.10)$$

$$\alpha_{RF} = c_{0,0}+c_{1,0}+c_{2,0}+c_{3,0} \quad (9.11)$$

Equation (9.10) is a huge improvement over (9.6) because it is relying on delaying baseband signals rather than RF signals. It should also be noted that the delay τ_{RF} does not actually need to be implemented as we can rely on the inherent $RFIN \rightarrow RFOUT$ group delay inside the linearizer.

Before moving to the hardware implementation, we will apply a last transformation to further simplify the implementation. Indeed, even though the output of $\sum_{p=0}^{3}\sum_{n=1}^{3} c_{p,2n}X^{2n}(\omega t - \tau_p)$ is a baseband signal, the input $X^{2n}(\omega t - \tau_p)$ is still an RF signal. We can rewrite this equation as follows:

$$\sum_{p=0}^{3}\sum_{n=1}^{3} c_{p,2n}X_p^{2n}(\omega t - \tau_p) = \sum_{p=0}^{3} X_p^2(c_{p,2} + X_p^2(c_{p,4} + X_p^2 c_{p,6})) \qquad (9.12)$$

where

$$X_p^2 = X^2(\omega t - \tau_p) = \Delta_p X^2(\omega t) \qquad (9.13)$$

Equation (9.12) is now a sum of delayed baseband signals derived from the term $\boldsymbol{X_p^2(\omega t)}$. As can be seen, this equation can be implemented using a single RF squarer to generate $X^2(\omega t)$, immediately followed by a set of baseband analog delays.

9.3 RFPD ARCHITECTURE

In this section, we will present some of the implementation details of Scintera's RFPD solution. The proper design partitioning is key to such a complex high-speed analog processing architecture. Indeed, the wrong partitioning can quickly lead to losses such as: signal degradation, high power consumption, linearity limitation, high noise figure, or bandwidth reduction. We therefore took great care into laying out the different block diagrams and selecting the right circuit topologies (Figure 9.5). Wherever possible, we used passive circuits instead of active ones. We also tried to reduce the number of active stages to the strict minimum. Finally, many of the functions were combined into single circuits. This work paid off and a clean and simple architecture came out of the process as we will see now.

But before going any further, I would like to show you a picture that illustrates perfectly one of the major benefits of this technology: the achievable form-factor. The PCB in Figure 9.6 is 30 mm × 60 mm in size and includes all the components necessary for the linearization of any PA! It contains input and output couplers, a 6-ns delay line (see Figure 9.3), matching networks, crystal, and voltage regulators.

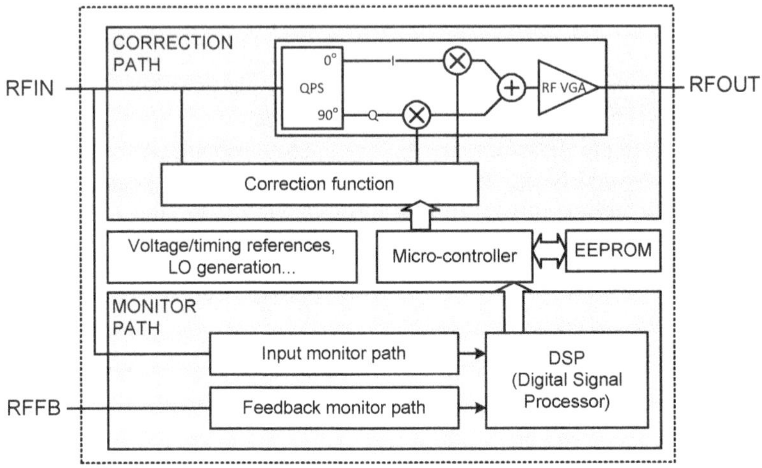

Figure 9.5 Block diagram.

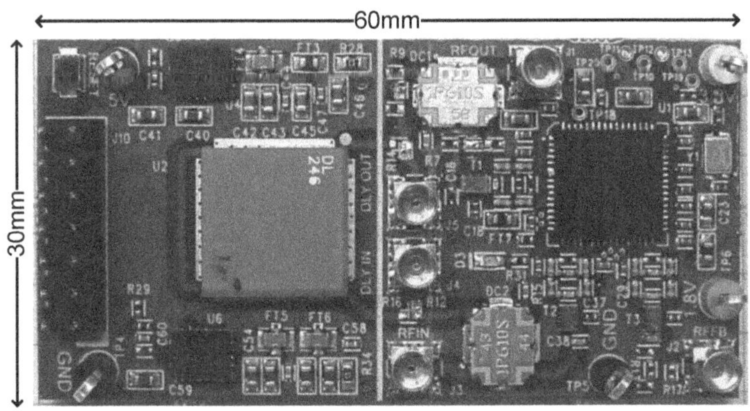

Figure 9.6 RFPD complete solution.

9.3.1 Analyzing and Processing Engine

As mentioned previously, the RFPD's feedback path is comparable to DPD (Figure 9.7). The feedback signal is down-converted, filtered, buffered, and converted to digital to be processed spectrally by a DSP. The originality of our approach lies

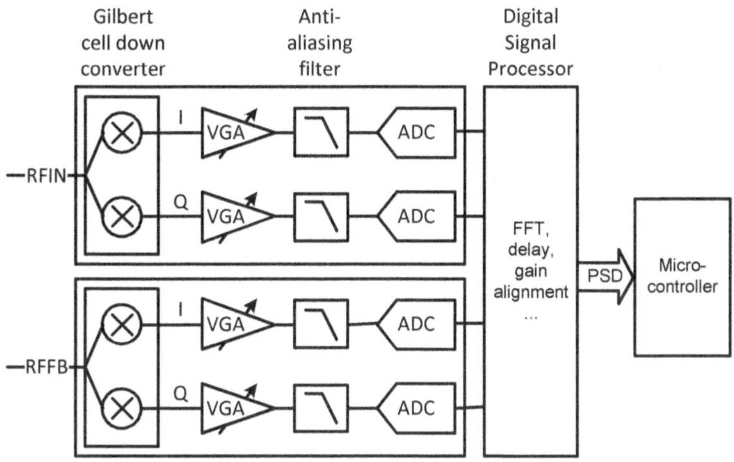

Figure 9.7 Adaptation paths.

in the full integration of this down-conversion chain. The integration of high speed ADCs and their associated low impedance buffers, digital output and clock drivers, in particular, leads to significant savings in both real-estate and power consumption (see Table 9.1 for a summary of these benefits).

For instance, most high-speed ADCs use LVDS signaling to transmit the conversion results. LVDS can be a real power hog as it requires a constant current equal to 3.5 mA/signal that is independent of the switching frequency. When powered from 3.3V, a 12-bit ADC will therefore dissipate 138 mW for the sole purpose of output data transmission ($3.5 \times 3.3 \times 12$).

9.3.2 Analog Data Path

Figure 9.8 depicts the top-level block diagram of SC1887/89/94 TX analog signal path. It is wideband and can virtually support any modern cellular standard with RF carrier anywhere between 160 MHz and 4.2 GHz.

The RF signal, RFIN, is fed to the RF signal processor (RFSP) and to the correction engine (CORR). The RFSP's first stage, called QPS, is an analog polyphase filter whose purpose is to extract the in-phase (I) and quadrature (Q) signals. Its RF outputs, I and Q, are multiplied by the corresponding I/Q signals generated by CORR, and added back together. The resulting pre-distortion signal is

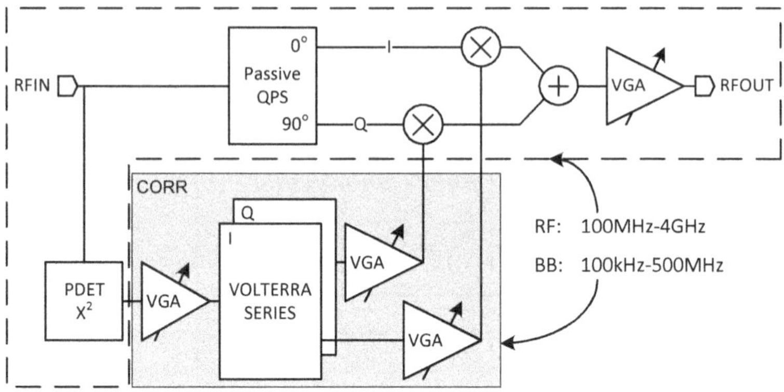

Figure 9.8 Signal Path.

then fed to the last stage that is the combination of an RF driver and a VGA. This VGA combination gives us some control over the power of the correction signal before it is sent out to the PA.

9.3.2.1 RF Squarer: PDET

From Section 9.2.3, we know that we first need to bring the RF signal down to baseband by squaring the incoming RF signal. This is the role of PDET (power detector or RF squarer).

The RF squarer plays a critical role in the performance of the system. It ought to convert the RF signal down to baseband (9.2) and drive the large capacitive load of the *CORR*. It needs to perform over the full RF frequency and input power ranges while, at the same time, keeping the signal integrity. Any signal degradation happening at this stage will indeed propagate through the signal path and degrade the final performance. It is interesting to note that the smallest gain variation at the input will translate into a large deviation at the output of the system due to the signal expansion inside the polynomial. A 1 dB variation at the input means 2 dB at the PDET output, 10 dB at the output of CORR, and 11 dB at the RF output. Table 9.2 presents a summary of the design target for PDET.

9.3.2.2 Correction Path

The correction block (CORR) is the heart of the system. Its function is to generate and apply nonlinear mathematical transformations to the incoming baseband signal.

Table 9.2
PDET Design Targets.

RF bandwidth	160 MHz to 4.2 GHz	
Video Bandwidth	100 kHz to 100 MHz	
Gain flatness within the VBW	< 0.1 dB	
Input dynamic range	> 10 dB	signal PAPR
Output dynamic range	> 20 dB	=2$\cdot PAPR$
Gain	20 dB	
Large driving capability	a few pF	needed to drive CORR

The SC1887's simplified Volterra series, (9.7), is made of four independent polynomials, or memory terms, connected in parallel (Figure 9.9). Each polynomial is identical but is fed by a time-shifted version of the PDET signal generated by the RF squarer.

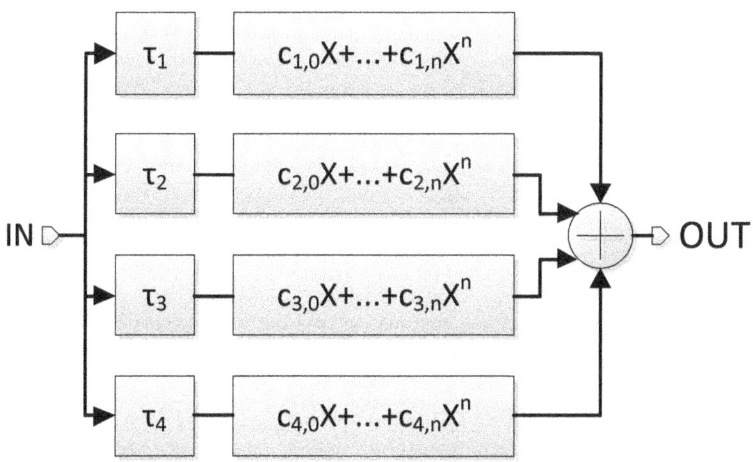

Figure 9.9 Work function.

Inside each memory kernel, the PDET signal is passed through a series of multipliers to generate the high-order harmonics of the RFIN signal $X(\omega t)$. Those harmonics are then multiplied by a series of coefficients $c_1, c_2, \ldots c_n$ and finally

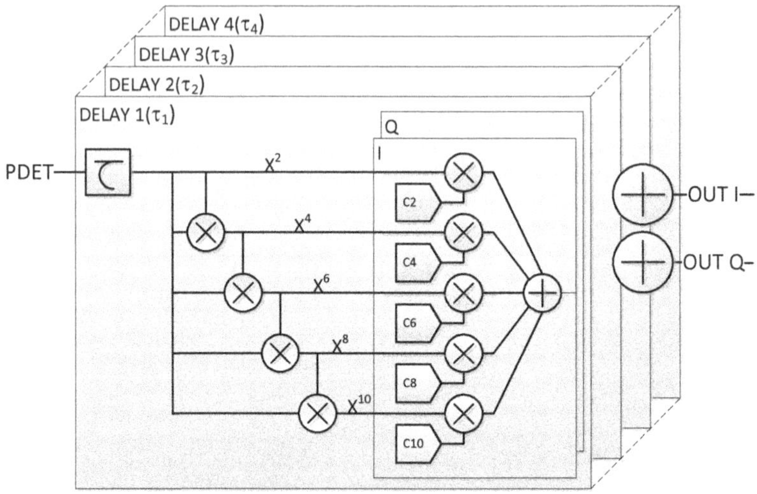

Figure 9.10 Memory kernels.

summed together to create a polynomial of the form: $c_1X + c_2X^2 + \ldots + c_nX^n$. This is shown in Figure 9.10. We will expand[1] on the resulting bandwidth expansion in Section 9.4.4.

Two independent Volterra series are actually generated, allowing us to create a complex pre-distortion signal and correct for AM-PM distortion as well as IQ imbalance resulting from analog impairments. Despite the CORR's complexity, there are no more than four active stages in series.

You might have noticed the absence of any dedicated RF filtering to remove the high frequency leakage resulting from the RF to BB down-conversion (see Section 9.2.3) . The reason is that each CORR stage, passive or active, introduces at least one low-frequency pole that limits the maximum signal BW. Consequently, the high-frequency harmonics are naturally attenuated and do not require any special attention.

1 I know... I know...☺

9.4 TECHNICAL CHALLENGES AND ANALOG LIMITATIONS

Implementing complex nonlinear signal processing in the analog domain is not such an easy task, as you might by now realize. Even after the work of simplification described in Section 9.2.4, implementing a Volterra series still requires a large number of analog signal-processing building blocks. The expansion of (9.12) speaks for itself:

$$\begin{aligned} H_{I|Q}(X) = & \sum_{n=1}^{3} c_{p,2n} X_p^{2n}(\omega t) \\ = & \quad X_0^2(\omega t)(c_{0,2} + X_0^2(\omega t)(c_{0,4} + X_0^2(\omega t)c_{0,6})) \\ & + X_1^2(\omega t)(c_{1,2} + X_1^2(\omega t)(c_{1,4} + X_1^2(\omega t)c_{1,6})) \\ & + X_2^2(\omega t)(c_{2,2} + X_2^2(\omega t)(c_{2,4} + X_2^2(\omega t)c_{2,6})) \\ & + X_3^2(\omega t)(c_{3,2} + X_3^2(\omega t)(c_{3,4} + X_3^2(\omega t)c_{3,6})) \end{aligned} \tag{9.14}$$

As is, implementing (9.14) requires 40 coefficient DACs, 56 analog multipliers, and 4 analog delays: see Figure 9.10.

In this section, we will review a few of the challenges specific to the implementation of this simplified Volterra series in the analog domain. While system limitations can be overcome by changing the architecture or topology of circuits, the laws of physics cannot be broken, and it is important to understand and accept their impact and limitations. We will take a quick look at the most fundamental challenges for our analog implementation: offsets, signal expansion – and compression, and peak- to-average power ratio.

9.4.1 Offsets and Their Impact on the Pre-distortion Signal

In Section 9.2.3, we purposely removed the DC term from the equations for greater clarity. Actually, as we will now see, the DC term plays a significant role in an analog implementation and cannot be simply discarded. In the following, we will review the different types of offset and describe their impact on the system.

9.4.1.1 Signal-Dependent Offsets

In the presence of any input signal, the output of any squarer exhibits a so-called signal dependent offset. It is called signal-dependent because it varies together with the input signal, and offset because of its DC nature. We can explain these phenomena quite easily by squaring a single tone having an amplitude A:

$$(A(t)\cos(\omega_0 t))^2 = \frac{A^2(t)}{2}(\cos(2\omega_0 t)+1) = \underbrace{\frac{A^2(t)\cos(2\omega_0 t)}{2}}_{Signal} + \underbrace{\frac{\boldsymbol{A^2(t)}}{\boldsymbol{2}}}_{DC_{Offset}} \tag{9.15}$$

As we see in (9.15), the term $A(t)^2/2$ is a time-varying *DC signal* that fluctuates together with the input signal.

Understanding the nature of this offset is critical, because it will propagate through the whole data path and have dramatic effects on the signal quality. To understand these effects, let us now take the output of (9.15) and square it to get the 4th-order harmonics:

$$(A(t)\cos(\omega_0 t))^4 = \frac{A^4(t)}{8}(\underbrace{\cos(4\omega_0 t)}_{Wanted} + \underbrace{4\cos(2\omega_0 t)}_{Cross-term} + \underbrace{3}_{DC}) \tag{9.16}$$

Equation (9.16) is composed of three terms:

- The *Wanted* signal is the 4th-order harmonic we are aiming for.
- The *Cross-term* is a 2nd-order leakage signal resulting from the DC component.
- The *DC* is the sum of the two successive squaring functions.

It is interesting, and somewhat unsettling, to note that the 4^{th}order wanted signal is four times smaller than the 2nd-order, and even three times smaller than the DC.

Equation (9.17) below shows what happens when we generate the 6th-order.

$$(A(t)\cos(\omega_0 t))^6 = \frac{A^6(t)}{32}(\underbrace{\cos(6\omega_0 t)}_{Wanted} + \underbrace{6\cos(4\omega_0 t) + 15\cos(2\omega_0 t)}_{Cross-terms} + \underbrace{10}_{DC}) \tag{9.17}$$

We could go on, and move on to the next multiplication but (take my word for it) this will not get any better! The higher the harmonics we create, the lower their amplitude, and the larger the resulting offset.

The consequences for the data path are various and beyond the scope of this chapter. We will only mention the obvious fact that the dynamic range and linearity will be greatly impacted because the DC terms easily overwhelm the useful signal and do not bring any value to the system.

9.4.1.2 Random Offsets

Random offsets come from the imperfect manufacturing of silicon structures and are inevitable. Doping and etching variations will result in devices with relatively poorly-controlled electrical characteristics. Transistors fabricated will therefore vary a lot from wafer to wafer and from device to device. Even when two devices are designed to match with each other and are laid out accordingly on a die, their threshold voltage can differ quite significantly.

Among them, the threshold voltage, V_T, will have a direct impact on the amount of offsets in differential structures using matched transistors. The V_T mismatch can be calculated using the well-known equation:

$$V_{offset}(mV/\sigma) = \frac{\alpha}{\sqrt{WL}}, \quad (\alpha \; in \; mV, \; W, L \; in \; \mu m) \tag{9.18}$$

In (9.18), W and L represent the width and length of the transistor, and α is a statistical parameter that is technology-dependent. What this equation tells us is that the V_T mismatch is a function of the area of the transistor and that making transistors bigger leads to better matching.

Random offsets do not only have different origins from signal-dependent ones, they also lead to different leakage products as shown below:

$$(A(t)\cos(\omega_0 t) + \beta)^2 = \frac{A^2(t)}{2}(\underbrace{cos(2\omega_0 t)}_{Wanted} + \underbrace{\frac{4\beta\cos(\omega_0 t)}{A(t)}}_{Cross-term} + 1 + \underbrace{\frac{2\beta^2}{A^2(t)}}_{DC_{Offset}}) \tag{9.19}$$

In (9.19), the random offset β was added to the single-tone signal. If we compare to (9.15), we notice the appearance of a new *odd-order* cross-term.

9.4.1.3 Offsets Cancellation?

What have we learned so far?

- The generated harmonics are not pure.
- Cross-products are created that are far greater than the desired harmonics.
- The generated terms in our Volterra series (see 9.6) are strongly correlated.
- Dynamic range and linearity are severely reduced by the level of unwanted DC.
- Signal-dependent offsets lead to *even-order* leakage.
- Random offsets lead to *odd-order* leakage.

Should we consider canceling offsets to get rid of all these issues? The short answer is: it depends!

Random offsets can actually be improved or corrected quite easily. For instance, if we target 1 mV offset for a pair of transistors, assuming $\alpha = 4mV/\sigma$, the needed area can be calculated as follows:

$$V_{offset}(mV) = 1mV = \frac{4mV3\sigma}{\sqrt{WL}} \rightarrow WL = 140\ \mu m^2 \tag{9.20}$$

Granted, this area might be somewhat large and the large capacitance resulting from such a large device could be prohibitive for circuits having large bandwidth. But relaxing the target specification from 1mV to 2mV would lead to an acceptable area of 35 μm^2. Of course, the target for acceptable random offsets is system dependent and 2 mV might still be unacceptable. For those cases, other techniques such as correction DACs are commonly employed in state-of-the-art ICs.

Let us assume now that we are able to keep the *random* offsets to an acceptable level and have a way of filtering the *signal-dependent* ones after each multiplication stage. What would our harmonics look like? Much better, as can be see in (9.21).

$$2^{nd} : \frac{A^2(t)}{2} cos(2\omega_0 t) \tag{9.21}$$

$$4^{th} : \frac{A^4(t)}{8} cos(4\omega_0 t) \tag{9.22}$$

$$6^{th} : \frac{A^6(t)}{32} (cos(2\omega_0 t) + cos(6\omega_0 t)) \tag{9.23}$$

As attractive as it looks, *signal-dependent* offsets cannot be compensated for as easily as *random* offsets. Indeed, the only way to compensate perfectly for them is to generate dynamic correction signals that vary together with the input signal and inject them after each multiplier. The practicality of such a solution is therefore doubtful as it would require an *a priori* knowledge of the incoming signal.

Another way of removing offsets is to use analog filtering between each stage. For instance, a simple high-pass filter can be used to remove the low-frequency content including both types of offsets. This method works, but, as expected with filtering, the settling behavior and cut-off frequencies might not be appropriate for all wireless standards. There is unfortunately no universal solution when it comes to offset compensation and we ended up using a combination of these techniques.

9.4.2 PAPR: Peak-to-Average Power Ratio

Multi-carrier applications, modern standards, and OFDM in particular, lead to very high PAPR. What is important to understand from a design standpoint is that the analog circuits within the linearizer have the same problem as the PA. They have to support a high dynamic range to accommodate the PAPR. Worse, because the linearizer has to create high-order correction terms for the polynomial, it will actually expand the PAPR. Table 9.3 demonstrates this expansion for a 10 dB PAPR signal.

Table 9.3

Effects of PAPR Inside the Nonlinear Processing Engine.

Signal	PAPR(dB)	PAPR	PEAK(V)	AVERAGE(V)
1st order	10	3.16	1	0.316
2nd order	20	10	1	0.1
4th order	40	100	1	0.01
6th order	60	1000	1	0.001

As we can see, when the order of the signal increases, the average signal drops at a very fast pace because every time we square the input signal, the output is an order of magnitude smaller than the input. If we want to compensate for these losses we will therefore have to add at least 20 dB gain after each squaring function. This corresponds to an aggregated gain of 80–100 dB for the whole signal path.

9.4.3 Signal Expansion and Compression

Galileo's square-cube law teaches us that certain forces will matter more or less depending on the magnitude of a particular physics quantity. Take a 1V sinusoid and square it. The output signal will be 1V. Now, do the same thing with 10V and 0.1V signals. The outputs will be 100V and 0.01V, respectively.

In the first case, the output of the multiplier expanded the signal: 100V>10V. In the second case, the signal compressed it: 0.01V<0.1V. What this means for us is that for larger input signals, the polynomial generator output will be rapidly overwhelmed while smaller signals will virtually vanish. Increasing the dynamic range and the resolution of the coefficient DACs can help dealing with the expansion.

The compression issue, on the other hand, can only be accommodated by amplifying the signal. A high gain can be achieved using a cascade of amplification stages with the drawback that each stage requires power and adds noise.

Unfortunately, signal expansion also impacts the bandwidth of the signal. When multiplied by itself, the negative portion of a sine wave becomes positive, effectively doubling the number of inflection points.

9.4.4 Bandwidth Expansion

We mentioned previously that a big advantage of RFPD over DPD is that the up-conversion chain does not need to accommodate the full linearization BW, leading to significant savings at the system level.

Figure 9.11 illustrates the BW expansion taking place inside a memory kernel. As you can see, each multiplication doubles the signal BW. By the time the signal reaches the 4th series multiplier, the BW has been multiplied by five times. For the targeted input VBW of 100 MHz, the high-order multipliers and adders (solid line in the figure) must ideally support a VBW in excess of 500 MHz.

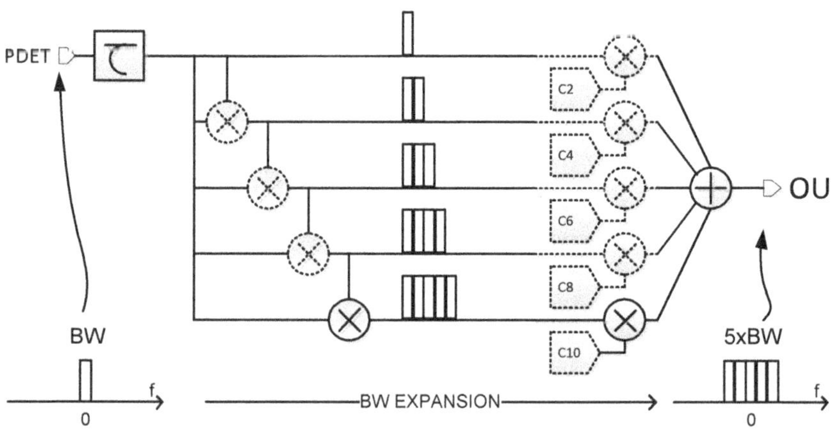

Figure 9.11 CORR: polynomial generation.

Adding the internal VBW requirement and the gain needed to compensate for the losses due to compression lead to a Unity Gain-BW product requirement of 5 GHz. This number does not account for further BW reduction coming from the cascade of active circuitry but is a good enough approximation for the purpose of this discussion. The conclusion is that feedback could not be used and we had to rely on open-loop circuitry for our correction engine.

9.4.5 Temperature Effects

Dealing with process, voltage, and temperature (PVT) variations is the essence of an analog designer's job.

Voltage variation can be mostly handled by architecture choice and has little impact on most of the important design parameters other than power consumption. The effect of process variation on offsets was handled in Section 9.4.1.2.

The electrical characteristics of a transistor on a chip will change a lot with temperature. For instance, the gain of an analog multiplier can easily change by a factor of 2.5 (8 dB) between −40 and 125°C (industrial temperature range). For reference, Table 9.4 shows how the temperature affects the harmonics' amplitude coming out of the polynomial generator.

Table 9.4

Gain Variation over Temperature.

Temperature (C)	GAIN	X (V)	X2 (V)	X3 (V)	X4 (V)
-40	25	0.1	0.25	0.6	1.5
25	15	0.1	0.15	0.22	0.34
125	10	0.1	0.1	0.1	0.1

As we can see, the signal amplitude after subsequent multiplications will vary greatly. Further non-idealities will actually make the situation even worse.

Our linearizer is an adaptive system. As such, it is continuously adapting the coefficients to compensate for changes in the system. As part of the adaptation algorithm, the device will duty cycle power-hungry circuits in order to save power. This algorithm puts a high burden on the analog blocks on the chip. When the device enters the 'sleep state', the on-chip temperature drops by a few tens of degrees Celsius. These rapid temperature changes are happening hundreds of times per second and should not impact the analog path for the PD signal to stay constant. Therefore, the gain settings throughout the chip need to be tightly regulated over temperature.

9.5 KEEPING THINGS UNDER CONTROL!

Handling both the complexity of the processing engine and analog limitation requires thorough planning. As we have seen in Section 9.4, numerous hurdles need to be overcome in order to create a cost-effective and high yield product.

Here is a summary of what we consider the biggest challenges from a design point of view:

- Wide RF bandwidth: 160 MHz to 4.2 GHz (3rd generation linearizer, SC1894).
- Wide video bandwidth: 100 kHz to 100 MHz → current wireless standards.
- Large input and output dynamic range → high PAPR signals and expansion.
- Good stability over PVT.

Signal normalization is our solution to these requirements. By continuously regulating the gain of the Volterra series components over PVT, the signal can be conditioned to accommodate the dynamic range limitation as well as the signal statistics.

9.5.1 Normalizing the CORR

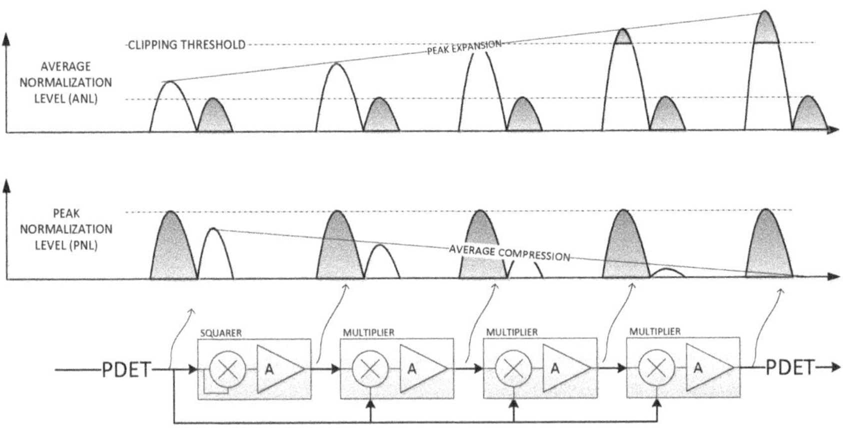

Figure 9.12 CORR normalization.

An effective way to handle the signal dynamic range is to normalize the signal at the input of the Volterra series. Doing so only puts a hard constraint on the incoming PDET signal. The normalization voltage should be selected based on the statistic of the input signal and expected signal expansion required by the PA to linearize. As a matter of fact, two types of normalization can be implemented: peak normalization level (PNL) and average normalization level (ANL). Figure 9.12 illustrates both concepts. Both methods can be implemented with the same hardware but lead to very different results.

The ANL is looking at the average power and hence enforces a constant power throughout the multiplier chain. The drawback of this method is that it is very difficult to ensure the integrity of the peaks. As you can see in Figure 9.12, the peaks will expand after each multiplication and eventually hit a *clipping threshold*. This clipping effect, due to the limited power supply, will limit our ability to correct high PAPR signals. This method is therefore better suited for signal with small PAPR and PAs that do not require higher-order large correction power.

The PNL, on the other hand, enforces a constant peak amplitude inside the harmonics generator. In this configuration, the peak integrity is guaranteed even for high-order distortion products. The drawback is a compression of the average signal that might disadvantage PAs with large small-signal distortion.

9.5.2 Nonlinear Operator Normalization

Normalization can only work if we can ensure a tight gain control for our adders and multipliers. In particular, the nonlinear operators are so sensitive to expansion or compression (Table 9.3), and experience such a great variation over PVT (Table 9.4), that it is critical to constrain their gain to a given target. This can be achieved by implementing following equations:

$$V_{out} = A \cdot V_{in1} \cdot V_{in2} \tag{9.24}$$

$$\text{for } V_{out} = V_{in1} = V_{in2} = V_{norm} \rightarrow V_{norm} = A \cdot V_{norm} \cdot V_{norm} \tag{9.25}$$

$$A = \frac{1}{V_{norm}} \tag{9.26}$$

Implementing (9.26) can be done using the circuit topology depicted in Figure 9.13. This well-known technique is called *replica biasing*. It consists in making a copy of an active circuit called replica. This replica is included in an off-line regulation loop that controls a specific parameter of the circuit. The output signal of the control loop is connected to both replica and active circuits. In this configuration, the gain of the squarer is regulated to enforce the normalization voltage V_{norm}. The accuracy of this technique is ultimately limited by the matching requirement and the fact that a DC signal is used for the replica instead of a dynamic signal. Frequency-varying parameters therefore cannot be compensated as effectively.

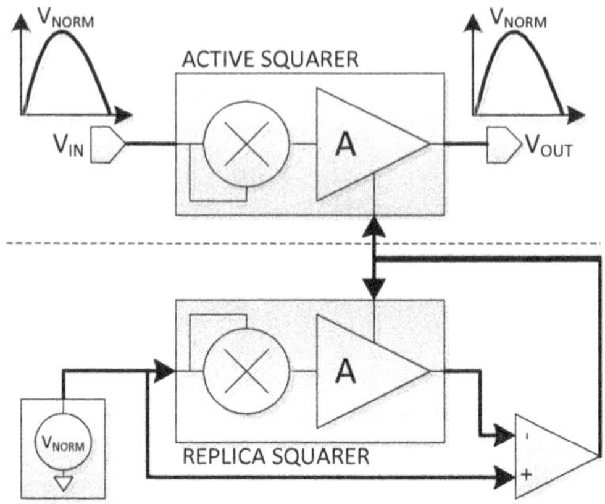

Figure 9.13 Multiplier normalization using replica biasing.

9.6 MEASUREMENT RESULTS

Before concluding this chapter, we would like to share a few typical measurements taken with Scintera's first three generations of linearizers. The SC1887 [2] was primarily developed for WCDMA signals. Its performance exceeded our expectation and paved the way to the further development of RFPD.

The SC1889, Scintera's second generation device, was developed to handle signals whose average power varies dynamically, such as WiMAX, HSDPA, and LTE.

With the third-generation device, SC1894, we decided to offer our customers a device covering an extended RF range (160 MHz – 4.2 GHz) as well as a larger linearization BW. The extended BW enables customers to have a multi-standard platform while the larger linearization BW is needed to support the growing demand for carrier aggregation.

It is impossible to predict the final performance after linearization because of the many variables involved in the design of a power amplifier. Nonetheless, most PAs will see their ACLR1 improve by 20 to 25 dB. Most importantly, efficiency can be improved up to four times, thus enabling proportional decreases in the overall system power consumption.

2 1887 was the year Volterra published his famous Volterra series

9.6.1 SC1887 and SC1889 Performance Comparison

In Figure 9.14, we are showing a Doherty PA being linearized by (a) SC1887, and (b) SC1889. We are operating at 47dBm output power and 0 dB back-off. The test signal is a 2-carrier WCDMA with a 6.5 dB PAPR. The SC1887 is achieving an ACLR1/2 correction of 21.7/16.8 dB, while the SC1889 outperforms it by about 3 dB, at 24.4/20.7 dB. The main reason for this performance boost is the introduction of an 11th-degree polynomial in SC1889, as opposed to a 7th-degree for SC1887 [5].

9.6.2 SC1894

Figures 9.15(a) to 9.17(b) show typical measurements taken with SC1894 for WCDMA and LTE waveforms. The same Doherty PA with 53.57 dBm PSAT was used for all these measurements. For these different cases, typical correction of 20 to 26 dB ACLR1 improvements were achieved. The operating points (0.5 to 1.5 dB back-off) were chosen to get performance close to -50 dBc ACLR1. This corresponds to real-world conditions where a 5-dB margin above the 45 dBc ACLR1 specification is typically required.

9.6.3 Carrier Aggregation

Before concluding this section, we would like to show an example of broadband carrier aggregation. In Figure 9.18, we are comparing the performance of SC1889 and SC1884 for a 20 MHz LTE and a 10 MHz LTE signals spaced by 45 MHz, resulting in an aggregated signal BW of 55 MHz. In-band and out-of-band, SC1894 is 3–4 dB better with respect to ACLR and meets an out-of-band specification of -50dBc over 150 MHz bandwidth. In-band, 10-dB improvement is achieved. The PA efficiency was measured at 35%.

9.7 CONCLUSION

While being a new technology, RFPD has proven to be a solid alternative to traditional DPD solutions. It has been quickly and widely accepted because of its unique benefits over traditional linearization techniques. Scintera's design is scalable, robust, and can easily be adapted to many applications. Originally developed for 3G standards, the same technology is now being used successfully for microwave and DVB-T applications. The simplicity of integration, low power consumption, and

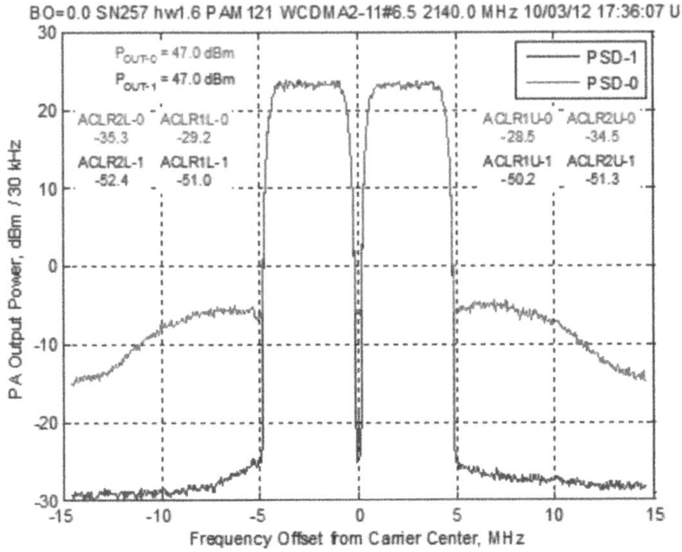

(a) SC1887

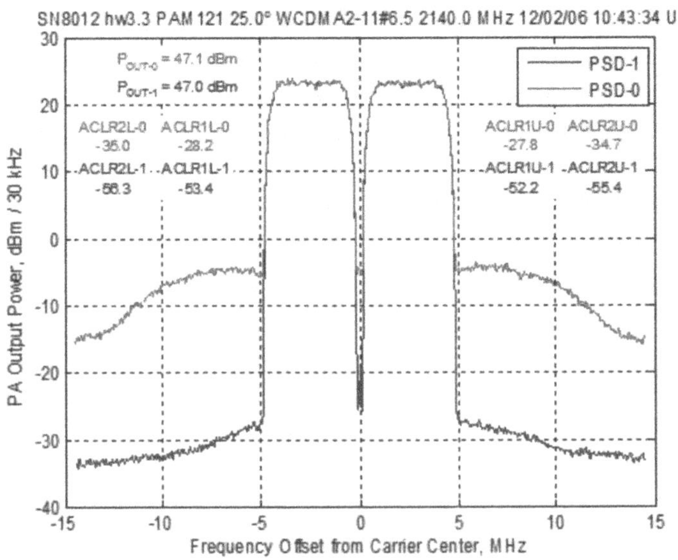

(b) SC1889

Figure 9.14 Linearization of a 2-carrier WCDMA signal using (a) SC1887 and (b) SC1889, at 2.14 GHz.

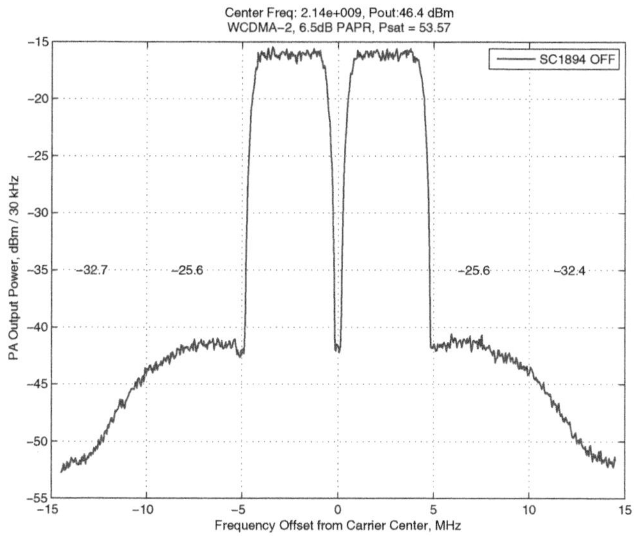

(a) Uncorrected

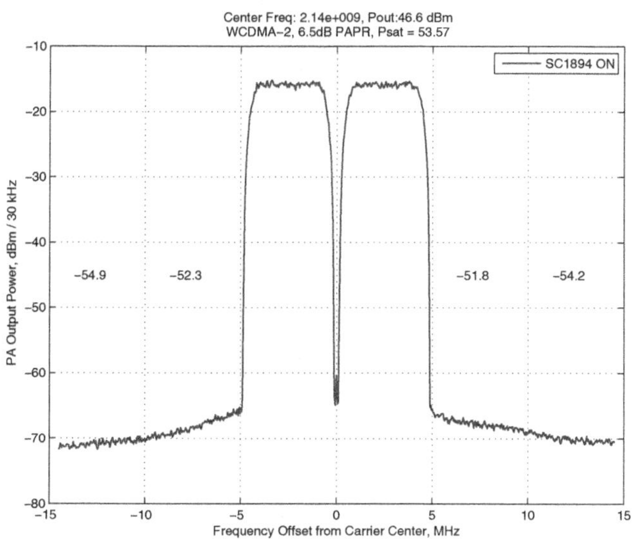

(b) Corrected

Figure 9.15 SC1894, 2-carrier WCDMA (a) uncorrected, and (b) corrected.

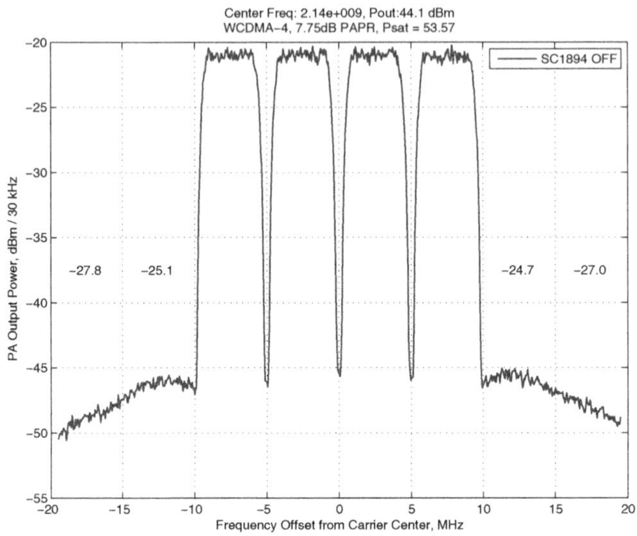

(a) Uncorrected

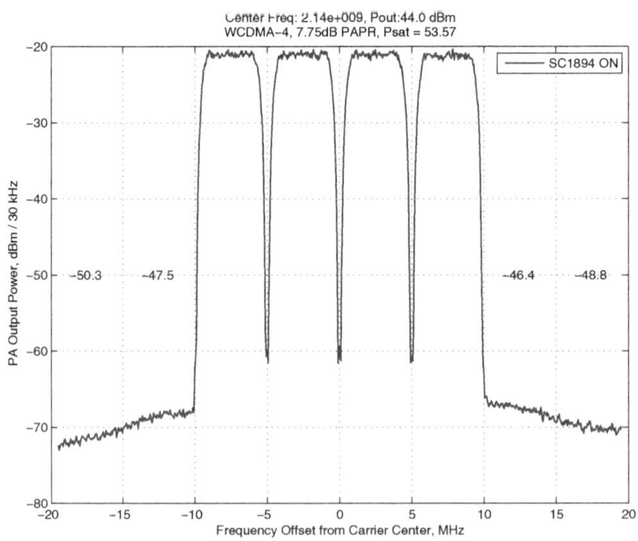

(b) Corrected

Figure 9.16 SC1894, 4-carrier WCDMA (a) uncorrected, and (b) corrected.

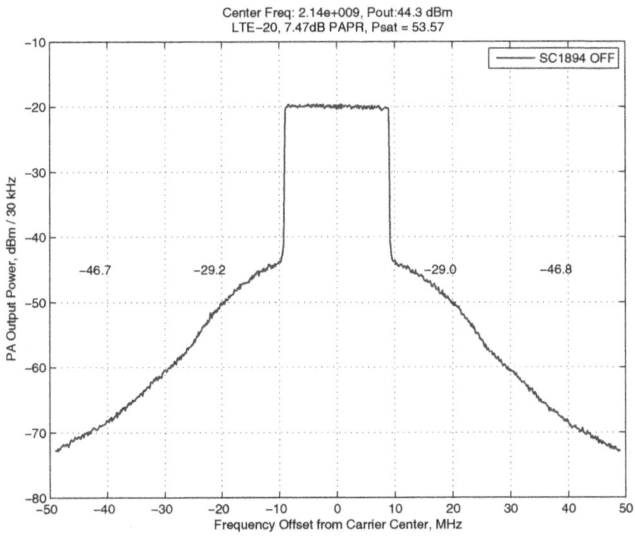

(a) Uncorrected

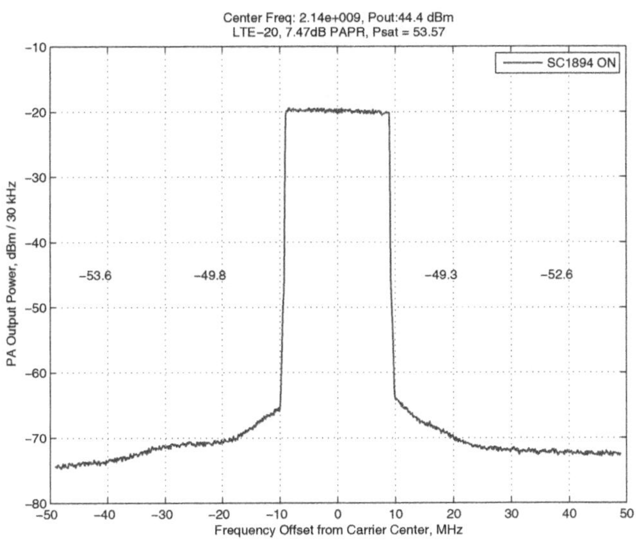

(b) Corrected

Figure 9.17 SC1894, 20 MHz LTE (a) uncorrected, and (b) corrected.

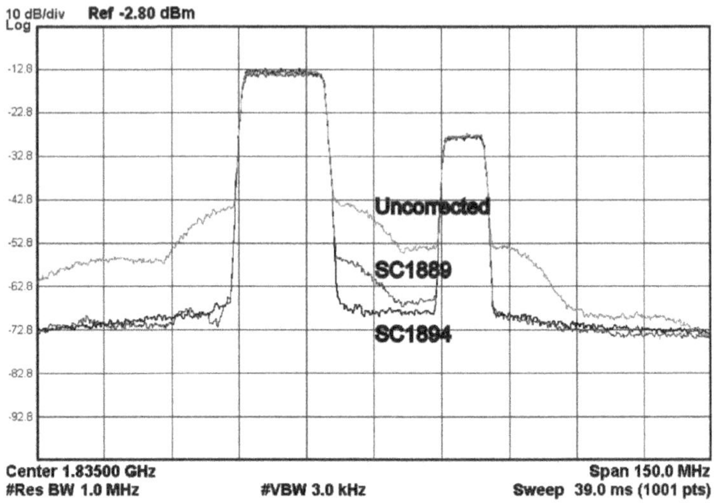

Figure 9.18 SC1889 and SC1894, test signal is 20 and 10 MHz LTE, 7.5 dB PAR, with 35% PA efficiency.

small footprint, make it a very attractive solution to implement fully adaptive RF pre-distortion in any system.

While it is now a maturing technology, RFPD is still novel and we are continuously learning and improving the technology. The third generation has now been released with emphasis on higher signal bandwidth and carrier aggregation, and we are already working on the fourth generation targeted primarily at the emerging Small-cell market, capitalizing on the unique advantages of this great alternative to PA linearization.

References

[1] Y. S. Lee, "A transistor-based high-order analog predistorter with memory correction for WCDMA applications," *IEEE Radio & Wireless Symp. Dig.*, New Orleans, LA, Jan. 2010.

[2] S.-Y. Lee, Y.-S. Lee, and Y.-H. Jeong, "Fully automated analog predistortion power amplifier in WCDMA applications," 35th *European Microwave Conf. Dig.*, vol. 2, Paris, France, Oct. 2005.

[3] E. Westesson and L. Sundstrom, "Low-power complex polynomial predistorter circuit in CMOS of RF power amplifier linearization," 27th *European Solid State Circuits Conf. ESSCIRC*, Villach, Austria, Sept. 2001, pp. 486–489.

[4] E. Westesson and L. Sundstrom, "Low-power complex polynomial predistorter chip in CMOS for baseband or IF linearization of RF power amplifiers," *IEEE Int. Symp. on Circuits and Systems ISCAS*, vol. 1, Orlando, FL, May 1999, pp. 206–209.

[5] F. Roger, "A 200mW 100MHz-to-4GHz 11th-Order complex analog memory polynomial predistorter for wireless infrastructure RF amplifiers," *IEEE Int. Solid State Circuits Conf. ISSCC*, San Francisco, CA, Feb. 2013.

Chapter 10

Frequency Domain Models

As microwave engineers, we tend to work almost exclusively in the frequency domain. In our amplifier designs, we are usually concerned with such factors as matching networks, filters, bandwidths, Q-factor, frequency response, and so on. We use tools such as S-parameters, and we rely on frequency-domain instrumentation such as spectrum analyzers and network analyzers for characterizing the performance of our amplifiers and other microwave components. Naturally, this leads us to look for frequency-domain nonlinear models that we can use in harmonic-balance simulators for circuit and system design.

In the harmonic-balance simulation, convergence is reached when the error voltage at each harmonic is minimized, under steady-state conditions. In essence, we are finding the unknown Fourier coefficients of the admittance matrix describing the network that relates the output and stimulus signals. In general, for a linear two-port network, if we take the input signal $u_p(t)$ and output signal $y_q(t)$, where $p, q = 1, 2$, we can write their Fourier expansions,

$$u_p(t) = \sum_{k=-K}^{K} U_p(\omega_k) e^{j\omega_k t} \tag{10.1}$$

$$y_q(t) = \sum_{k=-K}^{K} Y_q(\omega_k) e^{j\omega_k t} \tag{10.2}$$

and then express the output in terms of the Fourier coefficients in the frequency domain, through

$$Y_q(\omega_k) = \sum_{p=1}^{2} C_{qp}(\omega_k) U_p(\omega_k) \tag{10.3}$$

If we define the input signal as the small-signal incident wave, $A_p(\omega_k)$, and the output signal as the small-signal reflected or transmitted wave, $B_q(\omega_k)$, then the coefficients relating the input and output are the linear scattering or S-parameters, defined as:

$$s_{qp}(\omega_k) = \left. \frac{\partial Y_q(\omega_k)}{\partial U_p(\omega_k)} \right|_{U_0} \tag{10.4}$$

where U_0 is the bias point at which the system is placed for the input/output measurement.

This yields the well-known linear relationships for the two-port

$$\begin{aligned} B_1(\omega_k) &= s_{11} A_1(\omega_k) + s_{12} A_2(\omega_k) \\ B_2(\omega_k) &= s_{21} A_1(\omega_k) + s_{22} A_2(\omega_k) \end{aligned} \tag{10.5}$$

If we measure or otherwise determine the S-parameters over the range of possible bias or operating points, we have a bias-dependent linear model, but this is *not* a nonlinear model.

The extension of this simple linear form to a general nonlinear formulation is not possible, as there is no known way to treat a general nonlinearity on the frequency domain [1]. Instead, there have been a number of simplifications and approximations made in order to construct well-defined frequency domain nonlinear models that have specific ranges of applicability.

10.1 FREQUENCY DOMAIN NONLINEAR MODELS

Simple nonlinear system-level models have been around in the EDA simulation tools for many years. For example, in Agilent ADS there are two simple system-level amplifier models, the 'Amplifier2' or 'S2D' model, and the 'P2D' model, which are easy to use, offer different capabilities, and are somewhat complementary to each other. A more modern nonlinear model is the X-Parameters™ model developed by Agilent Technologies, which we shall describe in more detail below.

10.1.1 "Amplifier2" Model

Actually, the Amplifier2 model is a time-domain model, and strictly shouldn't be in this chapter, but as it is a ready-to-use nonlinear model for an amplifier, we shall consider it here. It can operate in all simulator modes. It is a polynomial model, with separate polynomials for the magnitude (AM-to-AM) and phase (AM-to-PM) responses of the amplifier to the input power level. The user can choose from a number of parameters such as the 1-dB gain compression point, saturated power, and third-order intercept point to determine the degree of polynomial for the magnitude characteristic. The model is carefully constructed to avoid causality problems that could arise from having different magnitude and phase behaviors. Alternatively, the user can provide AM-to-AM and AM-to-PM curves for the model to use. Using measured data enables realistic power drive-up curves to be introduced into the model, including effects such as gain expansion, and double-compression curves seen in some (poorly-designed) Doherty PAs.

The Amplifier2 model will generate harmonics, and is very fast in simulation. But the Amplifier2 model is essentially a single-frequency model: there is no explicit input parameter for the signal frequency, and so the model behavior is the same for signals at any input frequency. This introduces a cascadability problem: the harmonics generated by the first model are treated equally by the second model in the cascade, as there is no frequency response information included in the model. This can be overcome by introducing filters at the input and output of the amplifier model: effectively creating a Wiener-Hammerstein model structure from the simulator's built-in components. The Amplifier2 model does not recognize DC, and does not pass DC through the input/output terminals. This means we cannot readily calculate the power consumed or the amplifier efficiency. We have to introduce some other components, for example, to read measured current as a function of input or output power from a datafile. The initially simple model is becoming more complicated.

The model also has no capability of including baseband effects: memory effects are therefore not included in this model, so for detailed power amplifier modeling for wireless communications systems, especially those including linearization, it is of limited utility.

10.1.2 "P2D" Model

The P2D model is a frequency-domain model, and operates in S-parameters, harmonic balance, and circuit envelope simulators, but not in transient simulation. The model is based on large-signal S-parameters (LSSP), which mimic the definition of S-parameters, but for large-signal operation. The LSSP are defined for the fundamental frequency only, so no harmonic information is produced by this model. This model is best suited for narrowband applications where the harmonics are not of interest. This is not such a problem for relatively narrowband amplifiers such as the wireless infrastructure PAs that we are interested in, as the nonlinear behavior is captured in the LSSP, and the intermodulation distortion behavior, for example, can be accurately captured. The P2D model does not include baseband memory effects or DC, again requiring the inclusion of additional components to enable a calculation of power amplifier efficiency.

10.1.3 X-Parameters™

It was against this backdrop of limitations of the existing system models that Agilent began the development of the X-parameters model, which has recently been introduced into their family of simulators. The X-parameters model was conceived as a general-purpose, broadband, nonlinear model for system-level simulation of nonlinear microwave components such as amplifiers, mixers, and so on [2]. Initially developed for modules with 50Ω input and output, it has since been extended to accommodate source- and load-pull conditions for application in power amplifier characterization and modeling. The X-parameters formulation was invented by Jan Verspecht, as a development of a describing function approach that he outlined in the Annex to his Ph.D. thesis [3]. The X-parameters format and equations are open and documented, and have been implemented in other EDA tools.

10.1.3.1 X-Parameters Formulation

The X-parameters are a mathematically correct extension of S-parameters into the large-signal regime of operation. They have been described clearly in an article by Verspecht and Root [4], and we present a summary below, in the context of an amplifier model.

The formulation of the X-parameters is essentially one of linearization about a single large tone: the drive signal, which for X-parameters is a sinusoid. In this respect, it is no different from the formulation of the Volterra series that was discussed in Chapter 6, although the drive signal for the Volterra series was the modulated signal. The X-parameters model, like the Volterra model, is a *local* model.

The linearization for X-parameters is performed in the frequency domain. A single large-signal tone at the input to the PA produces a large output signal at the fundamental, plus harmonics tones at the output, and harmonics reflected at the input by the nonlinearity of the PA input impedance: usually a capacitive reactance. This is shown schematically in Fig. 10.1.

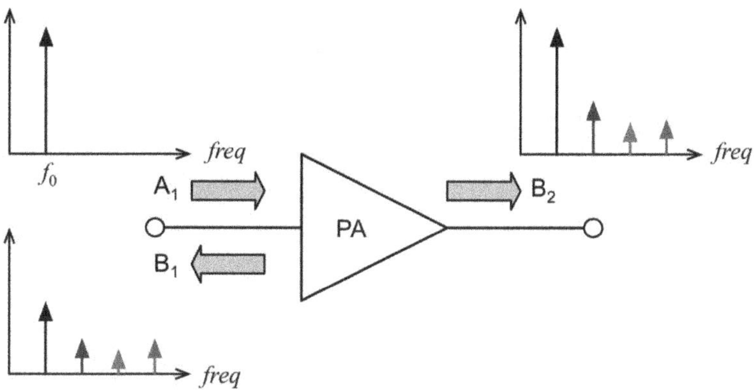

Figure 10.1 Power amplifier driven by a single-frequency tone. The nonlinear PA generates harmonics at the output, and also reflects (scatters) some harmonic energy at the input as a result of the nonlinear input capacitance of the transistor.

Under small-signal conditions, we would write the forward and reflected scattering waves as linear equations, in terms of S-parameters,

$$\begin{aligned} b_1 &= s_{11}a_1 \\ b_2 &= s_{21}a_1 \end{aligned} \tag{10.6}$$

where we assume there is no input signal at the output, port 2. With X-parameters, we now have a nonlinear formulation, and the vector of outputs $\mathbf{B}$ is a function of the input signal(s), $\mathbf{A}$

$$\mathbf{B} = \mathrm{F}(\mathbf{A}) \tag{10.7}$$

In the general case, the output wave of harmonic number m at any port p, B_{pm}, is a nonlinear function of all of the input waves and their harmonics at all of the ports:

$$B_{pm} = \mathrm{F}_{pm}\left(A_{11}, A_{12}, ..., A_{1M}, A_{21}, A_{22}, ..., A_{2M}, ..., A_{P1}, ...A_{PM}\right) \tag{10.8}$$

where M is the maximum number of harmonics, and P is the maximum number of ports, which will be two for a power amplifier. The drive signal is the fundamental at port 1, A_{11}.

In practice, we will need to measure these signals A_{pm}, B_{pm} accurately in terms of their magnitude *and* phase in order to be able to reconstruct the waveforms incident upon and scattered from the PA. To do this, we will need a phase reference to which we can refer the phases of the harmonics. We choose the phase of the drive signal A_{11} as the reference, using the following definition.

We assume that the power amplifier is a time-invariant system, and so the function F_{pm} is also time invariant. This means that any arbitrary time delay or phase applied at the input will result in the same delay or phase in the fundamental of the output signal. All harmonics will be multiples of this phase. Using this time invariance property, if we assume that the input signal to the system has an arbitrary phase, $e^{j\theta}$, we get

$$B_{pm}e^{jm\theta} = \mathrm{F}_{pm}\left(A_{11}e^{j\theta}, A_{12}e^{j2\theta}, ..., A_{1M}e^{jM\theta}, A_{21}e^{j\theta}, ..., A_{P1}e^{j\theta}, ...\right) \tag{10.9}$$

Recognizing that the input signal A_{11} is

$$[A_{11} = |A_{11}| \exp\{j\varphi(A_{11})\} \tag{10.10}$$

the key insight to the phase normalization is to make θ the inverse of the phase of A_{11}. If we write $\Phi = \exp\{j\varphi(A_{11})\}$, then $\Phi^{-1} = e^{j\theta}$ and we have

$$B_{pm} = \mathrm{F}_{pm}\left(|A_{11}|, A_{12}\Phi^{-2}, ..., A_{1M}\Phi^{-M}, A_{21}\Phi^{-1}, ..., A_{P1}\Phi^{-1}, ...\right)\Phi^{+m} \tag{10.11}$$

The more obvious choice of making the phase of A_{11} equal to zero actually leads to a much more complicated and less intuitive notation.

Now that we have a means of describing any output harmonic at any port, we are left with the task of identifying the nonlinear function F that relates the output to the inputs. For the X-parameters model, we do this by applying the drive signal A_{11} to establish the nonlinear state of the PA, and then applying only a single harmonic tone at one port, and measuring the resulting responses B at every harmonic and every port. In this way we have determined the input-output relationships between the outputs $\mathbf{B}$ and the input A_{pm}. This procedure is repeated for all harmonics of interest, including the fundamental, and at every port. The applied harmonic tone is at a very low amplitude compared with the drive tone, and acts only as a perturbing signal. We assume that the resulting output signals are also small, and the outputs from different harmonic stimuli at a given port and harmonic can be simply added together vectorially. This is the principle of *harmonic superposition*, illustrated in Figure 10.2.

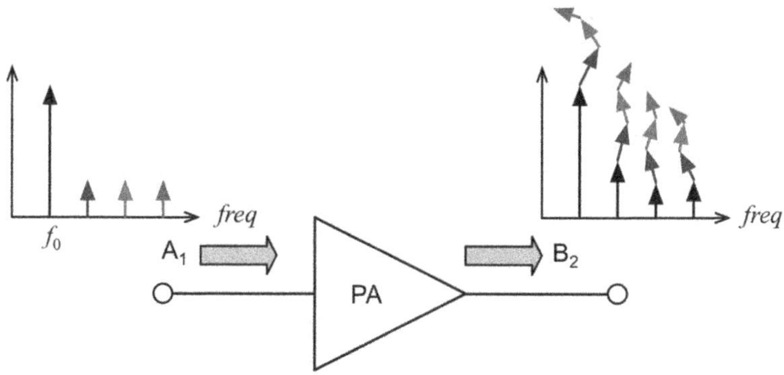

Figure 10.2 The power amplifier driven by a large single-frequency tone, generating an output signal and harmonic distortion products. An additional small tone at one harmonic, say the second, is also applied, resulting in output signals at fundamental and all harmonics, because of the PA nonlinearity. These additional output signals are small perturbations on the existing output, and can be added linearly. This small additional harmonic tone is switched off, and a different harmonic perturbing tone is applied (third harmonic), again resulting in output signals at fundamental and all harmonics. This procedure is repeated for all harmonics and ports.

The linearization of the expression for B_{pm} against all of the input components in the function F yields

$$\begin{aligned} B_{pm} &= \mathrm{F}_{pm}\left(\left|A_{11}\right|, 0, 0, ..., 0\right)\Phi^{+m} \\ &+ \sum_{q=1}^{P}\sum_{n=1}^{M} G_{pq,mn}\left(\left|A_{11}\right|\right)\Phi^{+m}\mathrm{Re}\left(A_{qn}\Phi^{-n}\right) \\ &+ \sum_{q=1}^{P}\sum_{n=1}^{M} H_{pq,mn}\left(\left|A_{11}\right|\right)\Phi^{+m}\mathrm{Im}\left(A_{qn}\Phi^{-n}\right) \end{aligned} \tag{10.12}$$

where the intermediate functions G, H are the sensitivities of the function F to the real and imaginary parts of the argument $A_{qn}\Phi^{-n}$, respectively. There are two things to note about these expressions. First, we have introduced some complicated subscripts: the first component is the output port, the second is the input port, so far, just like S-parameters, and the third and fourth components are the harmonic number of the output and input signals, respectively. The second point is that the linearization is limited to the first derivative of the function, or the first component in the Taylor series. While this could seem to be a limitation, it is quite an elegant solution: the simulator will calculate or use the first derivative in the Jacobian of the admittance matrix describing the circuit, and so this formalism fits in neatly with the way the simulator works.

The X-parameters description is then obtained by substituting the real and imaginary part of the arguments of the input signals by a linear combination of the input signals and their conjugates, to get

$$\begin{aligned} B_{pm} &= \sum_{q=1}^{P}\sum_{n=1}^{M} X^{S}_{pq,mn}\left(|A_{11}|\right)\Phi^{+m-n}A_{qn} \\ &+ \sum_{q=1}^{P}\sum_{n=1}^{M} X^{T}_{pq,mn}\left(|A_{11}|\right)\Phi^{+m+n}A^{*}_{qn} \end{aligned} \quad (10.13)$$

where the coefficients X^S, X^T are the X-parameters, relating the output at a given port and harmonic to all the harmonic inputs at all other ports, for the given drive level A_{11}. The X-parameters model is a table of these X^S, X^T coefficients, and can have multiple dimensions to include variations of the X-parameters with input power, fundamental frequency, bias conditions, temperature, and load-pull impedances.

A classic example of the value of X-parameters is given by the following example. Let's consider the output wave B_{21}, and restrict ourselves to fundamental signals only:

$$\begin{aligned} B_{21} &= X^{S}_{21,11}\left(|A_{11}|\right)A_{11} + X^{S}_{22,11}\left(|A_{11}|\right)A_{21} \\ &+ X^{T}_{22,11}\left(|A_{11}|\right)\Phi^{2}A^{*}_{21} \end{aligned} \quad (10.14)$$

There is also a contribution to B_{21} from a signal incident at port 2, possibly from an impedance mismatch. If we compare this expression with the small signal equivalent, and dropping the harmonic index for simplicity:

$$B_2 = S_{21}A_1 + S_{22}A_2 \quad (10.15)$$

we can associate the $X^S_{21,11}$ term with S_{21}, and $X^S_{22,11}$ with S_{22}, but we have an additional contribution to the S_{22} term from the X^T parameter. This shows that the extension from small signal to large signal parameters is more complicated than expected, with the additional X^T terms being necessary to describe the large signal scattering waves correctly. This additional contribution to the large-signal S_{22} gives the true 'hot-S_{22}' measure.

10.1.3.2 Measuring X-Parameters

While the X-parameter formulation outlined above is quite general, the first examples of measurements were made using a nonlinear vector network analyzer. This instrument is essentially a vector network analyzer (VNA) that is equipped with power and phase calibration in addition to the regular small-signal S-parameter calibration. Without going into details, these calibrations establish the measurement reference planes for the ratioed measurements, and for the absolute values of power and phase of the signal at fundamental and harmonics. In this way, the magnitude and phase of the harmonic components that are needed to calculate the X-parameters can be measured. The basic setup is shown in Figure 10.3.

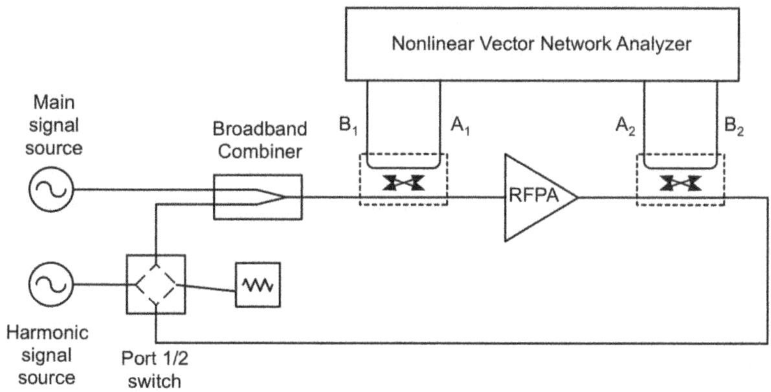

Figure 10.3 Schematic diagram of the X-parameters measurement system, showing the main and harmonic signal sources, switch, and the nonlinear VNA instrument used for measurement of the signal power and phase of fundamental and harmonic tones.

The main signal source provides the large signal single-tone drive to place the PA in the nonlinear regime of operation. The harmonic signal source provides the small amplitude signals at each harmonic in turn, and its output is switched between Ports 1 and 2 of the PA. The nonlinear VNA measures the responses at fundamental and each harmonic. The couplers can be external, though at low powers they are usually internal to the VNA. This is a 50-Ω measurement system, where the reference impedance is provided by the instrument. As an example of the nonlinear VNA, the Agilent PNA-X VNA comes with suitable features and options for X-parameters measurement: four ports are required, to enable connection of the phase reference source that ensures all of the VNA's receivers are phase locked, enabling the accurate measurement of absolute phase; the instrument is equipped with two controllable sources, for main and harmonic generation, and internal switching to route the signals correctly, under software control. This makes it quite straightforward to measure X-parameters of a device-under-test (DUT), especially at low power, and examples of such measurements and simulation models include power amplifiers for cellphone handsets [5].

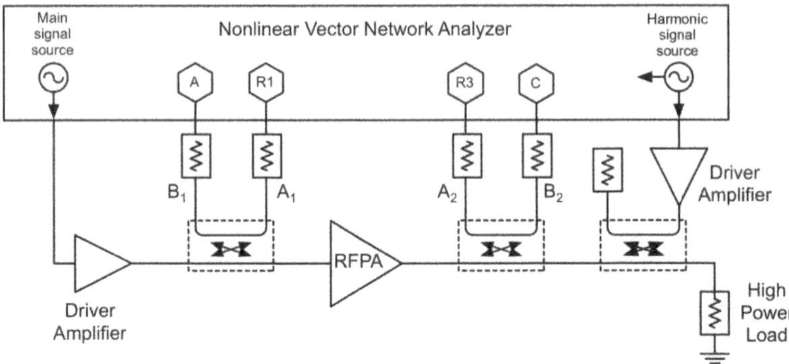

Figure 10.4 Schematic diagram of the X-parameters measurement system, extended for high-power measurements. External couplers and reflectometers, and high-power broadband low-distortion driver amplifiers are needed. We also use plenty of padding in front of the VNA's expensive receivers.

To extend this setup to higher powers, for measurement of wireless infrastructure PAs, for example, requires the addition of several external components, as shown in Figure 10.4 [6]. The reflectometers must be external to the VNA, to handle the higher power, and we place some attenuation between the coupled signals and the input to the VNA, to protect the instrument. The level of attenuation is carefully chosen to permit the maximum dynamic range while preventing overload of the receivers. The couplers must be broadband to allow for the capture of several harmonics. Similarly, the driver amplifiers must be broadband, and also very low harmonic distortion to prevent the injection of extra harmonic information not generated by the PA under test. The input driver amplifier must have enough power to drive the PA into the nonlinear operating regime. The output or reverse injection driver amplifier also needs to be of high power: to overcome the coupling factor of the reflectometer, and to provide sufficient signal level to be detected in the presence of the large output signal from the PA under test. An external high-power load is required to dump the power generated by the PA. An alternative setup has the reverse injection driver connected through a high-power attenuator rather than a coupler: this still requires the driver to have enough power to overcome the loss in the attenuator, and has the potential disadvantage that the attenuator resistance may be modulated thermally by the high power signals incident upon it, generating additional harmonic distortion products detected by the instrument, but not arising from the PA under test. While in principle this is still a 50-Ω system, high-power components tend not to be constructed to the exacting levels of precision as the small-signal counterparts, presenting VSWR that is not exactly unity. This is manifest as a measured system impedance that is different from 50Ω, by possibly an ohm or two. The drawback with this is that in validating the X-parameters model against the measured data in the simulator, a 50-Ω reference impedance can not be assumed: the measured system impedance must be used. The X-parameters table file does report the system impedance, so this data is available to the user.

The high-power setup can be further extended for PA characterization, by inclusion of load tuners to control the impedances presented at the input and output of the PA or power transistor: load-pull X-parameters [7]. The tuners would normally be placed as close as possible to the PA connections, to obtain the maximum possible impedance range, and get as close to the edge of the Smith chart as possible, which is essential to provide the appropriate impedances for high-power transistors. If ultra-low loss and broadband couplers are available, the tuner can be placed outside of the reflectometers, enabling accurate measurement of the fundamental and harmonic tones at the ports of the PA, and, incidentally, of the presented impedances, too [8].

10.1.4 S-Functions Model

The S-Functions model was developed by NMDG bv (now part of National Instruments), and is based on a describing function approach very similar to X-parameters. The S-Functions description uses a large-signal operating point (LSOP) to establish the nonlinear state of the device under test. The LSOP depends not only on the RF drive signal level, but also the DC conditions, and can include a second drive signal at port 2. In this way, the DC bias conditions become an explicit part of the model. The nonlinear behavior of the model is determined using injection of harmonic tones at all ports of the DUT, measuring the responses, and using the principle of harmonic superposition to build up the table of interactions. This is very similar to the way the X-parameters model is constructed, though there are detail differences in the construction of the table.

S-Functions models can be extracted using commercial equipment, based on 4-port Rohde & Schwarz VNA augmented by NMDG hardware to provide the absolute calibrations of power and phase. The S-Functions models are available in the AWR Microwave Office simulation environment.

10.1.5 The 'Cardiff Model'

This model was developed at Cardiff University as a mechanism for getting large amounts of active harmonic load-pull data into the microwave circuit simulator.

The group at Cardiff University have extensive experience in the measurement of high-power RF transistors under harmonic load-pull conditions, and the aim was to enable this data to be used efficiently in a simulator for PA design.

This model extends the X-parameters formalism by recognizing the inherent limitation of the linearization around the large-signal tone using only the first term in the Taylor series expansion. The Cardiff model uses higher-order terms in the Taylor series description of the X^S, X^T functions, and by using this approach a model can be generated that can cover a wide range of the Smith Chart impedance plane from relatively modest series of measurements. The added benefit is that this load-pull data can be imported easily into the simulator using the Cardiff model, enabling the PA design space to be investigated in simulation. Model structures for the Cardiff model are available for Agilent ADS and AWR Microwave Office.

10.2 CONCLUDING REMARKS

The models described in this chapter are all based on frequency-domain measurements. These measurements are made using VNAs, and are performed using periodic CW signals, typically single-tone sinusoids. This is very much in our comfort zone as microwave engineers. The drawback with this approach is that no transient or time-dependent effects can be measured, at least, not simply. This means that it is not possible to include memory effects into these models directly: the model structure does not inherently support memory effects.

There have been a number of attempts to include memory effects into the X-parameters type of formalism. Verspecht *et al* [9] added a memory effects module that was successful in describing bias circuit memory effects, although this addition has not (yet) been included in the official released version of the model. More recently, Soury *et al* [10] described how memory effects could be included into the X-parameters model by introducing a function to modify the single large tone drive. The assumption is that the harmonic interactions are already small enough (harmonic superposition) that the memory behavior will be negligible. The memory behavior is characterized and modeled using the modified Volterra Series approach, as described in the context of PA modeling in Chapter 3 of [11]. The drive tone is then modified by the Volterra series model. This approach would seem to give the advantage of having a wide band model, based on the X-parameters, that has memory capability included by means of a local (to the drive tone) Volterra Series model.

Using a single large-signal tone to place the PA in the nonlinear state, about which the linearized model is extracted, means that it is difficult to choose the power of this drive signal to correspond with a typical use case for the RF power amplifier. Modern wireless communications signals have a sizable peak-to-average power ratio (PAPR), and the average PA power will be several dB below the compression or saturation points. Using the drive tone at this power level will place the PA in the correct thermal environment, dissipating the right amount of heat, but we will be far from the nonlinear regime of the amplifier, underestimating the nonlinear behavior. Conversely, using a higher drive power will probe the nonlinear region, but will increase the dissipated power and temperature of the PA. Pulse techniques can be used to overcome this difficulty, choosing a duty cycle to give the average power while the signal amplitude is in the nonlinear region.

RF power amplifiers are relatively narrow bandwidth devices, of tens to hundreds of MHz bandwidth, and typically not extending over several octaves of frequency, as we would like for a complete X-parameters-type of characterization. This means we are limited to a fundamental-only characterization of the PA. This is performed at a single frequency, or, preferably, a number of frequencies across the band. A practical limitation currently with the measurement instrumentation is that the minimum frequency spacing between the tones is 10 MHz, determined by the phase reference, which is greater than the channel bandwidth of several common modulation formats.

The single-frequency fundamental-only X-parameters-type of characterization needs to be done at several power levels to get a model of the PA nonlinearity. This is essentially an instantaneous (memoryless) AM-to-AM/AM-to-PM characterization of the PA, although the resulting model is defined only at the measured power levels, and points between the data are obtained by interpolation rather than by fitting a continuous function. The interpolation is not a part of the X-parameters model, which is a set of tables, but is instead left to the simulator, and is either linear or cubic spline, either of which limit the order of nonlinearity that can be modeled over a wide power range. This is illustrated in Figure 10.5, which compares an X-parameters model and a simple memoryless polynomial. The X-parameters model is derived from simulations made at a single frequency, over a range of input powers to the PA; the polynomial model is derived from modulated AM-AM and AM-PM data. The test signal used here is a 20 MHz LTE signal. The linear interpolation between the measured power indices can be clearly seen. There is also some small artificial memory effect seen in the X-parameters model. This is because the model was extracted at a single frequency only, and it is applying constant gain and phase across the 20 MHz signal bandwidth, instead of a linear phase characteristic (although it is clearly impossible to create a linear phase characteristic from only one frequency point). This can be avoided by extracting several X-parameters models

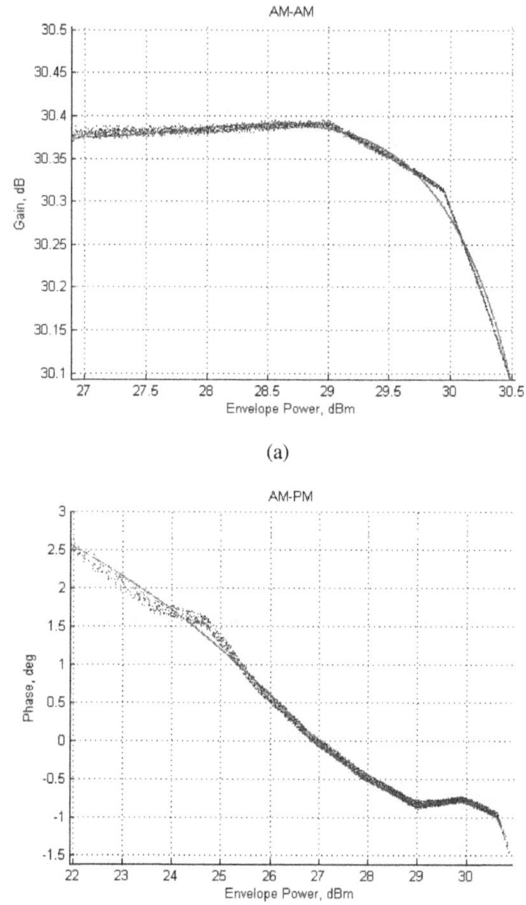

(a)

(b)

Figure 10.5 The AM-to-AM and AM-to-PM responses using a memoryless polynomial and a X-parameters model extracted at several power levels and at a single frequency. The test signal is 20 MHz LTE.

across the signal bandwidth, although the frequency step limitations mentioned above are still a consideration.

The value of the X-parameters-type frequency-domain models lies in their ability to predict behavior over a wide frequency range, and over a wide range of load terminations. This makes them valuable as a load-pull model that the designer can use to import the load-pull data of the device into the simulator, and investigate the design space of the transistor in the frequency domain: matching networks and harmonic loads, for example. Notwithstanding the recent developments in 'Dynamic X-parameters' [12], for modeling of the PA at the system level using wideband modulated signals, smooth nonlinear behavior and inclusion of memory effects over a wide range of signal levels and operating conditions are essential, especially when considering the use of the PA in the context of a linearized transmitter using DPD. In these circumstances, a nonlinear dynamical model, of the type described in Chapters 6 and 7, is, in my view, essential.

References

[1] J. C. Pedro, "A glimpse on behavioral modeling technology for microwave transistors," *IEEE 14th Wireless and Microwave Technology Conf.*, Orlando, FL, Apr. 2013.

[2] D. E. Root, J. Verspecht, D. Sharrit, J. Wood, and A. Cognata, "Broad-band poly-harmonic distortion (PHD) behavioral models from fast automated simulations and large-signal vectorial network measurements," *IEEE Trans. Microw. Theory Tech.*, Vol. 53, No. 11, pp. 3656–3664, Nov. 2005.

[3] J. Verspecht, "Describing functions can better model hard nonlinearities in the frequency domain than the Volterra theory," Ph.D. dissertation, Vrije Universiteit Brussel, Nov. 1995.

[4] J. Verspecht and D. E. Root, "Polyharmonic distortion modeling," *IEEE Microwave Mag.*, Vol. 7, No. 3, pp. 44–57, June 2006.

[5] J. M. Horn, J. Verspecht, D. Gunyan, L. Betts, D. E. Root, and J. Eriksson, "X-parameter measurement and simulation of a GSM handset amplifier," *14th European Microwave Integrated Circuit Conf. EuMIC*, Amsterdam, Holland, Oct. 2008, pp. 135–138.

[6] J. Wood and G. F. Collins, "Investigation of X-parameters measurements on a 100W Doherty power amplifier," *75th ARFTG Conf.*, Anaheim, CA, May 2010, pp. 1–7.

[7] G. Simpson, J. M. Horn, D. Gunyan, and D. E. Root, "Load-pull + NVNA = enhanced X-parameters for PA designs with high mismatch and technology-independent large-signal device models," *72nd ARFTG Conf.*, Portland, OR, Nov. 2008, pp. 88–91.

[8] V. Teppati and A. Ferrero, "A new class of nonuniform, broadband, nonsymmetrical rectangular coaxial-to-microstrip directional couplers for high power applications," *IEEE Microw. Wireless Compon. Lett.*, Vol. 13, No. 4, pp. 152–154, Apr. 2003.

[9] J. Verspecht, J. M. Horn, and D. E. Root, "A simplified extension of x-parameters to describe memory effects for wideband modulated signals," *75th ARFTG Conf.*, Anaheim, CA, May 2010, pp. 1–6.

[10] A. Soury and E. Ngoya, "Handling long-term memory effects in X-parameter model," *IEEE MTT-S Int. Microwave Symp. Dig.*, Montreal, Canada, June 2012.

[11] J. Wood and D. E. Root, *Fundamentals of Nonlinear Behavioral Modeling for RF and Microwave Design*. Norwood, MA: Artech House, 2005.

[12] J. Verspecht, D. E. Root, and T. Nielsen, "Digital predistortion technique for generating distortion-less high-power modulated signals," *82nd ARFTG Conf.*, Columbus, OH, Nov. 2013, pp. 1–6.

About the Authors

John Wood

John Wood received B.Sc. and Ph.D. degrees in Electrical and Electronic Engineering from the University of Leeds, United Kingdom. He is currently a Senior Principal Member of the Technical Staff in the RF Solutions Division at Maxim Integrated, Inc., San Jose, California, where he is working on system simulation of wireless communications RFICs and algorithm development for digital predistortion and other signal impairment corrections. He was formerly a Distinguished Member of the Technical Staff responsible for RF System and Device Modeling in the RF Division of Freescale Semiconductor, Inc., Tempe, Arizona. His areas of expertise include the development of compact device models and behavioral models for RF power transistors and ICs, and linearization and pre-distortion of high-power amplifiers. To enable and support these modeling requirements, he has been instrumental in the definition of high power pulsed I-V-RF test systems, for connectorized and on-wafer applications, and in the development of large-signal network analyzer (LSNA), loadpull, and envelope measurement techniques. From 1997–2005 he worked in the Microwave Technology Center of Agilent Technologies in Santa Rosa, California, where his research work has included the investigation, characterization, and development of large-signal and bias-dependent linear FET models for millimeter-wave applications, and nonlinear behavioral modeling using LSNA measurements and nonlinear system identification techniques. Between 1983 and 1997 he was a Professor in the Department of Electronics at the University of York, United Kingdom, where his research and teaching interests covered semiconductor devices, RF and microwave circuits, IC design, and semiconductor device modeling.

He has organized, co-organized, and presented at many workshops at IMS and RWS in recent years; he was on the Steering Committee for IMS 2006, and has been a member of the IMS Technical Program Committee since 2008. He has been a member of the ARFTG Executive Committee since 2007, and was the Technical Program Chair for the 70th & 75th ARFTG Conferences (2007, 2010), and the General Chair for the 78th ARFTG Conference in Fall, 2011. He was Technical Program Chair for the IEEE Power Amplifier Symposium 2008 and 2010, and was General Chair in 2009 and 2011. He is a regular reviewer for *IEEE Transactions on Microwave Theory & Techniques*, on *Electron Devices*, and on *Circuits & Systems*. He is author or co-author of over 140 papers and articles in the fields of microwave device and system modeling and characterization, and microwave device technology. He is the co-author of *Modeling and Characterization of RF and Microwave Power FETs* (Cambridge, 2007), and co-editor of *Fundamentals of Nonlinear Behavioral Modeling for RF and Microwave Design* (Artech House, 2005). From 2012–2014, he was the Editor-in-Chief of the *IEEE Microwave Magazine*; the magazine distributed to all members of the IEEE Microwave Theory & Techniques Society (MTT-S); he has been an MTT-S Distinguished Microwave Lecturer for 2012 to 2014. He received the ARFTG Technology Award in 2007. He is a Fellow of the IEEE.

Gayle Collins

Gayle Collins is currently a Senior Member of the Technical Staff and the Principal Staff Research Engineer at MaXentric Technologies, La Jolla, California. She has extensive R&D experience in the communications and instrumentation industries. She is currently working on microwave and millimeter-wave power amplifier design in MMIC technologies using power combining techniques. She has worked in the design and advanced development teams of Freescale Semiconductor and RF Micro Devices, designing power amplifiers for wireless infrastructure applications in discrete and RFIC technologies. She has successfully designed Doherty amplifiers pushing the performance envelope of the processes she has worked with, and obtained the highest figure of merit for a power amplifier (PA) design. Ms. Collins has also worked in the Microwave Technology Center of Agilent Technologies in Santa Rosa, California, where she worked closely with the circuit design and fabrication engineers and scientists to improve the process and to identify any potential failures in the MMICs. Her areas of research interest include the nonlinear analysis of RF and Microwave PAs, high power PA design, and filter synthesis for broadband matching; and her areas of expertise include circuit analysis and synthesis, and system identification.

Ms. Collins holds a Master of Science in Electrical Engineering from Arizona State University. Her thesis is entitled "Robust Linearization and Stability of RF

Power Amplifiers," in which she demonstrated the viability of robust control theory in the design of RFPA systems in the face of process variability.

Ms. Collins is a senior member of the IEEE, and she is active in the MTT Society as well as the ARFTG measurements group. She is a member of the IMS Steering Committees for 2013, 2015, and 2020. She has been the Technical Program Chair for the ARFTG/Power Amplifier Symposium in Fall 2011, and Local Organizer for ARFTG in Fall 2012. She has presented at several IMS Workshops in recent years, and was an invited panel speaker at the 32nd IEEE Compound Semiconductor IC Symposium (CSICS). She regularly reviews papers, and she has been a member of the IMS Technical Program Review Committee since 2013; the ARFTG TPC since 2006; and the PA Symposium since 2009.

R. Neil Braithwaite

R. Neil Braithwaite received his B.Sc. degree in electrical engineering from the University of Calgary, Canada, in 1985, and M.Sc. and Ph.D. degrees from the University of British Columbia, Canada, in 1989 and 1992, respectively.

From 1992 to 1995, he conducted postdoctoral work at the University of California, Riverside. He has worked for Computing Devices Company (acquired by General Dynamics) in Ottawa, Canada, from 1985 to 1987, Nortel in Calgary, Canada, from 1995 to 1997, and Agilent Laboratories in Palo Alto, California, from 1997 to 2002. From 2002 to 2013, he worked for Powerwave Technologies in Santa Ana, California. Since June 2013, he has been working for Qualcomm Technologies Inc. in San Diego, California. His recent research interest is in the field of wireless communications. Research in linearization of RF power amplifiers includes digital pre-distortion (DPD), analog pre-distortion, feedforward compensation, vector feedback, and RF feedback. Other wireless communications work includes crest factor reduction (CFR) of signals, LTE relays, LTE PHY layer, echo cancellation in RF repeaters, and distributed antenna systems (DAS). Algorithm work includes adaptive control, interference cancellation, learning systems, and signal processing. His past research includes image processing, computer vision, autonomous vehicle control, pattern recognition, and underwater acoustic processing.

He is the author of 27 U.S. patents, 11 journal publications, 4 workshop presentations, 41 conference papers, and 4 book chapters. He has been a frequent contributor to the International Microwave Symposium (IMS), European Microwave Conference, ARFTG Microwave Measurement Conference, Radio Wireless Symposium, Vehicular Technology Conference (VTC), and the Power Amplifier (PA) Symposium. He has been a member of the IEEE since 2005 as well as a reviewer for several journals including *IEEE Transactions on Microwave Theory and Techniques* and *IEEE Microwave Magazine*.

Frédéric Roger

Frédéric Roger received the Ingnieur degree from ESIEE, Paris, France, in 1996. He was a Research Engineer at Collge de France (CNRS), Paris, from 1996 to 1999, where he worked on data acquisition electronics used in particle physics experiments. In 1999, he joined Inneon, Munich, Germany, to work on GSM ICs. From 2001 to 2006, he was Director of Engineering with Xignal, Munich, Germany, where he worked on VDSL, 10-Gb/s ICs, and SONET timing devices. From 2007 to 2013, he was with Scintera Technologies, Sunnyvale, California, becoming Director of Analog, RFIC Design and Test Development in 2011. He was instrumental in developing the first commercial RF Pre-distorter IC. At Scintera, he successfully led the Analog and RF development of four generations of linearizer ICs. Since 2014, he has been with Genapsys, Redwood City, California, where he is working on next-generation DNA sequencers based on semiconductor devices, aimed at revolutionizing the future of research and testing of genetic disease, cancer, and microbes.

Index

Recent Titles in the Artech House Microwave Library

Behavioral Modeling and Linearization of RF Power Amplifiers, John Wood

Chipless RFID Reader Architecture, Nemai Chandra Karmakar, Prasanna Kalansuriya, Randika Koswatta, and Rubayet E-Azim

Design of Linear RF Outphasing Power Amplifiers, Xuejun Zhang, Lawrence E. Larson, and Peter M. Asbeck

Design Methodology for RF CMOS Phase Locked Loops, Carlos Quemada, Guillermo Bistué, and Iñigo Adin

Design of CMOS Operational Amplifiers, Rasoul Dehghani

Design of RF and Microwave Amplifiers and Oscillators, Second Edition, Pieter L. D. Abrie

Digital Filter Design Solutions, Jolyon M. De Freitas

Discrete Oscillator Design Linear, Nonlinear, Transient, and Noise Domains, Randall W. Rhea

Distortion in RF Power Amplifiers, Joel Vuolevi and Timo Rahkonen

EMPLAN: Electromagnetic Analysis of Printed Structures in Planarly Layered Media, Software and User's Manual, Noyan Kinayman and M. I. Aksun

An Engineer's Guide to Automated Testing of High-Speed Interfaces, José Moreira and Hubert Werkmann

Essentials of RF and Microwave Grounding, Eric Holzman

FAST: Fast Amplifier Synthesis Tool—Software and User's Guide, Dale D. Henkes

Feedforward Linear Power Amplifiers, Nick Pothecary

Filter Synthesis Using Genesys S/Filter, Randall W. Rhea

Foundations of Oscillator Circuit Design, Guillermo Gonzalez

Frequency Synthesizers: Concept to Product, Alexander Chenakin

Fundamentals of Nonlinear Behavioral Modeling for RF and Microwave Design, John Wood and David E. Root, editors

Generalized Filter Design by Computer Optimization, Djuradj Budimir

Handbook of Dielectric and Thermal Properties of Materials at Microwave Frequencies, Vyacheslav V. Komarov

Handbook of RF, Microwave, and Millimeter-Wave Components, Leonid A. Belov, Sergey M. Smolskiy, and Victor N. Kochemasov

High-Linearity RF Amplifier Design, Peter B. Kenington

High-Speed Circuit Board Signal Integrity, Stephen C. Thierauf

Integrated Microwave Front-Ends with Avionics Applications, Leo G. Maloratsky

Intermodulation Distortion in Microwave and Wireless Circuits, José Carlos Pedro and Nuno Borges Carvalho

Introduction to Modeling HBTs, Matthias Rudolph

Introduction to RF Design Using EM Simulators, Hiroaki Kogure, Yoshie Kogure, and James C. Rautio

Klystrons, Traveling Wave Tubes, Magnetrons, Crossed-Field Amplifiers, and Gyrotrons, A. S. Gilmour, Jr.

Lumped Elements for RF and Microwave Circuits, Inder Bahl

Lumped Element Quadrature Hybrids, David Andrews

Microstrip Lines and Slotlines, Third Edition, Ramesh Garg, Inder Bahl, and Maurizio Bozzi

Microwave Circuit Modeling Using Electromagnetic Field Simulation, Daniel G. Swanson, Jr. and Wolfgang J. R. Hoefer

Microwave Component Mechanics, Harri Eskelinen and Pekka Eskelinen

Microwave Differential Circuit Design Using Mixed-Mode S-Parameters, William R. Eisenstadt, Robert Stengel, and Bruce M. Thompson

Microwave Engineers' Handbook, Two Volumes, Theodore Saad, editor

Microwave Filters, Impedance-Matching Networks, and Coupling Structures, George L. Matthaei, Leo Young, and E. M. T. Jones

Microwave Materials and Fabrication Techniques, Second Edition, Thomas S. Laverghetta

Microwave Materials for Wireless Applications, David B. Cruickshank

Microwave Mixer Technology and Applications, Bert Henderson and Edmar Camargo

Microwave Mixers, Second Edition, Stephen A. Maas

Microwave Network Design Using the Scattering Matrix, Janusz A. Dobrowolski

Microwave Radio Transmission Design Guide, Second Edition, Trevor Manning

Microwave Transmission Line Circuits, William T. Joines, W. Devereux Palmer, and Jennifer T. Bernhard

Microwaves and Wireless Simplified, Third Edition, Thomas S. Laverghetta

Modern Microwave Circuits, Noyan Kinayman and M. I. Aksun

Modern Microwave Measurements and Techniques, Second Edition, Thomas S. Laverghetta

Neural Networks for RF and Microwave Design, Q. J. Zhang and K. C. Gupta

Noise in Linear and Nonlinear Circuits, Stephen A. Maas

Nonlinear Microwave and RF Circuits, Second Edition, Stephen A. Maas

Q Factor Measurements Using MATLAB®, Darko Kajfez

QMATCH: Lumped-Element Impedance Matching, Software and User's Guide, Pieter L. D. Abrie

Passive RF Component Technology: Materials, Techniques, and Applications, Guoan Wang and Bo Pan, editors

Practical Analog and Digital Filter Design, Les Thede

Practical Microstrip Design and Applications, Günter Kompa

Practical Microwave Circuits, Stephen Maas

Practical RF Circuit Design for Modern Wireless Systems, Volume I: Passive Circuits and Systems, Les Besser and Rowan Gilmore

Practical RF Circuit Design for Modern Wireless Systems, Volume II: Active Circuits and Systems, Rowan Gilmore and Les Besser

Production Testing of RF and System-on-a-Chip Devices for Wireless Communications, Keith B. Schaub and Joe Kelly

Radio Frequency Integrated Circuit Design, Second Edition, John W. M. Rogers and Calvin Plett

RF Bulk Acoustic Wave Filters for Communications, Ken-ya Hashimoto

RF Design Guide: Systems, Circuits, and Equations, Peter Vizmuller

RF Linear Accelerators for Medical and Industrial Applications, Samy Hanna

RF Measurements of Die and Packages, Scott A. Wartenberg

The RF and Microwave Circuit Design Handbook, Stephen A. Maas

RF and Microwave Coupled-Line Circuits, Rajesh Mongia, Inder Bahl, and Prakash Bhartia

RF and Microwave Oscillator Design, Michal Odyniec, editor

RF Power Amplifiers for Wireless Communications, Second Edition, Steve C. Cripps

RF Systems, Components, and Circuits Handbook, Ferril A. Losee

The Six-Port Technique with Microwave and Wireless Applications, Fadhel M. Ghannouchi and Abbas Mohammadi

Solid-State Microwave High-Power Amplifiers, Franco Sechi and Marina Bujatti

Stability Analysis of Nonlinear Microwave Circuits, Almudena Suárez and Raymond Quéré

Substrate Noise Coupling in Analog/RF Circuits, Stephane Bronckers, Geert Van der Plas, Gerd Vandersteen, and Yves Rolain

System-in-Package RF Design and Applications, Michael P. Gaynor

TRAVIS 2.0: Transmission Line Visualization Software and User's Guide, Version 2.0, Robert G. Kaires and Barton T. Hickman

Understanding Microwave Heating Cavities, Tse V. Chow Ting Chan and Howard C. Reader

Understanding Quartz Crystals and Oscillators, Ramón M. Cerda